D0857771

GLOBAL ENVIRONMENTAL
HISTORY

GLOBAL ENVIRONMENTAL HISTORY

I. G. SIMMONS

THE UNIVERSITY OF CHICAGO PRESS
Chicago and London

I. G. Simmons is professor of geography at the University of Durham, a fellow of the British Academy, and a member of Academia Europaea. He is also the author of *Interpreting Nature*, *Environmental History: A Concise Introduction*, *An Environmental History of Great Britain*, and *Changing the Face of the Earth*.

The University of Chicago Press, Chicago 60637
First published in the UK by Edinburgh University Press
© I. G. Simmons, 2008
All rights reserved. Published 2008
Printed in the United Kingdom

17 16 15 14 13 12 11 10 09 08   1 2 3 4 5

Library of Congress Cataloging-in-Publication Data

Simmons, I. G. (Ian Gordon), 1937–
    Global environmental history / I. G. Simmons.
        p. cm.
    Includes index.
    ISBN-13: 978-0-226-75810-7 (cloth : alk. paper)
    ISBN-10: 0-226-75810-9 (cloth : alk. paper) 1. Human ecology—History. 2. Nature—Effect of human beings on—History. I. Title.
    GF13 .S56 2008
    304.2—dc22
                                                              2007050673

♾ The paper used in this publication meets the minimum requirements of the American National Standard for Information Sciences–Permanence of Paper for Printed Library Materials, ANSI Z39.48-1992.

Contents

TYPOGRAPHIC NOTE

In the body of the text, words which are defined and explained in the Glossary are printed in **bold face**. Any other typographical enhancements are for local emphasis only.

Tables

Figures

Preface

This book completes a trio of planned works at different spatial scales: that of the country (Britain), an internal landscape type (moorlands) and now the whole globe.* The timescale has been the same in all of them: the last 10,000 years. When people ask, 'what are you writing?', and you tell them, then the usual reaction is one of amusement, qualified by a nod in the direction of the poor old fellow's age. They may well be right but, inspired by some other attempts at 'big' history, I wanted to try. As Chapter 1 shows, I want to move the writing of environmental history further in the direction of inclusiveness. I believe that the natural sciences are very important but they are not the whole story because they sit in the type of social framework analysed by the social sciences and the humanities. Hence there is reference to a wide variety of work in this volume. Beyond that, I have no methodological ambitions: I do not think that there is a 'right' way to write environmental histories.

Any book has to be selective: it would be impossible to mention even every outstanding example of the processes that have been chronicled, and so those included comprise both the obvious and the eccentric. Some cannot be ignored, while others result from trawls through the literature or, increasingly, a period of surfing the net. The last is influential in one particular way: I have not (as in my other books) included a plethora of numerical tables and graphs. All the information in them is always badly out of date by the time a book actually appears, and readers will find it easier to go to a website and call up the latest data. Some sites are specified, others not, but appropriate government departments, the United Nations Environment Programme (UNEP), and bodies such as the World Resources Institute, the Population Reference Bureau, and the World Wide Fund for Nature (WWF) will provide necessary numbers and graphics.

Another initial point to make is that this is a book of history and not prognosis. I have tried wherever possible to end the narrative at the year 2000 though, in Chapter 5, this gets to be more or less impossible because so many trends simply carry on at the point where they have been discussed quite recently. If there is anything to be carried forward then it is the suggestion that

---

* *An Environmental History of Great Britain from 10,000 years ago to the Present*, Edinburgh University Press, 2001; *The Moorlands of England and Wales. An Environmental History 8000 BC–AD 2000*, Edinburgh University Press, 2003.

major changes have involved technological developments (agriculture, the use of fossil fuels) and that the future will as likely be driven by an equivalent change as by the more modest requirements of environmentalists. But any future seems likely to have to respect the laws of physics and the biogeochemistry of the planet: a revived potential for the ideas of environmental determinism, perhaps. In line with my other books, then, I have used human access to energy sources as a periodisation device. This has its disadvantages in terms of asynchrony and accusations of technological determinism but has the up-side of connecting with lively debates at the present time for I do not believe that history is culturally irrelevant, only that it may not be an accurate guide to the future. In social terms, increased access has allowed social differentiation and so cultural fragmentation has resonated in our attitudes to nature. It is not so simple, of course: for the world has long been subject to coalescence by both natural and cultural processes of diffusion, if that last word can be decently applied to electronic communication as well as medieval trade.

Even though three-score years and ten is now reckoned to be no age at all in western countries, it is always possible that I may not write any more books. So this is a good time to acknowledge all the generous encouragement and help that I have had from so many people over a forty-year career in academia: colleagues and friends at Durham, Peter Haggett in Bristol, academics in sister universities in several countries (with special thanks to Aberdeen for the honorary doctorate and to the ACLS for a postdoctoral fellowship at Berkeley), my teachers and postgraduate supervisors at UCL, quiet neighbours, GPs, and cats. There are too many names to mention individually but not a day goes by without thinking of one of you. From Berkeley Square to George Square, John Davey has always been a constant source of discriminating encouragement. It is a source of great pleasure to me that the book will be published in the USA by the University of Chicago Press, since it was their *Man's Role in Changing the Face of the Earth* (ed. W. L. Thomas) in 1956 that more than any other book turned me to this kind of topic.

My offspring, Catherine and David, are also a great inspiration in several ways, and grandchildren are just the sheer pleasure needed to offset some of the things our species has been responsible for perpetrating. All my books have been written while married to Carol and so she is present within all of them. If the publishers allowed watermarks in books, then her name and picture would be visible on every page.

I. G. Simmons
Durham, October 2007

Prologue

Mustering the marks

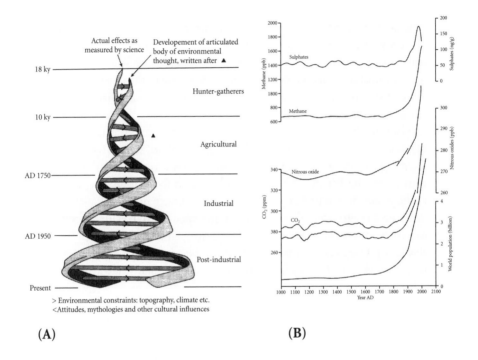

(A)   **(B)**

(A)   A simple model of human–non-human interactions on Earth in the last 12,000 years, based on the double helix conception of the DNA molecule. Here, the base pairs represent the influence of the natural on the cultural and vice versa. They should perhaps have different widths according to their strengths at various times, but the size of the diagram does not make that visually effective. In this version, the gyre of the helix is very roughly proportional to the size of the human population, with the downturns pointing out that population growth, while apparently inexorable, can be affected by plagues and pandemics. For greater accuracy, the diagram should be cut off at about the level of the label 'Post-industrial' but that would fail to convey any sense of vulnerability. But this model is mostly a guide to the structure of the material in this book, rather than a direct help to understanding the world.

(B)   A set of graphs for the period AD 1000–2000. The lowest curve is a numerical indication of the size of the gyre in (A) and the other curves reflect human activities. Carbon dioxide (CO_2) has a very high profile and represents the take-off of fossil-fuel use in the growth of industrialism, a curve echoed by methane which is a more effective 'greenhouse gas' than CO_2 by a factor of twenty-three. It is emitted from human activities that involve anaerobic digestion such as landfills, and the stomachs of cattle. Nitrous oxides, which are emitted by many forms of transport, are also greenhouse gases and fall out from the atmosphere as part of acid precipitation. Like the DNA gyre, these are both cause and effect. The growth in human numbers and the changes in economy increase the quantities of gases emitted to the atmosphere but the curves also symbolise cultural attitudes in which growth in wealth and throughput of resources are regarded as normal and every effort is made to sustain rates of growth rather than level out the curves. One task for environmental history is to chronicle and explain the strength of the interactions between the human and non-human worlds in terms of their mutual effects and the creation of hybrid forms.

This initial section of the book is basically an overview of what will be developed later in the text and may therefore allow potential readers to tell if it is the book they are looking for. It contains in brief many of the ideas and themes that are treated at greater length but obviously loses many of the nuances and caveats that pepper the longer chapters. But, in the spirit of the 'executive summary', it contains a compressed version of what follows: it musters together the essence of the printed text ('the marks').

An approach to a complex history

As a foundation, this narrative emphasises the empirical evidence for change in the last 10–12,000 years.[1] It is not confined to the material world, however, for it is also concerned with humans' ideas about the planet and their place on it. This inevitably means noticing the debates about the status of knowledge: how do we know what we think we know? This discussion of ideas *per se* is in Chapter 1, and readers can pass it by if they want the (relatively) simple epoch-by-epoch story. But, even then, there is no escape from discussions of the ideas formulated by various societies together with our recent interpretations of their perceptions and cognitions. There is also an attempt to draw out some abstract themes that carry across the whole timespan of the last 10,000-odd years (with even earlier roots) and which apply to society–nature interactions. These crystallise around notions of fragmentation and individualisation in society on the one hand, and coalescence and uniformity on the other; they are then examined for their impact on the human environment.

There is as well a stance in terms of definitions. A distinction is made between *worldwide*, in which a material entity is found throughout the world but in discrete patches and mostly on the land surfaces (e.g. soil erosion or Sky TV), and *global*, which is used only when there is the involvement of all the -spheres of the planet, including the upper atmosphere in its capability of diffusing uniformly the gases which it receives more regionally. Global phenomena are thus mostly relatively recent when brought about by humans, though natural climate change (as one example) has always been effective. This brings us up against the modelling of the 'greenhouse effect' and, while this must be included, the book is not about prognostication and is indeed a bit sceptical of the view that environmental history has a great deal to tell us about our future.

States of change

The world has been in a state of flux since the height of the last glacial maximum of the Pleistocene (1.8 million to 11,500 years ago); the post-glacial climate is sometimes said to be unusually stable but there have been notable fluctuations: a widespread 'optimum' in the mid-Holocene, sudden descents into cold phases and long periods of intense drought. Recurrent phenomena like the El Niño/La Niña variations in Pacific sea temperatures have experienced measurable fluctuations in frequency and intensity. Yet most of these second-order changes have not been uniform across the planet: there are regional differences in their incidence. There has been a continuous response by living organisms whose populations have grown or fallen and which have changed their distributions. New land surfaces have been colonised, and most human habitats have acquired a characteristic flora and fauna, including micro-organisms. None of the scientific investigations into the last 10,000 years has indicated a stable state of nature.

In addition to these transitions, human societies have changed their ways of life. From a population that was 100 per cent hunter-gatherer (or 'gatherer-hunter' or 'forager' – equivalent terms) and based on food collection from the wild, agriculture became dominant after about 8000 BC, though leaving large marginal areas for the hunters and gatherers. The solar-based agricultural economies persisted until after the mid-eighteenth century when the industrial economies then burgeoning in Europe and North America began to have a strong impact upon them. Although such agriculture has persisted until very recently, it can be argued that a fossil-fuel based industry was the world's major economy until about 1950, when it was intensified to a different level of interaction with the rest of the globe. All these changes (each of which is labelled an era) have been accompanied by a rise in the human population from a few million in 10,000 BC to just over 6,000 million (6 bn) in AD 2000. The main difference between the beginning and the end of this sequence has been a transition from patchy and temporary impacts upon the energy and material flows of the ecosystems inhabited by humans to a partial obliteration of the natural world in a series of very large conurbations together with a considerable degree of alteration of the terrain devoted to agriculture, grasslands and forests. Further, the effects wrought by carbon-based industrial activity upon the oceans and atmosphere have made *Homo sapiens* a species with a truly global reach.[2]

Parallel to this history devoted to alterations in the material world, there are the shifts in ideas about the kind of world we talk about and of the human place in it. There may have been a degree of commonality in most hunter-gatherers' world views as they adapted to circumstances over which they often had only a small degree of control. Agriculture seems to have produced many different interpretations of humans' place and role in the world but industrialisation brought about more uniformity as technologies powered by steam emplaced conquest, colonialisation and the spread of genetically uniform crops. Then, since 1950, there is the phenomenon called 'globalisation' in which instant communication and rapid transport have allowed an intensification and acceleration of most forms of interaction between humans and between humans and the non-human world of the globe: the 'post-industrial' economy. Both the last two eras have spawned countercultures which exist as islands in time as well as space.

Nobody can now imagine that these are stories in isolation from one another. They intertwine and are connected by strands of material flow and of meaning in which separation of either is very difficult. The quantity of food on a plate in the United States, for example, has more to do with the symbolism of plenty and achievement than with what is needed for healthy nutrition. A possible visualisation of these relationships might be the kind of DNA-style double helix, as presented above. Such imagery does not produce explanations and, in this case, it is only an aid to grasping the structure of the thinking behind the book. In fact the approach of this present volume is largely descriptive and even where, at the end, some 'why' questions are approached, it is in the knowledge that there are deeper levels of understanding that need another set of enquiries.

Perspectives

Even without humanity, the world would have changed and be changing. Humans have, though, produced many alterations which are very different from those of a 'natural' kind. Although the roots were much earlier, the period since 1950 has been the most extensive, the most intensive and the most measured. These features tend to overshadow the fact that each era has had its origins in an earlier phase but, once established, the later epoch dominates the scene. Equally, every subsequent era was not predictable by its predecessors, each of which would have declared itself to be the only way of living. Yet all of them were superseded by changes in the harnessing of energy and the application of that energy through technologies which move within a matrix of social attitudes. Hunter-gatherers, pre-industrial agriculturalists and hydrocarbon-based industry alike would have believed at the time of their zenith that they expected to go on for ever.[3]

NOTES

1. There is a number of textbooks which supply long-range and worldwide accounts of the development and activities of human societies. For 'prehistory', see C. Scarre (ed.) *The Human Past. World Prehistory and the Development of Human Societies*, London: Thames and Hudson, 2005 (784 pages); for later times there is R. Tignor et al., *Worlds Together, Worlds Apart. A History of the Modern World from the Mongol Empire to the Present*, New York: W. W. Norton, 2002 (462 pages + 49 pages of Index); more modestly there is P. Atkins, I. Simmons and B. Roberts, *People, Land and Time. An Historical Introduction to the Relations between Landscape, Culture and Environment*, London: Arnold, 1998 (a mere 286 pages). There is something of a gap between the chronological coverage of the first two, not filled by any comparable work. For really long-range history (the last 4.5 billion years), see D. Christian, *Maps of Time*, Berkeley, Los Angeles and London: University of California Press, 2005.
2. Overviews with an environmental emphasis include J. Diamond, *Guns, Germs and Steel. A Short History of Everybody for the Last 13,000 Years*, London: Chatto and Windus, 1997, Vintage 2005; J. R. McNeill and W. H. McNeill, *The Human Web. A Bird's Eye View of World History*, New York and London: W. W. Norton, 2003.
3. Doubts about the long-term availability of coal were expressed in the first quarter of the twentieth century, but nobody acted as if they were real. Hence my use of the verb 'believe'.

CHAPTER ONE

Resonances

FIGURE 1.1 *Kleine Orgel (small organ) at St Jacobskirche in Lübeck.*
Photograph by Wilhelm Castelli.

This is the Kleine Orgel (small organ) at St Jacobskirche in Lübeck,
dating from 1467 to 1636. There are some major sections such as the
upper set of pipes (the 'Hauptwerk') and a separate and lower set of pipes

1

which almost look like an separate instrument. As a metaphor, these might represent major sets of disciplines like the natural sciences and the humanities–social sciences. Authors may elect to play on one of these sets of pipes or may try to use both, sandwiched between the weight of one and the sharp ends of the other. Each separate pipe has its own sounding note and harmonic resonances, rather like many academic fields where each has their own special sounds: think of all the 'environmental' fields: economics, engineering, sciences, ethics, restoration, let alone the many other uses of the word. The player's seat might also symbolise a society's attitudes: is it best to have a score, to which adherence is compulsory, or is it a better survival technique to have a simple theme (such as basic needs) and improvise on it, with chance and contingency playing a full role?

Further, this is firstly an internal sound. When we represent environment in words and pictures we are talking to ourselves. When we use a bulldozer, the case is altered, though we are saying something about ourselves as well.

SOME ASSUMPTIONS

In bringing into one focus the whole of the world over a long period of time, certain assumptions are essential. Many of these are simply assumed within many societies while others are debated within the scholarly community. But without them, it is either impossible to write about humanity and its surroundings, or else the reader is left without knowing what the author takes for granted. So here are a few suppositions that will not be tested again in this book.

The first is that there is indeed a material world. A long tradition exists in western thinking that everything is only a product of Mind, either that of humans or of God. By contrast, in the present book it is assumed that the material world exists and that, for example, if humans vanished from the Earth, there would still be other animals, plants, rocks and water. This does not preclude the further assumption, also implicit in these pages, that the materiality of the globe is too complex and too dynamic for humans to know everything about, especially given their own limitations of brain capacity and sensory equipment. With Giovanni Battista Vico (1668–1744) we might argue that the human-made or 'social' world is something we have a chance of understanding but the 'natural' world is the outcome of processes of which we have only partial understanding.[1] It is not difficult to sympathise with the biologist J. B. S. Haldane (1892–1964) who remarked that 'my suspicion is the universe is not only queerer than we suppose but queerer than we *can* suppose'.[2] The present volume endorses the concept of the reality of a material world which, despite all our efforts to frame it culturally, may present its own limits in its own way.

Another basic notion is that humans act differently from other species in a number of ways. The more fundamental religionists favour the divergence as testimony of a divine mandate; their environmentalist equivalents are more likely to see it as evidence of a drive to destruction. Within such a spectrum, the scale and persistence of the human desire to control are relevant themes. Where the non-human world is concerned, this is most popularly summed up in the phrase 'the conquest of nature'. In the frequently adopted dualism of freedom and necessity, freedom usually implies the overcoming of nature; when there is disagreement over which bits are to be subjugated, it often involves the overpowering of other people first. In part, such processes acknowledge that 'humanity' does not exist as a single entity but in the form of humans (as individuals and as groups) driven by often conflicting needs, demands and illusions.[3] Thus, environmental history is made by individuals, by groups small and large, societies, nations and international agencies: there is much work to be done in investigating the scales of both conception and execution of environmental changes.[4]

In finding a workable language for this book, the terms 'human' and 'environment' or 'nature' are difficult enough, even without finding labels for the apparently hybrid forms which are emerging under the aegis of technologies such as micro-electronics and biotechnology. A vocabulary is necessary, however, and preferably one which (for the present purpose) does not have to resort all the time to quotation marks. *Human* will therefore be used to denote the genus *Homo*, including its present single species; *nature* will be used of the entire other material components of the cosmos; *environment* will refer to those elements of nature which are in an ecological relation with humans, that is, where there exists a possible transfer of energy and materials between them. *Culture* is the learned behaviour of humans which is transmitted down the generations. All of these can be the subject of non-material model-making in the human brain. (A number of other terms will be defined or glossed as they first occur.)

The behaviour of humans seems to be an interaction of the genetically determined and the culturally learned, with one class of behaviour most remarkably developed in humans being governed, as Charles Darwin said, 'by that short but imperious word "ought"'.[5] Social restrictions on feeding and reproduction are common in many species but the human unfolding of this trait has led to ideas of morality which are applied to standards of conduct[6] towards environment as well as to other fields of the life-world. A more developed, self-conscious form of morality is labelled ethics, and there is a whole academic field of 'environmental ethics'.[7] These constraints of right and wrong underlie many of the human actions upon nature (and the absence of others) which form the bulk of this volume. A few of the ingredients of any deliberations about environmental ethics might be:

- Humans have shown moral behaviour for as long as evidence exists. The boundaries of moral responsibility shift and, in the west, they have stopped for a long time at a species barrier between humans and other forms of life.
- Humans want incompatible things from their surroundings: they want material resources (of which there are inescapable minima) but also the company of other species and, often, intangible features such as beauty, which is identified relative to particular cultures.
- Humans have the power of understanding what is happening (albeit imperfectly) and using that power to regulate. At the same time, we have the imaginative power to know what we are missing. Much of this is tied up with purpose-centred thinking which, when compared with other animals, humans deploy in abundance.
- Developed, reflexive ethics has many approaches of the *-ism* and similar types: sentientism, ecofeminism, the land ethic, normative, deep ecology are examples that do not exhaust the roll-call even if the reading list is totally daunting.

Different mixes of these ingredients have produced different results over time. Two main categories are:

- An ethic for the use of the environment: the world is a set of resources for humanity to employ, though there may be limitations on that use, such as ensuring their perpetuation ('sustainability') or securing equitable distribution between various groups of people ('justice'). Terms such as 'utilitarian' and 'instrumental' are often applied to this view.
- An ethic which applies to the whole of nature, including humans, and which does not regard *Homo sapiens* as the culmination of evolution but as one species among many. The world is not our oyster but a place we share with the oysters: all species and ecosystems have an intrinsic value. The most developed form is called Deep Ecology. Terms such as 'impractical' and 'emotional' are often heard and sometimes written down.[8]

The ability to understand (even if only partially) means that simplified models of the relations of humanity and the cosmos are many and varied, and a few are mentioned here to give a flavour of the historical variety of them.

- The earth is senescent, having been occupied and degraded for a long time but there was once a Golden Age when humans and nature were in harmony; originated in Classical Greece, still around today.

- The notion of the sublime: that humans must relate to something bigger than ourselves, such as Nature or God. The poetry of William Wordsworth (1770–1850) is often seen as emblematic of the power of Nature to convey moral imperatives.
- The idea of progress and especially the eventual perfectibility of humankind. All human history is seen as progressing towards a better state, not without stumbling along the way.
- The adoption of Prometheus as a mythic icon. Stealing fire from the gods was just the first step in gaining tools that allowed mastery over nature; there will be a technological fix for everything. (What happened to him as a result is not usually mentioned by its advocates.)
- The idea that humans are on earth to divert or even thwart universal processes. Thomas Henry Huxley (1825–95: 'Darwin's bulldog') was a great advocate of the position that societies' actions were about the 'frustration' of the flows of the cosmos.
- The opposite view that humanity needs always to align itself with the flows of the cosmos and disturb them as little as possible: the Tao and Deep Ecology meet here.
- A model in which life on the planet works to maximise the conditions for its survival, and contrary human actions will eventually result in the demise of the species: the Gaia hypothesis. Nobody much discusses the likelihood that *Homo sapiens* is destined to have a short (if fiery) existence in evolutionary terms.
- The suggestion that the biophysics of the planet imposes limits to human actions. In the past, these may have been at local scales and surmountable by technology. Now, they are being seen at a global level, impelled by population growth and carbon-based economies. Usually labelled environmental determinism.
- Fatalism: *que sera, sera*, or 'God will decide, not us', or a no-model model in which history is simply 'one damn thing after another'.
- That very few, if any, things are free of ambiguity. Human actions produce up-sides and down-sides to almost everything. Mines are not pretty but silver and salt mines provided the riches that eventually fed Bach and Mozart, respectively.
- That the basic building block is always the notion of the Self as opposed to another, and that binary pairs are a common outcome. The two components are rarely of equal standing in human eyes.

There seems to be a number of common threads among these models. The first is that environmental ethics are necessary: in general, there is a need for 'oughts' since random behaviour is not acceptable. The second is that there is a concern for the future which aligns with the predictive disposition of the natural sciences. They look first for cosmic order and then transfer that idea of order (as in pattern, law, structure, construction, mechanism) in carrying out out human intentions.[9] Purpose usually involves control over nature such that it is transformed into environment, and over environment so that it becomes resources. And possibly over other humans so that they do not have access to those resources.

BASIC DEMOGRAPHY

No historical account can ignore the growth in the human population. The outline of humanity's major increase in numbers and the spread of the species from its origins in sub-Saharan Africa to most parts of the globe's land surface is well known. The term 'population growth' is usually used, and charts the rise of the numbers of the species *Homo sapiens* from perhaps 4 million in 10,000 BC (12 ky) to 6,000 million (or six billion) in 1999 and 6.5 billion in 2006.[10] Growth rates have not been constant: the fifth millennium BC saw a gain of 50 per cent, followed by 100 per cent in each millennium after that so that the total was 100 million in 500 BC. By contrast, the second century AD was the time of a slowing down of growth. This first cycle was largely a consequence of the invention and spread of agriculture, which released controls on the densities and growth rates of gatherer-hunters, and it mostly took place in Europe, mainland Asia and North Africa. In the tenth century AD another cycle of growth began, running its course until a slackening-off around 1400, after major disease epidemics. From 200 million in AD 400, a peak of 360 million was reached in 1300, with a fall to 350 million in 1400. Europe and China were the dominant contributors to this era of growth. The necessary concomitant of this phase was the improvement of food production within 'Malthusian limits', that is, environmentally produced upper boundaries of energy and protein gain, although other interpreters prefer explanations based on social and political structures. A third cycle can be postulated in which Europe leads population growth from the fifteenth century onwards, with the world total going from the 350 million of 1400 to the 6,000 million of 1999. In this stage Africa and the Americas add to the growth and, indeed, have some of the fastest rates of growth; in absolute numbers China is still a major builder. After the fifteenth century, the intercontinental exchange of food plants allows more intensive crop growth in many regions, and after the eighteenth century, any economic enterprise is liable to be subsidised by fossil fuel energy.

Interruptions to the apparently inexorable growth of human numbers have tended to be short-lived. Disasters such as major earthquakes, wars and famines have been locally or regionally significant for a time but births have generally made up for the lost numbers. The exception seems to have been the plague, erupting westwards from time to time, losing demographic power only in the seventeenth century. Its environmental relations are not obvious if vectors such as the rat and dense habitation are excluded. Many chronic diseases are more unambiguously linked to environment: malaria is one, and others belong to the suite of 'development diseases', as where irrigation projects spread the incidence of schistosomiasis. As human populations press more and more against the wild, then zoonoses are more likely to be transferred into humans, and viruses in particular may then show a remarkable ability to undergo mutations, just as other organisms, such as malarial mosquitoes can develop resistance to pesticides. Mutation has been a feature of the virus causing HIV/AIDS in humans, which emerged in central Africa in about 1959, with the syndrome getting its name in 1982. This disease has resonances with other major epidemics in human history: it is transmitted via sexual contact like syphilis, affects children and young adults as does smallpox and has a long incubation period like tuberculosis. In 1999, infection rates in sub-Saharan Africa were 80/1,000, in the Caribbean 20/1,000, in south and south-east Asia 7/1,000 and North America 6/1,000. South Africa and Zimbabwe had 30 per cent infection rates. Populations continue to grow in these areas, though at slower rates than hitherto; the effects, however, are concentrated on children, because many orphans are created, and in the working population which lacks a proportion of young adults. Here, as elsewhere, poverty is part of the complex.

Demography and demographic history have a distinct set of social contexts. For example, much interpretation has been underlain by a demographic transition theory which makes the assumption that the falls in fertility in the west since the nineteenth century will be echoed in lower-income economies as they get less poor.[11] Many funding bodies, too, were about 'population control' in search of relatively painless ways of reducing growth rates in the south. In general, until recent years, demography could be said to be strong on mathematics and weak on interface with social theory; its interface with environmental work was generally confined to the Malthusian assertions of environmentalists who were convinced that population growth in all types of economies was the root cause of degradational environmental change. Now that rates of fertility are actually declining in most parts of the world,[12] the great surge of Malthusian environmentalism in the 1960–80 period can be interpreted as either (a) having been totally irrelevant scaremongering or (b) a brilliantly triumphant piece of consciousness-raising with positive consequences.[13]

Until the Industrial Revolution, plants and animals, wind and water were the only sources of energy accessible to humans. One calculation suggests that 314 square kilometres (km^2) used as gatherer-hunters' territory would support three people in the Arctic, eleven in semi-desert, fifty-four in grassland and 136 in subtropical savanna. These numbers were exploded with the coming of agriculture, often by a factor of 100, though not in the Arctic where agriculture has never been successfully established. By the first quarter of the nineteenth century the worldwide energy availability had increased by sixfold. Thus, above-ground environmental constraints were obviously highly relevant until the coming of fossil-fuel energy (either as power in, for example, steam form or embedded in materials such as fertilisers) but thereafter began to fade as clear-cut and immediate sources of limits.

MATERIAL LINKAGES IN HUMAN–ENVIRONMENT
RELATIONSHIPS

The cosmos is a material entity with flows of energy: matter can be seen as energy at rest. Humans, too, are made of materials and are fuelled by energy intake. Humans tap into the material stocks and the energy streams in order both to survive (as do other living things) and uniquely to advance cultural ambitions. For our species, the use and control of energy is the key to much of our use of planetary materials (which we label as 'resources') and to our manipulation of the materials of nature.[14] Inevitably, the more people there are, the greater the volume of usage but the relationship has become more exponential than linear since so many people have commanded much higher levels than those needed simply for survival and reproduction.

Radiant energy from the sun can be fixed to a chemical form, oxidised to provide heat and electromagnetism, and then transformed into kinetic energy of the pushing and shoving variety. Formally defined as the capacity to do work, most forms of energy gradually lose that capacity as they are transformed, ending up as heat which is radiated to space. The measure of the loss of the capacity to do work is called **entropy** and a defining quality of living organisms is that they can temporarily defy entropy while building up complexity and undergoing **evolution**. The starting point for practical considerations is solar radiation which is fixed by green plants in the process known as **photosynthesis**. Globally, photosynthesis is not very efficient if looked at with an engineering cast of mind. The solar radiation reaching outer layer of the Earth's atmosphere is about 5,500,000 exajoules per year (EJ/yr) and global net photosynthesis reaches 2,000 EJ (1 EJ = 10^{18}J), so the efficiency is about 0.3 per cent. The mass of animals which feed on plants (which is most of them) is about 200 EJ, about 1 per cent of all the phytomass. (For comparison, worldwide fossil fuel production is 300 EJ.) A historical view of the relentless rise of humanity is given in the statistic that in AD 1900 the biomasses of humans and wild vertebrates were equal but that, by 2000, there was a difference of an order of magnitude, and further that domesticated vertebrates exceeded wild species by twenty times. Humans' interest in the energy content of plants and animals was for millennia in recently captured energy coming from the last year in most plants, a bit longer in animals and longer still with wood for fuel, but there was a massive change when, in the eighteenth century, fossil photosynthesis became widely usable in the form of coal: the timescale of interest was now geological as well as biological.

The availability of energy is fundamental for human access to materials, including the supply of more of it. There is firstly somatic energy: that of the body itself, which can be expended in, for example, running after prey animals or walking to look for plant foods, and which can be maximised by channeling, as in using the spear or the bow and arrow. Then there is extrasomatic energy in which other energies are harnessed to human ends, such as the use of draught animals, for instance, or with the energy of fossil fuels directed through technology. In both cases, energy becomes a 'binding resource' in the sense that without adequate somatic energy intake, we die and that a 'modern' lifestyle is possible only, when extrasomatic energy is available, usually in large quantities. Energy surplus is therefore an important goal of many societies in order to devote time to non-subsistence activities. Some gatherer-hunters were said to be nourished on a few hours' gathering and hunting per week, large wheat surpluses fuelled the building of the Egyptian pyramids, and a coal mine produces many times more energy than is needed to dig it. Manifestly, the idea of surplus has an objective and measurable component (figures at the local health club for calorie intake versus exercise levels might be an example) but also has a social and cultural component: the quantity of 'surplus' energy used in packaging in western economies is a matter of corporate policy rather than necessity.

The idea of toothpaste in tubes inside cartons is a reminder that goods and services all have an embedded energy content, that is, that energy has been used to make and transport them (energy intensity, EI) and that some of that energy may still reside in the materials (energy density). Worldwide plant mass is a store of 10,000 EJ at any one time, a 100-tonne wagon of coal contains 2 terajoules (TJ), a barrel of crude oil 6 gigajoules (GJ), a bottle of white wine 3 megajoules (MJ), and a raindrop on a blade of grass 4 microjoules (μJ; $\mu = 10^{-6}$). In comparable energy density terms, each kilogram of crude oil averages about about 43 MJ, natural gas 35 MJ, coal 23 MJ, air-dried wood 14 MJ, cereal grains 15 MJ, lean meat 7 MJ, fish 6 MJ, potatoes 4 MJ, vegetables 1 MJ, and human faeces 2 MJ. The energy intensities of materials vary according to their methods of production and the technologies used, and so vary through time and place. Aluminium, for example, has a high cost at 227–342 MJ/kg, compared with iron at 20–35 MJ, and steel at 20–50 MJ, and with water at <1 MJ. Comparisons can be made for the energy intensities of, for example, water and sewage at 17,000 kilocalories per dollar of 'product' (note the different units from previous data), through railways at 15 kcal/US$, hotels at 11 kcal, education and medicine at 8 kcal and radio/television at 4 kcal, a position it shares with finance and insurance.[15] It is valuable to know that, in an industrial economy such as that of Australia, energy consumption by final use (if calculated to include indirect and embedded energy) has been dominated by households, which account for 53 per cent. By contrast energy embedded in capital formation (infrastructural constructions like buildings, roads and pipelines) was only 11 per cent. Most such calculations have an arbitrary cut-off point in the calculations: the energy costs of a slice of white sliced bread can, at the extreme, be taken back to, for example, the energy cost of digging the iron ore that made the digger that dug the phosphate that fed the worker that drove the tractor . . . and so on. A comparable horizon has to be fixed for each process. In an overall historical perspective, two major trends can be singled out here: the proportion of energy use represented by food fell as societies industrialised and, within industrial economies, energy intensities declined in the later twentieth century as energy costs rose and technologies became more efficient.

The history of access to energy can be seen as a set of additive stages in which an economy adds new sources while not relinquishing all the older technologies, even though they may become diminished in importance. Gatherer-hunters are reliant on solar power as biomass which has mostly been recently photosynthesised, with the exception of wood used in fuels and tool-making and dry plant matter which is fired in the landscape. A few bones of longer-lived animals were added to the fuel and construction repertoires in the Palaeolithic of Eurasia. The whole system collects energy from large areas as people move around. With the coming of agriculture, solar energy is still crucial but it is collected over smaller spaces (such as in fields and in herds) and so denser populations of humans can be supported. This era also sees the use of falling water and wind energy in machines like mills and boats. The tapping of fossil fuels on a large scale is a major move along the intensity gradient since the energy densities of coal, oil and natural gas are many times those of wood and other plant materials. Falling water plus concrete allowed the installation of large-scale hydropower units in climatically and topographically suitable places. Electricity entered the mix in the nineteenth century although it was initially nowhere near as important as it is today where the post-industrial economy comes very quickly to a halt when it fails (be in a supermarket when the tills get no power, let alone an underground

TABLE 1.1 Gross energy (E) expended by humans in history

Period	Number of years	E in 10^{18} J	E/yr in J	Notes
50,000–8000 BC	42,000	2.5	6	Largely hunter-gatherers
8000 BC–AD 1	8,000	506	16	Mostly solar agriculture, some irrigated
AD 1–1750	1,750	1,400	8×10^{18}	Solar powered but increasingly efficient
1750–1950	200	360	18×10^{18}	Major years of industrial economies powered by fossil fuels
1950–2002	52	647	124×10^{18}	High population plus fossil fuels and other energy sources

The table uses data from a Population Research Bureau (www.prb.org) table of the number of people who have ever lived and then multiplies that number by a representative figure for energy consumption at each stage, bearing in mind that, after 1750, there are large disparities between the populations of industrial countries and the great majority of people in developing nations whose commercial energy consumption is small. Nevertheless, poor people's environmental impact can be very great.

train or an airport). Thus, the new sources of the period since about 1950, such as nuclear power and 'alternative' energies like wind turbines and photovoltaic cells, mostly generate electricity though a few heat water systems directly. Uranium is a form of intensification: fissioning 1 kg of U^{235} releases 8.2 TJ of energy which is about 2.7×10^5 more than the same amount of coal. The long-term trend is therefore towards the conversion of ever larger amounts of more concentrated forms of energy.

Using estimated data, the quantitative energy use of different types of historic economies can be outlined. Measurements in 10^3 kcal/person/day suggests a level of 5.0 for gatherer-hunters, of which about 3.0 are needed for bodily metabolism, and some of the rest comes as fire at the hearth and, often, in the landscape. But there is little energy consumed otherwise. Early dryland agriculture pushes the figure up to 12.0 so that there is a surplus that allows many activities other than subsistence to be pursued and population densities to rise. An advanced form of this type of economy, with better water control and informed breeding of plants and animals can reach 25.0 and so allow a bigger leisured class. When fossil fuels enter the energy mix then an immediate increase to about 77 kcal/day is possible, and in a fully 'post-industrial' economy, where electricity is a major source of energy for all consumers, then 230 is an representative figure.[16] If we think of these numbers as surrogates for interaction with the environment, then the magnitude of the more recent changes (in effect since 1950) immediately falls into place in space and time. What Table 1.1 shows is that the cumulative amount of energy expended by humans is not only very great but that the last 250 years have seen the dispersal of almost as much energy as the preceding 1750 and about twice as much as the whole history of gatherer-hunters who have occupied about 90 per cent of our species's evolutionary time. Nonetheless, the quantity of energy expended during the period of solar-powered agriculture is by no means negligible. Since the coming of fossil fuels, the amount of energy available to those in rich countries vastly exceeds the availability to the poorer nations and thus average figures mean little. Some emissions increase with income (sulphur dioxide [SO_2], for example) to a certain level and then begin to fall. Poor people, however, exert very strong influences upon water, soils and vegetation and can bring about change in very short times. At the other end of the scale, the EI of industrial nations began to fall markedly after the oil-price shocks of the 1970s. It is all more complex than a simple graph of commercial energy consumption against gross domestic product (GDP). Access to, and control over, energy sources allow humans to extend their reach in all kinds of ways. Few are more important than transferring materials from one place to another on every conceivable scale, including outer space. To take just one instance, about 85 per cent of the infrastructure of cities consists of mined products. There is a secondary level, too: industrial manufacture of, for example, anti-malarial drugs enables people to do otherwise impossible things like forest clearance or waging war. What is clear in a historic perspective is the parallel trend of loss of biodiversity and energy use.[17]

In sum, energy transformations provide a way of carrying thoughts into action. In a world without humans, or in an early Holocene one with very few, material transfers were, of course, taking place. The natural rates of sediment formation, movement, and dissolved chemical content have been much studied, as have the human alterations of the processes. Soils are formed at rates that average about 0.25 millimetres per year (mm/yr) in farmed areas, with a maximum global rate of 0.8 mm/yr. In the opposite process, weathering strips about 0.1 mm/yr from the Earth's crust (the maximum being about 10 mm/yr in the Himalayas), and the material is moved by water, wind and ice. Most of it is deposited in valleys, at any rate temporarily, and perhaps only one-tenth of the annually eroded material reaches the oceans. Within the human-caused movement of materials, the balance of deliberate and accidental transfers seems to be about equal at present; the rates are higher than that of population growth and rates have risen by three- to ten-fold in the years since 1920. In aggregate quantities, about $2–3 \times 10^{12}$ tonnes per year (t/yr) of soil and rock are moved by the mining and processing of minerals, which amounts to some 0.2 per cent of the Earth's surface. Data on these relocations can be produced in great quantities and with increasing precision for recent years; in historical contexts we need to remember that many are extrapolations from analogous situations of today and fewer are from direct measurements of deposited and dated sediments. A list of the ways in which humans change the ways in which sediments and dissolved chemicals move towards the sea could apply to most eras from the beginning of agriculture onwards, and possibly even before in a few limited places. They nearly all changed gear upwards many-fold with the coming of steam power although a few decelerated natural processes. While the outlines of the historic progressions are clear, the detail needs cautious interpretation.[18] In a similar way, a natural world had species extinction and evolution, migrations on many spatial and temporal scales; gradual long-term processes might (as with material transport) be over-ridden by catastrophic events such as rapid climate change, volcanic eruptions, major storms, and earthquakes.

Quantitative calculations of the costs and benefits of energy and materials availability may well not tell the whole story. Many data for energy use focus on commercial energy and so omit local sources of biomass fuels: an inhabitant of a remote part of the Himalayas might have a high quality of life, provided enough wood is accessible, just as backwoods survivalists in Montana rely on National Forests. Yet, for most of the high-income world, the flood of energy supply since the eighteenth century has spawned the mythologies of cornucopia which dominate many human–environment relationships today. Yet there is no escaping the second law of thermodynamics – at every energy transformation useful energy is transformed into heat which eventually finds its way back out to space.

Talking to ourselves

The sociologist Niklas Luhmann reminded us that we cannot talk to nature, only to ourselves.[19] His main conclusion was that we had come to simplify the complexity of it all and so we labelled a number of channels; environmental science and sciences; environmental economics, politics, ethics, religion and sociology, to name only some. In each, we seem to strive for a zero–one or binary resolution as in economic or uneconomic (a killer phrase for many a project), right or wrong, and perhaps good or evil. This model is very useful, and helpful metaphors can be constructed from it as in the analogy between the on-off sounds in each separate pipe of a church organ (as at the head of this chapter) and whether they make for harmony or dissonance. But there seem to be other subjects of conversation which cut across these channels and inform those parts of our culture which deal with the environment. There are three categories of knowledge, for instance, that cannot be ignored in any deliberations: facts, values and myths.

A factual approach to environmental history is an accepted route. The information in it may be derived from work in the natural sciences or the positivist social sciences where that term includes historiography.[20] Especially where the natural science component is strong, there tends to be an 'inevitable' conclusion: 'humanity is undermining its own resource base', perhaps, or 'technological ingenuity has always provided solutions when their need became evident'. The philosophically aware see in these latter statements the operation of value. An environmental history which contained no implicit statement of values would be almost impossible, but some approaches to both past and present are firmly based on humans rather than on nature: religious views with a hierarchy of god(s)–humans–animals–plants–rocks allow dominance of nature by humans, albeit usually with some constraints though not usually ones with the direr immediate penalties for transgression. Another historiography could bring forward all those instances where natural phenomena have controlled human affairs (at disasters and hinge-points as well as long-term and gradual processes) with the value content that Mother Nature is not to be gainsaid: in its extreme form this is called environmental determinism. Then there is the use of myths to sustain living among difficulties and contradictions. 'Myth' is used here to mean a condensed and vivid (poetic, indeed) story which encapsulates a story of events thought to be undoubtedly true and which is just as relevant today as when first formulated. Thus the myths about the expulsion from the Garden of Eden can be used to explain why poverty and degradation are inevitable (for some),[21] just as any account of Prometheus stealing fire from the gods can be seen as an inspiring vision for technological advance.

'Facts' are crucial to the models made of human–environment relations. In some people's perfect world, the positivist sciences sit outside society. Yet practically nobody would now prosecute the view that the human mind can be a clean slate upon which the senses (probably aided by technology) simply record the world around us, rather like burning files on to a newly unwrapped compact disc. Eminent scientists, such as P. B. Medawar, have quoted Kant (1724–1804) and Nietzsche (1844–1900) in support of the way in which cognition is constructed out of many more elements than perception. Kant, for instance, thought that experience was itself a form of knowledge that informed understanding, and Nietzsche goes further in asserting that everything that reaches consciousness is utterly and completely adjusted, simplified, schematised and interpreted. He anticipated recent studies of the sociology and philosophy of science by a hundred years in saying that there is a transition phase (or in some cases a dissonance) between the reality of the material world and its description by humans.[22]

Friedrich Nietzsche is often credited with the crystallisation of the type of thought that spread from France in the 1970s and 1980s, in which any 'grand narrative' was eschewed in favour of local and contingent truths; everything was valid only in its particular historical context. The most articulate proponent was Michel Foucault (1926–84) who thought that all knowledge was split into discourses that were specific to time and place, and that they in turn were at the service of increased production or, just as likely, harnessed as sources of political power. Knowledge and power are then temporarily bound together since 'truth' in any culture is a product of forces which work to legitimise some ideas and repress others.[23] The application of this notion to environmental attitudes is obvious: think only of the European debates over the introduction of genetically modified (GM) crops or the assertions that climatic change science in the United States has been subject to McCarthy-like attack. Cynics point out that, if all truths are relative, then the statement must apply to itself and so there is a further layer of uncertainty. This chimes with Medawar's view that a scientist tells stories, albeit ones which are always read with scepticism. They all agree that there is no ultimate, self-validating viewpoint from which all other limited perspectives would suddenly fall into place. But while total objectivity may be impossible, it may also mean an openness to the needs of others or even a way to feel a way into the experience of others without any excess of self-interest. The closeness of the material and the moral is shown when we are dependent on others for survival but are grateful for it.[24]

The cultural strand of understanding human–other relationships has contained the category usually labelled 'philosophy' since Classical times.[25] Though often anathema to the practically minded, some of its notions invariably underlie lived experience and form a part of inter-human communications. Immanuel Kant posed the question 'how should I live?' as central to moral philosophy and so environmentally minded thinkers have quarried his work, and that of many others, in search of a set of more or less universal abstractions that might guide us in everyday life. The idea of abstraction is important: historiography and the natural sciences maintain a roughly constant level of abstraction whereas philosophy constantly seeks to excavate one more level of irreducible meaning. Word-based philosophers, such as Plato, Bacon, Descartes, Rousseau and Heidegger, have all been targeted for their possible application to human–environment relations. In North America, more practical men like Thoreau and Aldo Leopold have spawned huge industries of abstract commentary which might well have surprised these adepts of axe and gun. 'Has there been any impact on impact?' is a question implicit throughout the empirical matter of the present volume.

The tension between the local–relative interpretations of the world and the 'grand narratives' has not prevented the believers in the latter from presenting their case. One type of master narrative is that of the natural sciences. The natural sciences' findings are at the heart of the grand construction known as the 'Gaia hypothesis', conceived by James Lovelock (1919–).[26] This inverts the usual sequence in which climate demarcates the distribution and nature of life for one in which life as a whole produces the global climate. It is argued that the global temperature and the gaseous composition of the atmosphere, and the salinity of the oceans would be different without the evolution of life forms. Initially, the hypothesis was rejected by the scientific community because no mechanisms could be found for linkages to produce the predicted effects. As is often the case, evidence turned up once people started looking for it; in this case a major discovery was that plankton in the oceans produced an aerosol (dimethyl sulphide) which initiated cloud formation which, in turn, engendered rainfall over the continents. Gaia can thus be seen as an holistic expression of a complex system of feedback loops that have the apparent 'purpose' of keeping the planet safe for life. This latter idea is a piece of teleology still unacceptable to many scientists, especially to neo-Darwinian proponents of organic evolution; it is highly acceptable to many environmentalists since all forms of life, and not necessarily the human species, are the benefactors.[27] It is also becoming more acceptable in 'mainstream' science under the label 'earth system science' without any hint of goddesses or teleology. The Gaia concept has also chimed very well with 'New Age' types of spirituality and, at one time, was a good component of various types of advertising.

A few other evaluations of the world aspire to findings of a unitary character: the energy-use history outlined in a previous section of this chapter is one such. Another was the attempt to provide a monetary value for the 'work' of the world's ecosystems and for natural capital. This concluded that an annual global gross national product (GNP) of 18 trillion (18×10^{12}) US dollars was far exceeded by a 'natural' value of US\$ 33 trillion.[28] (Although designed to show how valuable nature is to human societies, the findings were also a challenge to those minded to convert the one category into the other.)

A socially based grand narrative can be founded on the idea that technology is basically determinative: that it drives social change along before it. This attitude is exemplified by the Indian former Prime Minister, Jawaharlal Nehru (in office 1947–64), who tried to combine science and technology with the concept of planned development with a view to achieving a socialist pattern of society. In 1958 he obtained the Indian Parliament's acclamation for the statement that, 'The key to national prosperity, apart from the spirit of the people, lies, in the modern age, in the effective combination of three factors: technology, raw materials and capital, of which the first is perhaps the most important . . .' and in which 'dams and laboratories became temples of modern India.'[29] He would perhaps not have wanted the loss of social control which comes about from the restructuring inherent in high technology. This includes the need to facilitate control from a single centre (for example, the railways, air traffic control), the replacement of religion in the hierarchical order of authority, the formation of large organisations with their own social patterns, and the ways in which technologically based organisations tend to dominate the socio-political influences that are supposed to control them. Technology, it can be argued,[30] creates a new way of building order, almost like a new form of life, and social choices are introduced only after that fact. The mechanical clock may have been the key to many social revolutions and certainly to industrialisation.[31] Put more informally, Robert Oppenheimer (1904–67, the 'father of the atomic bomb') is quoted as saying, 'When you see something that is technically sweet, you go ahead and do it and you only argue about what to do about it after you have had your technical success.'[32] Historically, technological determinism may have been held at bay by particular sources of authority but, when technology itself is the source of authority, then control is virtually impossible. It is inconceivable to exclude the natural world from the penumbra of effects.

The connections between the world and ideas about it seem strong: in a time when the fragmentation of societies and the distancing of individuals from one another are remarkably obvious, then 'separate discourses' seem to be a set of ideas whose correspondence with the material world is good. Yet there can be a double counterpoint. If it can be shown that some facets of human behaviour are transhistorical and transcultural, then there are spatial and temporal linkages. In these deliberations of the social-natural interactions, there may be two processes which seem to be found across time and space. The first is that societies may be prone to fragmentation, as when social classes emerge based on, for example, birth or wealth or on technologies of separation such as the mobile headset. The second is the opposite: a tendency to coalescence, as happens, for example, with trade or with access to instant electronic communication. There are analogous processes in the natural world: the formation of species in isolated places is a form of fragmentation just as their colonisation of new environments is a coalescence. Put the human and the natural together and it seems likely that humans will alter, extinguish and bring into being both processes. A set of ideas to deal with that situation is needed: perhaps a case for 'grand narratives' can be made which needs to formulate a framework in which to discuss an ongoing tension between the two. Discussion at this abstract level carries the danger of forgetting that governments very often lag behind social changes acceptable to their people but that corporations and individuals do not always respect the dicta of governments: both have environmental relevance.

LOCAL, REGIONAL, CONTINENTAL, HEMISPHERICAL, GLOBAL

In the absence of humans, the world had a number of flows and cycles that are largely local and unconnected with their 'outside'. A heavy rainstorm will cause some landslides that eventually contribute silt to a river which then deposits most of it on its floodplain in the next episode of over-bank flow. But the phenomenon is confined to the one river basin, albeit there may have been many other such events in other basins. An isolated mountain range may function as a biotic 'island' and some species (endemic species) evolve there which are unique to that island and cannot disperse across the lowland habitats to another mountain range. At a rather larger scale, the faunas of the great ocean basins evolved separately, and only when canals like those at Suez and Panama have been built has there been diffusion of species into the other ocean basins. A very few species of higher plants are cosmopolitan in the sense that they are found on all the continents except Antarctica: the common reed (*Phragmites australis*) is one and a plantain (*Plantago major*) another, with the latter having been spread by humans because it is a weed of paths and field margins.

There are processes which are worldwide but unconnected. All the major rivers contribute some silt to the oceans but the majority of it falls to the seafloor soon after emergence into the lower-energy environment: the plume does not necessarily join up with other such effluxions. On land, climatic variability may cause desertification but in different places at varying times; even a worldwide climatic change may not be synchronous everywhere. There are, too, similar species which are confined to one major region and cannot interbreed: flightless avian herbivores such as the emu, ostrich and rhea are examples from, respectively, Australia, Africa and South America. Moving to a larger scale, the natural world has had worldwide and connected phenomena. Warming one ocean produces effects in all the others, as is seen in the ENSO (El Niño-Southern Oscillation) phenomenon; the minerals that escape from geyser-like vents in the ocean floor will, in solution, find their way to other oceans.

There are also truly global cycles and flows, where changes have global consequences. The incidence of solar radiation upon the Earth's atmosphere is a major example and it is difficult to think of any corner of the planet that did not change in response to the major cooling of the Pleistocene ice age. Another great cycle is that of carbon, which is found in liquid, gaseous and solid phases. It is present in several 'pools', such as atmospheric carbon dioxide (CO_2), organic compounds in living matter and fossil forms of life such as coal and oil, and as calcium carbonate ($CaCO_3$) on the ocean floors. The importance of carbon is immense: it is needed for photosynthesis; it flows between human-dominated and nature-dominated processes, is at the heart of the economies of the industrial world, and is implicated in current anxieties such as global climatic change. Along with water, its flows are pivotal for the way in which the planet works, and for many of the human activities thereon.[33] The carbon cycle is mirrored by some other elements, such as nitrogen and sulphur and they all have in common an atmospheric pool which diffuses to a uniformity of concentration: even if a great deal of them is injected into the atmosphere by, for example, a volcanic eruption, it soon becomes spread very widely.[34] The impact of a comet or a truly massive volcanic eruption may have global consequences: the global cooling of about −3 °C in AD 536–45 resulted from a 'dry fog' or dust veil which could have come from either cause.[35]

The majority of human impacts on nature for many millennia were on land and in the near-shore area of the seas. Several were worldwide but none probably truly global (though a case has been made for the atmospheric consequences of gatherer-hunters' fire) until the effects of steam power began to be felt in the late eighteenth century.[36] Since then, the human mobilisation of materials has equalled that of the natural world in some instances, and the historical context of the higher concentrations of 'greenhouse' gases (especially carbon dioxide and methane) is well understood. Concentration is indeed a hallmark of human interactions with environment: at its simplest, a river can process the sewage output of a couple of pre-industrial farms along its banks, but a city of 500,000 people (or a few housed piggeries) is a different matter. Humans have introduced one entire novelty: the formulation of substances which are unknown in nature. Many chemicals used in industrial processes are of this type, as are many polymers in everyday use such as PVC and other plastics. The relevance here is that such compounds are unlikely to have the kinds of breakdown pathways which 'natural' materials have attracted in an evolutionary time-frame. Most complex molecules eventually form the substrate for weathering, plants, animals, fungi and bacteria that break them into simpler elements (which we often label 'rotting') but chemists have produced molecules which are very slow indeed to break apart, usually by design so that they are effective for longer. Poisons of the group that contains DDT are good examples, for not only does DDT break down very slowly but its successor compounds are sometimes more toxic than the DDT itself and have diffused into most of the world's ecosystems. The inventiveness of the chemical industry is now continually challenged to produce effective but life-friendly substances.

Climate is a phenomenon of all spatial scales and an essential element of environmental history. It is important, however, to avoid crude determinism in discussions of global climate. Many instances of human disaster seem to coincide with various phases of the ENSO phenomenon: the two retreats from Russia (Napoleon in 1812 and Hitler in 1942) coincide with the end of El Niño phases; but, as one commentator points out, a combination of unexpectedly poor weather and bad military judgements is probably involved.[37] The wide reach of El Niño and La Niña forces us to consider the 'big picture' in all kinds of environmental history as well as the local and the regional.[38] The idea of sudden environmental change, rather than a gradual transition from state A to state B is at the forefront of current climatic change models, and so historical examples and their possible causes are being re-evaluated. Apart from the obvious natural causes, such as earthquakes, tsunamis, and volcanic eruptions, phenomena like fireballs from extinct comets and oceanic degassing are getting attention. Earthquakes may disturb ocean-floor sediments that contain large quantities of hydrogen sulphide and methane which can cause morbidity and death in living creatures. So, one suggestion is that earthquakes in China in 1334 resulted in the recording in many places of dead fish, a red and yellow sky, and a 'corrupted atmosphere' which not only sounds like outgassing but preceded the outbreak of the Black Death. So the chance event may have widespread consequences when there is a disease pandemic or when the crops fail. Most agricultural societies are buffered against a year's failure and can adapt to a long-term drift in conditions but a few years' consecutive disasters are likely to obliterate a way of life.[39]

The overall impact of human societies on nature has been subject to attempts at measurement in recent years. Ideas include 'the ecological footprint' which is a measure of the area of productive land and sea that underlies the consumption of energy and materials and compares the footprint of human groups with the renewable capacity, showing that current consumption exceeds the planet's capacity to sustain such levels. The WWF has produced a 'living planet index' which shows the 'average change over time in the state of forest, freshwater, and marine ecosystems; it is basically an attempt to quantify the extent and severity of biodiversity loss'.[40] Time-depth is, however, restricted to thirty years or so because of the limitations of statistical data but space can sometimes be substituted for time: the Greater London area currently has a footprint which is 125 times larger than the actual area that it occupies. Like many grand narratives, these measurements have grand conclusions, usually of a pessimistic kind.

Two broad-scale features stand out: the first is that global ubiquity is now present in a cultural sense, albeit as the result of a long history of convergences. True, there are islands of resistance to the 'common culture' of western capitalism but the central elements of the knowledge of it are accessible to almost everyone, even if they are prevented from being an active participant. The second is that the world is in a physiological state (in terms of its species mix and gaseous levels) which has no past analogy: there never was a time when it was like this.[41] Such knowledges have consequences for how we think about the world as well as, perhaps, act.

A TRANSITION TO THE LATER CHAPTERS

Keeping in mind the opportunities and the constraints outlined in this chapter, we must now address the chronicle of human–nature relations during the Holocene. The technological span from gatherer-hunters to today's electronic world is immense and we shall use energy access as a periodisation. Each of the next four chapters will start with, and be dominated by, material relating to the ecology of that phase in its broadest meaning: the basic environmental relations of the phase, its demographic characteristics and the social properties which seem most relevant. There will then be an attempt to tease out any signs of the two long-term interactions of nature and society which can be realised as fragmentation and coalescence. To end with, the ways in which that phase can be represented, both by its inhabitants and by us today, are briefly highlighted before the outcomes of the era are laid out.

NOTES

1. L. Pompa, *Vico: a Study of the 'New Science'*, Cambridge University Press, 1975; I. Berlin, *Vico and Herder*, London: The Hogarth Press, 1976 and 1992. A thorough examination of realism and idealism in philosophical terms can be found in F. Mathews 'The real, the one and the many in ecological thought', in D. E. Cooper and J. Palmer (eds) *Spirit of the Environment*, London and New York: Routledge, 1998, 86–99.

2. J. B. S. Haldane, *Possible Worlds. And Other Essays*, London: Chatto & Windus 1927; reprinted as *Possible Worlds*, New Brunswick, NJ and London: Transaction Publishers, 2002. Haldane and his colourful wife Helen Spurway were often to be seen around UCL in my undergraduate days. There is a dissection of the view that 'nature' can only be socially constructed in the essays in M. Soulé and G. Lease (eds) *Reinventing Nature? Responses to Modern Deconstruction*, Washington DC and Covelo CA: Island Press, 1995.

3. M. Bookchin, *The Philosophy of Social Ecology*, Montreal and New York: Black Rose Books, 1990; J. Gray, *Straw Dogs. Thoughts on Humans and Other Animals*, London: Granta Books, 2002.

4. See for instance the attempts at a 'template' for knowledge integration present in B. Newell, C. L. Crumley, N. Hassan, E. F. Lambin, C. Pahl-Wostl, A. Underdal and R. Wasson, 'A conceptual template for integrative human-environment research', *Global Environmental Change* Part **A, 15**, 2005, 299–307.

5. In *The Descent of Man*, 1871, ch. IV.

6. The notion of 'good' and 'bad' as polar antagonisms is an example of western tendencies to divide everything into opposing twosomes. The idea of 'man' and 'the animals' is one example, as are 'economic' and 'uneconomic' and many other pairings. The emotional temperature of this dualism is raised when 'good' is opposed to 'evil'.

7. These include D. Jamieson (ed.) *A Companion to Environmental Philosophy*, Malden and Oxford: Blackwell, 2001; P. Singer (ed.) *A Companion to Ethics*, Malden and Oxford: Blackwell, 1991, including M. Midgley, 'The origins of ethics', pp. 3–13; M. Midgley, *The Ethical Primate. Humans, Freedom and Morality*, London and New York: Routledge, 1994. My own *Interpreting Nature. Cultural Constructions of the Environment*, London and New York: Routledge, 1993, ch. 5 has a more extended treatment, though several of the arguments have moved on by now.

8. The charge of 'emotion' is often levelled at anybody arguing for nature at times of conflicting views. It seems to me that (a) emotion is highly necessary: without it

nothing at all gets up any steam; and (b) is the love of making money somehow not emotional?

9. M. Midgley, *Science as Salvation. A Modern Myth and its Meaning*, London and New York: Routledge, 1992. She points out as well that these are all metaphors, 'but not optional, disposable metaphors'. (p. 10).

10. For the period up to 1975, a good source is C. McEvedy and R. Jones, *Atlas of World Population History*, Harmondsworth: Penguin Books, 1978; thereafter the numbers can be obtained from numerous website and commentaries thereon in those of e.g. United Nations Population Agency, the World Resources Institute and the World Bank. A standard interpretive history is M. Livi-Bacci, *A Concise History of World Population*, Oxford: Blackwell, 2001, 3rd edn (first published in Italian in 1989).

11. S. Greenhalgh, 'The social construction of population science: an intellectual, institutional, and political history of twentieth-century demography', *Comparative Studies in Society and History* **38**, 1996, 26–66. (*De facto*, it deals only with the United States.)

12. M. Connelly, 'Population control is history: new perspectives on the international campaign to limit population growth', *Comparative Studies in Society and History* **45**, 2003, 122–47.

13. Without doubt the most high-profile combination was the Stanford biologist P. R. Ehrlich and his book *The Population Bomb*, New York: Sierra Club/ Ballantine Books, 1968 and many subsequent editions and translations. A discussion of his work and publications is in I. G. Simmons, 'Paul Ehrlich 1932– , in J. Palmer (ed.) *Fifty Key Thinkers on the Environment*, London and New York: Routledge, 2001, 252–60. Ehrlich worked in an atmosphere in which the work of D. B. Luten was influential: it is collected in T. R. Vale, *Progress Against Growth*, New York and London: Guilford Press, 1986.

14. Most of the numbers and not a few of the ideas come from the work of V. Smil, especially his *General Energetics. Energy in the Biosphere and Civilization*, New York and Chichester: Wiley Interscience, 1991; *Energy in World History*, Boulder CO, 1994, and *The Earth's Biosphere. Evolution, Dynamics and Change*, Cambridge MA and London: MIT Press, 2002. His works with a wider scope such as *Global Ecology. Environmental Change and Social Flexibility*, London and New York: Routledge, 1993, are always worth reading. See also J.-C. Debeir, J.-P. Deléage and D. Hémery, *In the Servitude of Power. Energy and Civilization through the Ages*, London and New Jersey: Zed Books, 1991 [first published in French (Paris: Flammarion) as *Servitudes de la Puissance* in 1986].

15. These and other data for the period to the 1980s may be found in C. A. S. Hall, C. J. Cleveland and R. Kaufmann, *Energy and Resource Quality. The Ecology of the Economic Process*, New York: Wiley Interscience, 1985.

16. Plotted diagrammatically in my *Humanity and Environment. A Cultural Ecology*. Harlow: Longman, 1997, p. 151.

17. J. Goldemberg, *Energy, Environment and Development*, London; Earthscan, 1996; P. R. Ehrlich, 'Energy Use and Biodiversity Loss', *Philosophical Transactions of the Royal Society of London* **B 344**, 1994, 99–104.

18. I. Douglas, 'Sediment transfer and siltation', in B. L. Turner et al. (eds) *The Earth as Transformed by Human Action*, Cambridge: Cambridge University Press, 1990, 215–34. The pioneer work was R. L. Sherlock, *Man as a Geological Agent*, London: Witherby, 1922.

19. N. Luhmann op. cit. 1989; of course people talk to plants and animals but we might doubt whether this is actually communication. There is a good Gary Larson cartoon which contrasts what 'master' says with what Rover hears.

20. There is an excellent review of the scientific approach to environmental history in J. A. Dearing, R. W. Battarbe, R. Dikau, I. Larocque and F. Oldfield, 'Human–environment interactions: learning from the past', *Regional Environmental Change* **6**, 2006, 1–16.

21. See C. Merchant, 'Reinventing Eden: Western culture as a recovery narrative', in W. Cronon (ed.) *Uncommon Ground. Toward Reinventing Nature*, New York and London: W. W. Norton, 1995, 132–59. For her, there seems to be a very close identification between 'western' and 'North American' but if we look beyond that there is a very grand master narrative indeed.

22. Possibly the best book ever from a practising scientist on these topics: P. B. Medawar, *Pluto's Republic*, Oxford University Press, 1982; see also N. Smith, 'Nature at the millennium. Production and Re-enchantment', in B. Braun and N. Castree (eds) *Remaking Reality. Nature at the Millennium*, London and New York: Routledge, 1998, 271–85.

23. C. Norris, *The Contest of Faculties: Philosophy and Theory after Deconstruction*, London: Methuen, 1985; P. B. Medawar, *The Art of the Soluble: Creativity and Originality in Science*, Harmondsworth: Penguin Books, 1969. The positive value of postmodern thinking in environmental matters is explored in P. Quigley, 'Nature as dangerous space', in É. Darier (ed.) *Discourses of the Environment*, Oxford: Blackwell, 1999, 181–202.

24. T. Eagleton, *After Theory*, London: Allen Lane, 2003.

25. The literature is enormous but three possible starting places are D. Macaulay (ed.) *Minding Nature. The Philosophers of Ecology*, New York and London: Guildford Press, 1996; V. Pratt (with J. Howarth and E. Brady), *Environment and Philosophy*, London and New York: Routledge, 2000; D. Jamieson (ed.) *A Companion to Environmental Philosophy*, Oxford: Blackwell, 2001. The outstanding history of environmental ideas before the nineteenth century is still that of C. Glacken, *Traces on the Rhodian Shore. Nature and Culture in Western thought from Ancient Times to the End of the Eighteenth Century*, Berkeley and Los Angeles: University of California Press, 1967.

26. Gaia was the goddess of the Earth in Greek mythology. She now has an enormous literature at all levels of complexity and advocacy. Lovelock has written an autobiography, *Homage to Gaia*, Oxford University Press, 2000. In many ways, the founding document is his *Gaia. A New Look at Life on Earth*, Oxford University Press, 1979, but there is also his *The Ages of Gaia: A Biography of Our Living Earth*, New York: W. W. Norton, 1988; *Gaia: the Practical Science of Planetary Medicine*, London: Gaia Books, 1991, and M. Midgley, *Gaia: the Next Big Idea*, London: Demos, 2001. (Most of Lovelock's own books have later reprints and editions.)

27. Bunyard, P. (ed.) *Gaia in Action: Science of the Living Earth*, Edinburgh: Floris Books, 1996; the religious possibilities of Gaia as a metaphor even extend to its creator, for Lovelock has written 'For me, Gaia is a religious as well as a scientific concept, and in both spheres it is manageable . . . God and Gaia, theology and science, even physics and biology are not separate but a single way of thought.'

28. R. Costanza et al., 'The value of the world's ecosystem services and natural capital', *Nature* **387**, 1997, 253–60; S. L. Pimm provides a summary and commentary in the same issue, pp. 231–2.

29. Nehru is quoted in S. Visvanathan, 'A Celebration of Difference: Science and Democracy in India', *Science* **280**, no. 5360, 1998, 42–3; online at http://www.sciencemag.org/cgi/content/full/280/5360/42

30. This analysis follows J. Winner, *The Whale and the Reactor. A Search for Limits in an Age of High Technology*, Chicago: University of Chicago Press, 1986.

31. M. Kearney, *World View*, Novato CA: Chandler and Sharp, 1984.

32. *The Oxford Dictionary of Phrase, Saying and Quotation*, Oxford: OUP, 2002, 2nd edn, p. 451. Oppenheimer made the remark in 1954 during investigations into his security status.

33. V. Smil, *The Earth's Biosphere*, Cambridge MA and London: MIT Press, 2002, ch. 5.

34. *Idem, Cycles of Life, Civilization and the Biosphere*, New York: W. H. Freeman, 1997.

35. See among many R. B. Stothers and M. R. Rampino, 'Historic volcanism, European dry fogs and Greenland acid precipitation, 1500 BC to AD 1500', *Science* **222**, 1983 411–13; M. G. L. Baillie, 'Dendrochronology raises questions about the nature of the AD 536 dust-veil event', *The Holocene* **4**, 1994, 212–18; E. Rigby, M. Symonds and D. Ward-Thompson, 'A comet impact in AD 536?', *Astronomy and Geophysics* **45**, 2004, 1.23–1.26.

36. There is an excellent one-page summary by P. J. Crutzen and E. F. Stoermer, 'The "Anthropocene"' *IGBP Newsletter* **41**, 2000, 17–18, though there seems no need to introduce a term like 'Anthropocene' to anybody except geologists.

37. C. N. Caviedes, *El Niño in History*, Gainesville, FL: University Press of Florida, 2001. See also M. Davis, *Late Victorian Holocausts: El Niño Famines and the Making of the Third World*, New York: Verso, 2001; a semi-popular account is B. Fagan, *Floods, Famines and Emperors: El Niño and the Fate of Civilizations*, New York: Basic Books, 1999.

38. M. G. L. Baillie, 'A view from outside: recognising the big picture', *Quaternary Proceedings* **7**, 1999, 625–35.

39. *Idem*, 'Putting abrupt environmental change back into human history', in P. Slack (ed.) *Environments and Historical Change*, Oxford: Oxford University Press, 1999, 46–75; J. D. Post, 'The impact of climate on political, social, and economic change: a comment', *Journal of Interdisciplinary History* **10**, 1980, 719–23.

40. N. Chambers, C. Simmons and M. Wackernagel, *Sharing Nature's Interest*, London: Earthscan, 2000; WWF, *Living Planet Report 1998*, Gland: WWF International, 1998, p. 2.

41. J. Friedman, 'General historical and culturally specific properties of global systems', *Review* **15**, 1992, 335–72; C. Chase-Dunn, 'The historical evolution of world-systems', *Sociological Inquiry* **64**, 1994, 257–80; C. Chase-Dunn, S. Manning and T. D. Hall, 'Rise and fall: East-West synchrony and Indic exceptionalism reexamined', *Social Science History* **24**, 2000, 727–54; H. Haberl, S. Batterbury and E. Moran, 'Using and shaping the land: a long-term perspective', *Land Use Policy* **18**, 2001, 1–8.

CHAPTER TWO

The gatherer-hunters and their world

FIGURE 2.1 *Depiction of an owl in the Hillaire Chamber of La Grotte Chauvet-Pont-d'Arc.*
Photograph supplied by the French Ministry of Culture and Communication, Regional Direction for Cultural Affairs – Rhône-Alpes, Regional Department of Archaeology.

This an image of an owl from one of the most famous caves in France, La Grotte Chauvet-Pont-d'Arc in the Ardèche, 45 kilometres from Montélimar. Chauvet has several very large galleries with more than 300 paintings and engravings of rhinoceroses, felines, bears, owls and mammoths dating to between 31,000 and 29,000 years ago, that is, Palaeolithic in date. This is a finger painting on a scraped wall of an individual of *Asio otus*, the long-eared owl (Fr: *hibou moyen-duc*), with its head at 180 degrees to its body.

Beyond the breathtaking impression of the 'art', there are at least two questions of environmental significance here. The first, as with all such early depictions, is the simple one of 'what was it for?' There is no agreed answer (could there ever be?) but there is one likely certainty: that the people who fabricated the owl and the other animals had a vivid awareness of the presence and importance of the wild creatures in their lives. This sounds like a cliché, but contrast it with the likely situation today, where only a limited range of people could produce the equivalent. The second question relates to some of the speculative scholarship that inevitably surrounds such images, as it has at other famous sites in France and Spain, as well as rock art out in the open in, for example, Africa and Australia. This is concerned with the possibility of control: is the making of images one way of trying to exert influence over animals which might be food resources or potential enemies? Is the production of an image one of the stages (here, quite a long way off) towards domestication? Perhaps the inclusion of an owl argues against instrumental interpretations.

The environmental and cultural associations of owls are well known. In forager cultures, such as the Kwakiutl and Tlingit of the Pacific northwest of North America, the owl is a harbinger of danger and death, as in many later agricultural societies, with some remnants to the present day. The notion of the owl as a repository of wisdom derives mostly from Classical representations of the one-eyed goddess Athene (bringer of wisdom) who carried an owl on her blind-side shoulder, and it reported the state of people's souls as well as the weather. Owls thus showed that piercing the darkness was a transferable skill.

A guide to the cave (which is not open to the public) is at http://www.culture.gouv.fr/culture/arcnat/chauvet/en/index.html

'JOINT TENANTS OF THE WORLD'

John Donne's poems contain some environmental surprises, and his 'joint tenants of the world' were the sun, and man, who outlived the stag, raven and long-lived tree and so 'there is not now that mankind, which was then'. So might we conceive of cultural traditions which were entirely dependent on recently fixed solar energy in the form of plant and animal tissues, though our understanding of forensic archaeology leads us to think of such people as having rather short lives compared with most species of tree, though of the same order of magnitude as deer and raven.[1]

Beyond the life-span of the individual, the antiquity of human existence is constantly being revised as more archaeological evidence is found and as the techniques for determining age and genetics improve. As more and more bones yield measurable DNA, for example, a lineage becomes clearer, though not always one which accords with the archaeometric data. So the Pleistocene speciation of the genus *Homo* and its immediate precursors is not agreed upon, and any statement in print is hostage to the fortunes of bone-hunters and laboratory finances. A majority view, perhaps, is that Africa was the region of the evolution of *H. sapiens* by about 300,000 years ago (300 ky) and that the species spread from that continent. By the time of the Last Glacial Maximum (LGM) at about 22,000 to 20,000 years ago (in round figures 20,000 years before the present), *H. sapiens* has become the only member of its genus: *H. erectus* in the Far East and *H. neanderthalensis* in Europe became extinct, the one at about 90 ky and the latter at about 30 ky. The LGM may have been something of a population bottleneck for *Homo* but was followed by a major population expansion after about 16,000 years ago.

THE CULTURAL ECOLOGY OF GATHERER-HUNTERS

Until the early Holocene all humans practised the gatherer-hunter way of life so that, if our evolution as a genus was perhaps a million years ago, then over 90 per cent of our evolutionary history has been as gatherer-hunters. A species which was 100 per cent food collecting as it emerged from the Pleistocene, however, was quite quickly converted to agriculture so that by AD 1500, this life-way was restricted to uncultivable areas of the cool temperate, the Arctic and sub-arctic zones, and to areas not yet deemed agricultural, such as Australia and Argentina. In our own day the proportion of people with even a dominance of hunting and gathering in their culture is less than 0.001 per cent. In today's terminology, it has not been 'sustainable' though for reasons that may be more cultural than environmental.[2]

Evolution and dispersal

The complexities of the study of human evolution are marginal to most of this book. What is germane is that humans survived the unfavourable conditions of the various glacial maxima of the Pleistocene and, indeed, used the periods of low sea-level to colonise new lands, Tasmania included. The southern margin of an ice sheet filling the North Atlantic basin may even have allowed European peoples to work along it (in about 20–16.5 ky) to North America, much as the Innuit hunt the margins of the Arctic ice for seals and fish.[3] Certainly, the southern margins of the great Eurasian ice sheets were inhabited during the LGM by *H. sapiens* groups. Remains from 15 ky have been excavated at Mezhirich and Pushkari in Ukraine where the tundra provided enough plant and animal material for support as a seasonal occupation and the hunters may have been specialised in killing mammoths: up to 650 mammoth bones formed the skeleton of each 'house'.[4] Later in the Pleistocene, as the ice sheets waned, the mammoth was a favoured item of prey in tundra and open-steppe environments in Eurasia and in North America, to the extent that humans are often said to have been implicated in its extinction.[5]

The retreat of the massive ice sheets of the Pleistocene provided a sequence of new habitats that could be colonised by foraging groups of humans as well as by the wild fauna and flora. For none of them was it a stable world: ice-retreat is not necessarily orderly and cumulative, sea and lake levels were in flux, and faunal migration patterns were adjusting to new routes. It seems as if human migration from East Asia via the land bridge to Alaska, possibly during 25–14 ky, was periodically held up by the opening and closing of an ice-free corridor south-south-east from the upper Yukon river to the meeting of Alberta and Montana.[6] Other migrations brought people to Australia by boat certainly by 32 ky, though possibly by 60 ky. The climatic changes right at the end of the Pleistocene included a very dry period in the western United States area at 11,000 BP (before the present) which may have been implicated in the loss of mammal fauna described below.[7] Such was the success of human migrations that most land surfaces of the Earth not covered in year-round ice by the 10,000 BP mark were subject to the presence of *Homo sapiens*, though often this was seasonally transient in any one place. This was a world in transition: almost every feature of an ecosystem was changing, including many areas of low latitudes far from the reach of the Pleistocene ice. Humans failed to navigate to the remoter parts of the South Atlantic and to much of Polynesia (including New Zealand/Aotearoa) in this terminal Pleistocene–early Holocene phase of being outward-bound, for the oceans were the great barrier to movement. Nevertheless, from the margins of the shrinking Arctic ice to the interior of tropical forests, there were human societies, many of whom successfully adapted to rapid changes in climate and sea-level.[8]

This worldwide dispersal has meant that not all the people who have recently been identified as gatherer-hunters have had the same kind of history. There are groups whose whole existence has been a lineal descent from 'ancient' hunters of the early Holocene, but there are also people who have lived in various degrees of contact with non-hunting societies, and those who have been themselves herders or farmers in the past. Inevitably this led to different relations with the natural world. Most studies have concentrated upon the 'ancient-lineage' hunters who, in spite of their diversity of environments, seem to have certain cultural features in common. Their world view, for example, emphasises that, although resources may be owned, the land is held as common property. Nature is animated and mystical; the land is giving and sharing (compare with the modern concept of 'wilderness') and, indeed, there was a very early time when nature and humans were not separated but out of which matrix the various distinctions have crystallised. Yet this original time may persist into the present: in Australia the aboriginal song lines saturate the land with significance.

The energy relationships of gatherer-hunters

A term often used of this phase of human culture is 'foragers', which reasonably describes people who 'collect' food from the wild rather than produce it from domesticated species. A basic lineament of foraging is that people move to where the resources are found and thus the environment in all its manifestations is crucial in their lives. Such movements are made easier if there are few material possessions to be transported and if only one baby has to be carried.

All movement is an energy cost, however, unless food is encountered en route. This points up the core of gatherer-hunters' existence (as, indeed, at root our own) which is to maintain enough energy intake to survive and to reproduce. Most able people in the group, therefore, will contribute to energy gathering: the women, especially, often provide the essential ground-bass of calories in the form of plant materials, while men supply the improvised melody of animal meat and fat. The latter becomes more important towards the polar regions where cold seasons necessitate higher calorie intake, a need best satisfied by the provision of animal fat. There are social contexts: young men may hunt large animals to demonstrate their potential as fathers though, in more general terms, material possession routes to prosperity are disdained. The whole group may have reciprocality with other groups as insurance against famine or water shortage and the internal degree of sharing is likely to be high. Some groups move often so that resources garnered are consumed immediately; others move less (a few not at all) and rely more on storage of food which is especially important in environments with a seasonal shortage.

The spectrum of energy costs is wide though not especially complicated. The energy looked for is mostly 'this year's' solar energy in chemical form (seeds, nuts, fruit, animal tissue), though some previous years' capture such as older animals and the roots of perennial plants will also be sought out. So finding the food source is a first cost, as is subduing it if need be. Recent experience and stored knowledge are vital here and are an instance of a cultural filter where a taboo can be introduced. The gathering of plant foods is likely to give a net energy return of ten to twenty times at the camp gateway but will be quite trustworthy. In spite of its higher energy content, the figure for animals is about the same because reliability is lower, a daily success rate of 10 to 30 per cent being normal.[9] Coastal groups with access to whales as well as fish may net (though not literally) a surplus of 2,000 times. Transport costs are low because most foraging takes place within 5 kilometres of camp, and people may come to the food. The energy costs of processing and preparing food are not usually that high because either the materials are eaten quite quickly or are subjected to a preservation method not involving much human effort, such as sun-drying, smoking or burial. But some yams contain alkaloids which have to be leached out and so washing, slicing and boiling are essential. Provided populations are low, so that resources are not over-used, there can be an energy surplus that leaves time for dancing, gambling and sleep. Some energy surpluses go into periods of rapid population growth though these seem often to have been checked by disease outbreaks.

This situation for early Holocene groups is clearly the outcome of many millennia of cultural learning. Each band and tribe had its own stock of knowledge about how to survive though, in the rapidly changing environments of the early Holocene, the authority of traditional wisdom must have often been severely tested. The people of this era also built upon phases of cultural evolution when there were bursts of innovation that added to their repertoire of foraging skills. The Upper Palaeolithic (45–10 ky) seems to have been the period of invention of specialised stone tools, slings, the bow and arrow, harpoons, fish gorges, snares and pitfall traps, bone tools, and perhaps the boat. This 'explosion' of ability can be summarised in the statistic that the earliest stone tool-makers produced about 50 centimetres of cutting edge from 2 kilograms of raw material whereas, at 25 ky, some 350 centimetres could be made from the same amount of stone. Somewhere in this sequence, or soon afterwards, the blowpipe and the ability to add poisons to darts and arrows were appended.[10] Inhabiting hostile areas like Siberia and crossing short stretches of sea were now possible.

Between 12 and 10 ky comes another significant symbol of human–environment relations: the domestication of the dog (*Canis familiaris*). Many changes in animal habits and plant reproduction must have accompanied all human settlements, but here we have the example of deliberate manipulation of the genetics of a wild genus (in this case from the wolf family) to isolate genetically a species and its varieties.[11] In this case the object of attention was obedient to human command, lived off reject food, could seek out and chase other animals, was better at smell, might provide warmth at night, keep the children amused and, if need be, go on to the spit. Early examples of domestic dog come from a variety of places as far apart as northern England, Arctic Siberia and North America, suggesting that the same process occurred independently in widely separated cultures, and not that there was one instance whose example diffused via any medium of travel or trade. In contrast, DNA evidence has raised the prospect of a single origin of domestic breeds in East Asia at about 15 ky.[12]

'The first great force employed by man'

This was the title of the discussion of fire by the anthropologist Omer Stewart in the seminal volume *Man's Role in Changing the Face of the Earth* in 1956.[13] Much more is now known about fire in gatherer-hunters' times, and it has all confirmed and amplified the central importance of fire in that type of culture and, indeed, in subsequent economies as well.

Fire has a natural ecology. It occurs at the margins of volcanoes but the most widespread source is lightning fire. Lightning strikes which hit a suitable source of fuel may cause conflagrations to take hold and spread: tundra, grasslands, shrublands, all kinds of forest and even seasonally parched wetlands will burn. The spectacular fires of tree crowns in a gusty wind are complemented by a slow underground smouldering in peatlands. Plants and animals can become adapted to repeated patterns of fire (the fire regime). For animals, the adaptation is partly that of run or fly away unless they are predators: birds especially have a feast. In the case of plants, evolution has resulted in species with very thick barks, with fat leaves that protect reproductive organs, with seeds that do not germinate unless fired, or with cones that open only if subjected to a very high temperature: the pyrophytes. There are post-fire opportunists that can spread from abundant seed or fire-proof rhizomes into bare areas: bracken fern (*Pteridium aquilinum*) is one example, as is the aptly named fireweed, or rosebay willowherb, (*Chamerion angustifolium*). Plants that survive a fire may benefit from the reduced competition and also from the mineralisation of organic material: leaves of shrubs may have 300 to 700 per cent more protein than in the pre-burn state. Such nutritive levels, along with salts in the ash, attract many grazing and browsing animals, together with their predators.

Into this extra-human pattern is woven the millennia of human control of fire, from perhaps 800 ky onwards.[14] Our genus learned to preserve and transport fire, and to create it. It was probably first used at the hearth for light, warmth, cooking and to frighten away predators and, as such, doubtless was instrumental in forming social behaviour. Its possession was essential for moving out of the tropics, making possible the occupation of periglacial environments such as those of the Late Pleistocene mammoth-hunters. Even more important is the ability to use fire as a tool at landscape scale. This may be a short-term management device as when a fire produces fleeing and confused animals at its margins and in the aftermath a series of baked lizard dishes from underground 'ovens'. Animals may be driven towards hunters' spears or arrows, over cliffs or up narrow ravines, where the kill rate is much improved, even if rather indiscriminate. Fire is also useful for discouraging insects and for decontaminating a settlement area of, for example, fleas. In a longer-term perspective, regular fires can produce a landscape attractive to game animals, with perhaps high forest supplemented by zones of shrubs where the browse is at the right height for, for example, deer species, and also patches of salty ash are available. Edible pyrophytes are also encouraged. If we imagine such processes being co-evolutionary with the speciation of hominins in Africa, then ecological suites of humans–fire–plants–animals can readily be designated. This is a result of (a) one special property of fire: that it creates its own fuel supply as it progresses, by heating up otherwise non-flammable materials, and (b) one general property of resource exploitation: that once a regime is established then going back to an earlier stage often constitutes exceptional behaviour. Given the first quality, it is less surprising that most of the world's vegetation types can be set alight at one season or another. Given the second, some human groups came to depend upon fire, so knowledge about its management had to be preserved in the culture.[15] The role of fire goes even further. A regularly fired landscape acquires cultural significance because it is first of all claimed for a human group by being burned over: this is an act of possession-taking in the same way as later people might erect fencing. Spatially isolated fires might also mark out territorial bounds: *we* hunt here. Not surprisingly the landscape becomes symbolically valuable and the locus of myths, possibly dealing with the origins of the group or even of humankind itself. The potency of the material force is then interwoven with the authority of human custom.[16]

Even in wet areas, examples of the importance of human-produced fire can be seen. In the later Mesolithic cultures of upland Britain, 8500–5500 BP, evidence for forest recession is quite plentiful: the pollen of forest trees is partially replaced by species of the open ground; there are many deposits with charcoal in them, and woodland is overtaken by bog vegetation. The upper edge of the woodland was the main scene of disturbance. Such an **ecotone** is, of course, always the most susceptible to climatic change, but this zone is also prone to the kinds of disturbance in which fire is implicated. The clearest indications of the processes at work come from central Pennine sites.[17] At 6000 BP, a pollen-influx diagram for Robinson's Moss near Manchester proposes that the upper level of the 'lowland forest' was at 425 metres above sea-level and that of the 'upland forest' at 460 metres. This date, however, coincides with a second temporary retraction of the limit of the upland forest which is associated with evidence of fire. This is seen by Tallis and Switsur[18] as just one of a series of fires in which burning: 'probably prevented the upward spread of the component tree taxa . . . right from the time when upward forest expansion was just commencing in the early Flandrian'. [= early Holocene]

The principal woody species at the tree-line at 6000 BP was probably hazel (*Corylus avellana*). Despite the fires, the tree-line moved up slowly between 6800 BP and 5500 BP, which probably signifies a continuing response to climatic change. Given that exposure, rock type and slope are all additional variables, it is to be expected that the impacts of burning at the upper edge of woodland are highly variable from place to place. Between 8000 and 6000 BP on Dartmoor, hazel is dominant in the woodland community but its values appear to have been negatively influenced by fire after 7500 BP.[19]

In many investigations, estimates of areas involved in disturbance rely simply upon the degree to which certain pollen frequencies alter. The outstanding example of a more reliable data-set comes from Waun-Fignen-Felen on the Black Mountain of South Wales, where analysis of multiple profiles within a small basin has allowed the construction of a series of diagrams reconstructing the vegetation at 8000, 7500, 6500, 5700, 4700 and 3700 BP. The actual edges of the vegetation mosaic are not mapped but, at 8000 BP for example, the mixed woodland shows a burned-over opening which abuts a small shallow lake of about 200 by 200 metres in size which contains an 'early mesolithic' flint-knapping site. This site forms a nucleus for the spread of blanket bog to the north-west but, on other sides, continues to be set in mixed woodland until 4700 BP when mixed woodland appears to form islands in a sea of blanket bog (the lake being now covered with acid peats), a reverse of the position at 6500 BP. Fire is also implicated in the basal layers of the blanket peat and played a role in its inception.

This growth of blanket peat took place in the presence of humans. If inception happened shortly after a major climatic shift to wetter conditions, such as is postulated for the British Isles around 7500–7000 BP, then the role of climatic factors must be suspected, as permissive if not necessarily decisive.[20]

The accumulation of water is enhanced in woodlands by the removal (by any means) of trees. Deciduous trees act (a) as a shelter layer, intercepting precipitation and re-evaporating it from the canopy and trunk, and (b) as water pumps, removing water from the soils via their root systems and transpiration mechanisms. Experiments have shown that run-off increases by as much as 40 per cent after clear felling of deciduous forest. The pathways to peat accumulation are to some extent more complicated than allowed in this brief description but prior to them all is the removal of woodland and/or the presence of fire leading to waterlogging. Once that process has started, then the sequence of biochemical mechanisms will usually lead to peat formation.

The possibility of reversal is shown when wood layers are found in peat stratigraphy. These indicate that the peat was thin enough and dry enough to permit recolonisation by trees: birch (*Betula* spp.) and pine (*Pinus sylvestris*) are common species, for example, at Lady Clough Moor in the south Pennines, while birch is at Bonfield Gill Head on the North York Moors. Generally, recolonisation by trees coincides with lower levels of charcoal in the peat, suggesting that fewer fires as well as (or even rather than) any climatic shifts were a factor in their re-growth. To the populations of the later Mesolithic, though, heather moors accumulating more humus and becoming seasonally waterlogged (with underlying soils undergoing gleying [formation of sticky clays]) and invaded by wet-tolerant sedges and *Sphagnum* must have been a familiar part of their environment. There were also cotton-sedge mires which were wet year-round, *Sphagnum* bogs, open hazel and birch scrub with a variety of wet-tolerant ground flora species and a high proportion of dead trees, as well as wet mires in water-collecting sites, Fire was a factor in a wet climate and damp places. It may as well have been important in drier places because its use in the management of oaks (*Quercus* spp.) for acorns is a distinct possibility.[21]

If upland England and Wales seem an unlikely places for fire management by gatherer-hunters, Australia seems less so. In 1664, Abel Tasman sailed along the west coast and reported fire and smoke everywhere; one late-nineteenth-century European traveller was so struck by the amount of aboriginal burning that he wondered if the people '. . . lived on fire instead of water'.[22] The use of fire management of landscapes by a small population on a wide scale, with high frequency and with considerable effects on vegetation, is probably the world's most obviously fire-created landscape, though with savanna Africa a close second. (Europeans saw it as 'devastated' in Australia, less so in Africa.) Not only was it a particular human-managed ecology but an economy in the sense that rights of usage flowed from its deployment, and emotional ties to the land resulted.

The European incursion into Australia went into a landscape shaped by millennia of fire-use. Since 40 ky at least, humans have ranked with climate as the arbiters of change in the ecology of much of Australia. Reports from early visitors suggest that the firestick was an important accompaniment of travel by aborigines, and that routes across country might be marked by burned swaths. One investigation reported some 5,000 separate bush fires per year in an area of 30 square kilometres. Some areas were not burned so as to maintain a mosaic of vegetation patches, but a great number of plants became more accessible and better yielding to the gatherer economy. Examples include bracken fern (for its rhizome), wild tomatoes, wild banana, wild millet, and especially the cycads of the genera *Cycas* and *Macrozamia* where fire aided all stages of their growth and seed production. Fire cleared away the competing plants, raised the output by about eight times and encouraged the seeds to ripen simultaneously. Yet yam patches were kept free from burning. Firing also aided hunting, for thickets of spinifex could be set alight to flush out lizards, bandicoots and kangaroos; and the fire torch made night fishing possible. On the continental scale, burning has probably enabled the eucalypts to spread into areas climatically suitable for rain forest. In today's evaluations, fire is a maintainer of biodiversity.

The cultural relations of fire are well documented for Australia. For the aborigines, land with vegetation 'litter' on it was dirty – even disgraceful – and needed to be cleared up. Fire made the world habitable because it was the subject of predictive knowledge but it also rendered it understandable via a corpus of legends and myths, including those of the foundational Dreamtime. The spiritual world of the aborigine would collapse without it.

The absence of fire in post-European Australia has created problems, especially in fuel accumulations in the peri-urban bush, and burning has new friends in the conservation movement who see it as the perpetuator of biodiversity because a mosaic of habitats supports a variety of specials, especially mammals, and at least one species of tree (*Callitris intratropica*) is dependent on it in the monsoon areas.[23] So here is an example of the inseparable inter-linkages between culture and environment: without both, there is a large degree of dislocation of both spheres.

A third and final example is found in the moist evergreen forests of tropical Latin America, including Amazonia. The conventional scientific view of such forests has been that that they were not flammable, provided they were intact. Yet more recent research involving archaeological excavations, pollen and phytolith analysis, and flood plain stratigraphy, as well as off-shore coring, has revealed a picture of Late Pleistocene and Holocene vegetation change which involves fire.[24] The palaeoecological evidence allows the inference that not all the fires were 'natural' fires resulting from lightning, and that human occupance of the forest zone preceded the development of agriculture. Clearly, there were climatically drier periods when the frequent lightning strikes would have been a major cause of fire in forests with a susceptible fuel load.[25] In lowland Panama, for example, there was a dry period with a cooling of 5 °C in 16,000–11,000 BP. There is then evidence of human activity at 11,050 BP with charcoal and the pollen of sedges and *Helicona* species which are plants of forest openings and edges. By 8600 BP, the human modification of the forest was apparently systematic, with agriculture entering the region (in the form of maize) at 7000 BP. The alteration of the forest was characterised by repeated burning and the formation of small openings. In the Manaus region of the lowland Amazon basin of Brazil, foraging societies who used caves (such as Pedra Pintada) have left evidence of paintings, pigments, carbonised fruits and wood in those sites, dating from 11,000–8000 BP.[26] The mode of utilisation of lowland tropical forests seems to have centred upon the small clearing in which there was a form of horticulture even before the planting of full domesticates such as maize. This opening also acted as a magnet for mammals such as white-tailed deer, peccary and agouti, so that the term 'garden hunting' is sometimes used. The importance of the garden led some interpreters to suggest that hunting and gathering were never possible unless there was access to cultivation as well, though the very early finds of human presence rather suggest the opposite.[27]

There is no doubt that natural factors such as climatic drying and hurricanes can add to the fuel potential of lowland forests, and some tree species have fire adaptations so, at some time in the past, fires have been a component of the ecology of near-natural forests: in the *terra firme* of Brazil the return period was 390 to 1,540 years under those conditions. That humans increased the fire frequency in the millennia before agriculture now seems certain and there must have been concomitant effects upon the fauna. Anthropological studies in the 1980s have shown, for instance, that in Ecuador 230 inhabitants of three villages killed 3,165 birds, mammals and reptiles in less than one year; in Brazil, 8,850 kilograms of mammals were killed by one community in four months of 1978, of which 70 per cent by weight was one species of peccary (*Tayassu pecari*). Estimates suggest that, when hunting was done with spear and blowpipe, the off-take of animals would have been sustainable. Changing to the shotgun made the kill much more effective, with a yield of 1.6 kilograms per hour of meat as against 0.53 kilograms an hour by the older ways. So, while industrial-era technology has revolutionised hunting (and fishing, too, with headlamps as a nocturnal aid), the foraging period had an impact upon both landscape and fauna though, of course, nothing like the impact of European colonisation and then later national programmes of forest utilisation and conversion.[25] But, given many centuries of occupation by gatherer-hunters followed by horticulture and farming, the possibility exists that the Amazon of today's environmental concern is mostly a product of 300 years of depopulation since European incursions for, by 1650, the population had been reduced by 90 per cent and not many Europeans had penetrated by 1750. In essence, the Amazonia presented to early science was an early succession type of secondary woodland. Alexander von Humboldt's early nineteenth-century warning that not every tropical forest was primeval forest has been borne out.[29]

The close relationships of fire, culture and economy in all these examples allow us to think that the use of fire is as much a mental process as a technical one. It is uniquely human and universally human: at once it widens the gap between humans and other animal species but also links all humans in providing an element of control. Though it is often largely considered in the context of pre-industrial economies, it is not absent now because controlled combustion of fossil fuels is at the heart of today's world even in more remote zones. Further, there seems to be a linkage with climate still in the sense that major forest-fire outbreaks still happen regionally in dry years, which themselves seem to be products of ENSO cycles, with worldwide fires in 1982–83 and 1997–98. Into post-industrial times, the imaginatively named 'firewall' has become a major feature of the electronic age.

Management and impact

Gatherer-hunter populations were clearly involved with the manipulation of their environments. There seems to have been a scale of effects from a very transient presence (as with the tundra dwellers of the ice margins in the Palaeolithic) to an enduring source of makeover, as in Australia. This apparent spectrum needs further examination, not least because gatherer-hunters have been proposed as exterminators of genera of mammals as well as being thought to have been conservationists with a special closeness to the natural world. More detail must also acknowledge that they almost all lived on the land surface but often near lakes, rivers and the sea.[30] Thus, the focus on gatherer-hunters as humanisers of environments is on their capacity to alter the ecology of a region in a quasi-permanent fashion. A general framework for the different types of change imposed by human groups might include direct changes in animal and plant populations due to gathering or hunting. The effects were very variable in time and space but might result in the extirpation of a whole species at local, regional, continental or even global scales, depending on the abundance and distribution of that species. The use of fire pushed alterations in the direction of permanence because human groups came to rely on the spectrum of subsistence opportunities thus offered.

The most-discussed examples of extirpation come from North America and Australia, since Eurasia and Africa seem to have lost fewer genera and species in the wave of Late Pleistocene and early Holocene extinctions that have been documented. In North America soon after 11,000 BP two-thirds of the large (adult weight over 50 kilograms) mammal fauna disappeared: it included three genera of elephants, six genera of armadillos, ant-eaters and sloths, fifteen genera of ungulates, and many carnivores dependent upon those groups. This mass extinction more or less coincides with one accepted set of dates for human colonisation of what is now Canada and the United States via the Bering Strait land bridge and an ice-free corridor just east of the Cordillera. On the other hand, the rapid fluctuations in climate of the terminal Pleistocene provide a possible explanation. Yet earlier swings of a higher amplitude produced no such dramas of 'overkill'. But were there enough people to kill off all those animals? Both views have been extensively canvassed, with the anthropogenic camp bringing in as validation the extinction of the moa bird in New Zealand/Aotearoa, the megafauna of Madagascar and dwarf elephants in Java and Sulawesi because they all occurred soon after initial human colonisation of those islands. For the Americas the unresolved arguments can be reconciled with the idea that the climatic changes introduced considerable tension in the animal populations, which were then more easily wiped out by a new and socially adept preda-tor. On the other hand, the dominant Clovis culture seems to have been one of generalised foragers and certainly not the first humans into the relevant places. Because the controversy is between American scholars, however, there has to be a loser.[31] A version of the same debate has been conducted in Australia where, in general, the climatic determinism hypothesis is strongest, though one review of the data from all continents confirms that the advent of humans in the Late Pleistocene brought about some rapid extinctions and that the surviving fauna was suppressed to densities below those of pre-colonisation times.[32]

Different technologies might lead to divergent ecologies. The adoption of trapping technologies, as distinct from trying to shoot them with arrows or darts, has a distinct impact on many mammal populations. Skilled trapping reduces the energy expenditure of the families involved, compared with a hunt that might involve (especially before the adoption of firearms) chasing wounded beasts. The boreal forest Athapaskans (much studied in recent times) rely very heavily upon trapping: it is not only an efficient way of procuring many game species but has now become so tightly woven into their culture that men will do it even when there is no real need. In the tropical forests of Zaire, the Mbuti used nets to cull antelopes, with enough success eventually to lower the kill rate substantially. An indirect effect would result from any better technology which allowed greater success in animal hunting that then led to a higher fat intake and thence to lower the age of menarche and allow increased rates of human population growth, a process inferred for the eastern Canadian arctic in the Holocene and the European Middle–Upper Palaeolithic transition.[33] A similar impact type might produce a different ecology if the target populations had different behaviours. If all the pregnant females gather together and that band is then killed in its entirety for more than a few years in succession, then a decline in its abundance is probable unless there is colonisation from adjoining populations. By contrast, if adult males are the target for hunters who kill for status rather than for food, then, in many mammal species, the males on the margins of sexual success may move in and replace those taken out. Communal hunts seem to result in higher kill rates which may also have its resonances today. Such apparently profligate hunting has been documented unambiguously for the Great Plains of the United States. Many excavations of buffalo kills reveal that herds were stampeded (often with fire as an aid) into narrow canyons with a hide-walled pound at the end, or into mires within sand-dune complexes, or simply driven over cliffs. In one arroyo about 190 animals were killed in late June, *c.* 8200 BC, from a herd of about 200 to 300 beasts, of which 37 per cent were immature and 6 per cent juvenile. There were no foetuses, which allows the inference (a) that gravid females were spared the drive or (b) that the embryos were taken away to eat as a delicacy.[34]

There must be space, too, for different cultural attitudes. Clearly, the most important of these is the presence or absence of a 'conservationist' approach to animal numbers in which there is a voluntary restriction on off-take, whereas other groups regarded high levels of killing as essential to ensure the perpetuation of the target species: the more deaths, the more animal souls to be reincarnated. Pressure upon resources (in which rising population numbers might be implicated), however, is the most likely explanation for the instances of excavated human skeletons with signs of violent injuries and deaths. Examples have been found in Egypt (14,000–12,000 BP) and Sweden (7500 BP); rock paintings in Australia from the mid-Holocene have been interpreted to signify increasing levels of inter-group conflict.[35]

One outstanding example of the integration of a world of empirical knowledge and of spiritual guardianship is the Koyukon village in the boreal forest of Alaska studied by Richard Nelson. Human behaviour towards natural entities is based on spiritually based rules (including the treatment of usable and unusable parts and the avoidance of waste) even when the white man's technology is adopted. The result seems to be a model of what an ecologist would call sustained-yield practices.[36] By contrast, many other cultures were able to exterminate animals on a large scale and that, sometimes after contact with external traders, they did so, with the classic example being the penetration of the Hudson's Bay Company into what is now Canada. They traded in the valuable beaver pelts and contributed to a condition in which the beaver harvest in North America in the late nineteenth century was only about 10 per cent of its level a hundred years earlier because the animal itself had become scarce or locally extinct. One interesting question is the level of its population in, say, AD 1700, when it was abundant in the interior of the continent but less so in, for example, New England. Was this mostly due to conscious 'conservation' of beavers by the inland native people, using a variety of cultural mechanisms such as rotational trapping and moiety totemism, or was it rather a lesser ecological impact from a low population density, an inefficient extraction technology, and the lack of external trade contacts that would have provided a market for pelts?[37] In general, wetlands are manipulated less than dry ground, for obvious reasons. But if they dry out seasonally, they can be burned: early Mesolithic (*c.*9000 BP) people did so to lake-edge reed fringes in northern England, just as Indians behaved in prairie Canada within historic times. In central Australia, channels were dug in swamps to encourage the expansion of eel populations. In Lapland, the Saami (who were hunters of reindeer until they took up herding in the nineteenth century) avoided the extensive use of fire in case it consumed the precious pine woodlands. In some cultures, the great range of species available provided a steady flow of food without over-using any one species.[38] In all, we may suspect that low population densities made it much easier to live by foraging, and that only in specific circumstances were populations of animals and plants made regionally extinct.

One impression from many studies is that gatherer-hunters maintained a closely woven set of relations with the non-human world and that this was often, but not always, manifested as a form of respect. The foundation seems to be a metaphor based on kin relations in which unconditional giving is central, and often the environment is seen as a parent who never withholds. The relationship seems to have been fragile, however, so that the intrusion of agricultural economies, whether as, for example, farmers or as purchasers of furs, tore its fabric easily.[39] We are unlikely now ever to get enough data to understand why this was so, even if, indeed, there was one all-encompassing reason for it.

The diminution of foraging societies

There have been two major phases in which gatherer-hunters have disappeared. The first is quite obvious: the introduction of agriculture into Africa and Eurasia in the early and mid-Holocene took over many such economies. Many foragers became agriculturalists in the great millennia of agricultural expansion from core regions.[40] Near-recent evidence suggests that trading goods often precedes invasion, so that the new economy may have conquered by economic rather than military means. In some instances there will have been a fertile seedbed (as it were) because not all gatherer-hunters had a continuous history of foraging: many moved into and out of herding and farming.[41] Some African hunters, for example, were once herders who had lost their cattle, and the apparent loss in Newfoundland of the Beothuk Indians and their ancestral groups may owe much to their adaptation to other ways of life rather than to extinction.[42]

The second phase is the colonial era from about 1800 to 1945 when the spread of industrial economies was accompanied by genocide together with the loss of territory and autonomy. Perhaps 50 million tribal people were killed by colonial settlers and capitalists, and many of them were gatherer-hunters. They were also subject to introduced diseases, such as those that took off 75 per cent of the Yokut and Wintun people of California in 1830–33 or that contributed to the reduction in native Tasmanians from 5,000 to 111 in thirty years, with final extinction in 1876.[43] In an opposite case, the control of malaria after 1947 in Boksu and Tharu at the foot of the Indian Himalayas meant that agriculturalists from the Punjab could colonise the region.[44] Even when the European attitudes to hunting and gathering peoples (often lumped together with other tribal groups as 'savages') had become more protective, that way of life still diminished, not least because the thought-patterns of the nation-state demanded assimilation, via school and mission.[45] So what was once the only way of life for humans is now confined to a few enclaves in remote and marginal environments or is a barely recognisable form of it. Nevertheless, such peoples are still contributors to the making of environmental histories: the San bushmen of Botswana have been in conflict with the national government which wants them to move into new settlements and leave a Kalahari game reserve, a change bound to have ecological effects, not least if diamond mining is a successor land use.[46]

PROPER RESPECT: HUNTER-GATHERERS IN A COHESIVE WORLD

Nelson wrote of the Koyukon Indians that they lived in a world of eyes in which the surroundings were aware, sensate, personified. They could feel and be offended so that they must 'at every moment, be treated with respect'.[47] If, therefore, the ecological relationships of this world are to be explored, then the social characteristics of any group must be examined with regard to their ecological consequences: hunting and gathering is as much a system of thinking as a system of production. For most gatherer-hunters this separation of humans and nature does not exist.[48] At its most generalised, the cosmos can be said to be 'an organism at once real, living and sacred; it simultaneously reveals the modalities of being and of sacrality'.[49]

We find that the cultural and natural worlds are distinct but that they interpenetrate constantly; each depends upon the other and they must maintain a harmony. Any differentiation of the two worlds can be bridged by supernatural means and the proper ritual. Thus, the beings of the wild function as validators of what really matters and so the social and personal construction of self and self-efficacy is negotiated not only with humans but with the entire cosmos. There is, indeed, a cosmic economy of sharing which is managed locally by political regulation supported by an appeal to individual morality. Control of resources is vested in the whole group and administered by chiefs: any power that is exercised has to be validated by a tremendous knowledge of the entire ecosystem. The distribution of power is based on trust rather than on domination by rank and this trust extends to the whole cosmos, including the food sources which, for example, may be conceived as humans in temporary disguise that have chosen to don animal skins in order to offer themselves to their human brethren. Similarly, the hunters' weapons were means of knowledge of nature rather than tools of control.[50] The ecological outworkings, seen from the rationalist perspective of many scholars, often have what are viewed as conservationist principles: those of respect, of democratic decision-making, and of the avoidance of waste. But not always: some groups do not practise the conservationist tenets all the time and, in contact with other economies, have been ready to abandon their former ways of thought and action. If we are tempted to think of gatherer-hunters as maintainers of equilibria or 'the balance of nature' then we need to recall (a) that many lived in early Holocene environments subject to climatic and ecological change, and

(b) many of them moved in and out of agriculture and herding once these were in contact with the hunters, and very many stayed in those new economies. One attraction of newer ways was perhaps a greater sense of control since, in full gatherer-hunter mode, the powers that animated the universe were in the end to be served by humans and not vice versa.

Nevertheless, there are some empirical conclusions that can be drawn: hunters apparently need a knowledge of fewer species than agriculturalists and so their taxonomies are different: they recognise fewer taxa at the species level and none at varietal rank, whereas agricultural societies use both. Any explanation, however, is likely to be as fully cultural as instrumental. Pre-agricultural technologies often permitted long-distance travel, such as the southward occupation of North America, or perhaps continued migration into the British Isles after the drowning of Doggerland, but no examples of long-distance species transfer are known, apart from the dog.

BUYING THE LAND: FRAGMENTATION IN THE FORAGERS' WORLDS

The most famous lament of the breaking up of the gatherer-hunters' world is the speech supposed to have been given by Chief Seattle in January of 1854. 'How can you buy or sell the sky, the warmth of the land? The idea is strange to us.' It certainly conveys the notion of an ideological as well as a physical tearing although it is a fake which, in this form, dates from a film script of the 1970s.[51]

In the social arena, there is the image of the foragers as members of a co-operative society with a felt flow of identification with the cosmos. There were many commonalities of outlook and practice across the world as well as divergencies, such as the roles of men and women. Often regarded as strongly dichotomous, there are examples of men who gather and women who hunt. Likewise, the quantity of possessions is very variable. It may well rise when there is a long sedentary period during the year, as on the north-west coast of North America, though this is the heartland of the potlatch celebration where wealth is denoted by how much is given away in a party mood.

Ecologically, a fragmentation process seems to have occurred in the late Pleistocene and early Holocene with the extinction of many genera of megafauna, to which the label 'Pleistocene overkill' has been attached. This seems to have coincided with the first appearance of humans, especially in North America and possibly also in Australia. Overkill also occurs later in islands of the Pacific and Indian Oceans. They may have lacked megafauna (except for the birds of New Zealand), but their ground-nesting avifauna was especially vulnerable. This happened too when the first humans were agriculturalists. Eurasia and Africa seem not have suffered on the same scale though the mammoth's demise is common to both. At later times there seem to be the indiscriminate killings of the buffalo-jumps and other examples of the massacres of large numbers of animals without regard for their future numbers. If the animal population is large and the human population low or at least mobile, then there may not be any discernable long-term effects.

Though reciprocality is still important, examples can be seen of stricter land allocation. It has been argued both for near-recent Australia and Mesolithic Scandinavia that the presence of cemeteries is an indication not only of a sedentary phase in the yearly cycle but also of a land-claiming, in the way that was not true of leaving bodies to the elements. It is easy to see fire in this way, too, with the smoke signalling that this land is primarily 'our' territory. The domestication of the dog can be regarded in both ways: it is a coalescence of human social qualities plus the sensory capabilities of a wild animal but is also the detachment of wild creatures and their incorporation into the human realm.[52] The taming of wolves may have been a bigger jump than is generally thought since many groups had a sense of 'rightness' for wild organisms: plants may 'belong' in certain places and lose their virtue (in, for example, medicine) if moved.

REPRESENTING HUNTERS AND GATHERERS

In the spirit of asking 'what do we think we know?' there are two aspects to ideas of presenting information about these people and their worlds. Firstly, we need to think about how they represented themselves in various ways which we can now find intelligible, and secondly how we have formulated our ideas about them in ways likely to colour any accounts that are constructed.

Literacy came to gatherer-hunters only with contact with European and North American cultures. Although there are transcriptions of what they said about themselves and some pre-twentieth-century but post-contact writing, the main conveyance of any reflexive thought and accounts of their groups' history has been oral. This means that most were never set down in any form and also that any interpretation of any surviving tales and myths is doubly difficult: not only is there a cultural barrier to twenty-first-century 'western' readers but, as we all know, oral accounts of, for example, family history for a couple of generations tend to be subject to differences in emphasis: which great-great-uncle was *really* the black sheep?

A more widespread way for gatherer-hunters to express themselves has been in visual terms. The cave paintings of the south of France and northern Spain are well known and new examples are continually coming to light. Other forms of art, such as carved ivory figurines, are also known from the European Palaeolithic. Across much of Africa and also in Australia, there is 'rock art' in which depictions of life are made on rock surfaces in a variety of mediums: some in caves but many in the open air. Given the actions of weathering, some commentators have suggested that this was a virtually universal practice. Paintings are also found on bark, wood, bones and bodies. Carving and object-making may also have had a role which is apparently non-instrumental. Two observations need to preface any accounts of art: firstly that when the Altamira caves in Spain were first discovered in modern times, the paintings were assumed to be fakes: gatherer-hunters were too primitive to have produced such work. Though such attitudes are less common, a kind of lowest-level-possible attribution of 'primitive-religion-and-hunting-magic' purposes has suffused much writing about these artefacts. Secondly, the category of 'art' as we now see it may not be applicable directly to gatherer-hunter cultures. In a world that does not differentiate between humans and the rest of the cosmos, an apparent piece of 'art' may have had a functional role: what is crucial is that it had a meaning. In recent times, that meaning may extend to being a form of political action: we are here and we need you to take notice.[53] But in another commonality, much of the 'art' does depict environmental features, notably animals, sometimes with, and sometimes without, humans.[54] Even here, there are at least two ways of interpreting the marks that have been made: it can be seen literally in the sense that nobody can read art from the past and so a search for meaning is useless and the phenomenon has to be enjoyed in existentialist terms. But given a little ethnohistorical or ethnographic information, then the desire to understand these complex interweavings of the real and the non-real is strong. The range of depictions is very wide: astronomy, mixed human–animal creatures, economically important animals and their opposites, and possible states of altered human consciousness are all possible. A single motif may have multiple meanings, as we discover with today's equivalents. Perhaps we need to avoid the term 'art' and, thinking of 'painting' or 'engraving', allow for the disappearance of much of the outdoor material and, with Bahn, think that 'at least that's what it looks like to me, but what do I know?'[55]

The generous variety of interpretative possibilities for the art of the far past is exhibited in the many studies of the rock art of the caves of southern France and northern Spain. The majority of them were created in the Magdalenian phase of the Upper Palaeolithic, c.17–12 ky but examples of equally well-produced images of animals from the Chauvet caves of the Ardèche region east of the Massif Centrale have been dated to 32,000–29,000 BP, which attracts the label Aurignacian, a period previously known mostly for its carvings.[56] The meanings attributed to European Palaeolithic art are too diverse to be summarised here except to say that, quantitatively, most of the source material relates to the world of wild animals, although abstract patterns and humans are also present. The sense of connection with the world around the makers has been enhanced by the discovery that at c.22,000–21,000 ky in the Magdalenian and Solutrean, star patterns, such as the Milky Way and the Pleiades, have been painted in caves.[57] These, in one view, 'helped to organise the spatiotemporal structure of daily and spiritual life of Palaeolithic man'. Here, we can emphasise that the majority of this type of art (whatever that means) places humans within a frame of the rest of the universe; it is notable perhaps that, today, we organise our relationship to time with ever greater precision and frequency (I can see four clocks and three calendars from where I am now sitting) but images of nature are far less pervasive for most western people.[58]

The art of near-recent and extant gatherer-hunters is, of course, influenced by the world outside the producer groups. Some of it, indeed, is produced for a very affluent market as the price of Innuit soapstone carvings in the galleries of Toronto or Montreal will confirm. Yet its expression of environmental relationships is not necessarily invalidated by its commercial position: perhaps some buyers are attracted by the sense of connection that it gives them. Pictorial art is better at expressing relationships than sculpture, however, and the Innuit painter, sculptor and craftswoman, Kenojuak (1927–) from Cape Dorset on Baffin Island, who flourished in the 1970s, made many pictures of humans and animals in which birds have a central role, though bears and seals make appearances as well. Time after time, the sun, animals, occasionally plants and often humans are linked either by theriomorphy (animal–human hybrids) or, as it were, hold hands in the image. Once again, there is a non-separation of humans and the cosmos, even though there is in some of her work an endearing but not detracting sense of the domestic as well.[59] Art in the Canadian north is one of the ways in which the native peoples can project their world view as being distinct from the dominant industrial culture to the south.[60]

This leaves us to consider our own representation of foraging cultures. It says much that, when the Altamira cave paintings were discovered in Spain, they were considered to be a hoax. This illuminates the fact that hunter-gatherers have been subject to extremes of cultural appraisal by outsiders, including scholars. In the early twentieth century, for instance, they were frequently regarded as savages of a miserable and primitive kind (by Sigmund Freud, among many) whereas in the 1960s and 1970s there was rehabilitation on the grounds that they had shown the possibilities of human life without the inequalities of state and class societies, and ones whose attunement to the carrying capacity of their environments had lessons for the late twentieth century. There have also been national attitudes based on, most obviously, colonial relationships and in societies with a dominant and explicit ideology like the former USSR. Beyond these empirical matters, the ways in which history is understood and, most importantly, written are bound to affect the understanding and the representation of the peoples of hunter-gatherer economies, both recent and in the early Holocene.[61]

OUTCOMES

Reminders: gatherer-hunter people represent at least 90 per cent of human evolutionary history; they never had any significant impact on the oceans; in AD 1500 they still occupied large areas of Australia, North America and the southern cone of South America, along with regional patches of southern

Africa, Siberia and New Guinea. Elsewhere there were remnant groups in deep forests, those who had gone back to foraging after a period of agriculture, and some groups on coastal fringes like the Ainu of Hokkaido. Yet groups like the Saami and the San were foragers from the Upper Palaeolithic to the nineteenth century. Now, they are present in only a few places and are either being assimilated into industrial economies or trying to distance themselves from pervasive westernisation.

Hunter-gatherers in their ecosystems

If we try to see people of foraging cultures set within ecosystems which are in effect hybrid systems of culture and nature, then a number of salient characteristics can be identified. The first (and most important in terms of comparison with later types of economies) is the question of energy flow and the ways in which this can vary for gatherer-hunters.[62] Climatic change in the form of increases or decreases of solar energy are clearly important and were probably critical to the survival of some groups at the end of the Pleistocene. Viewed scientifically, the use of fire in the Late Pleistocene and early Holocene affected many ecosystem characteristics: species composition, rates and directions of plant succession, permanently altered communities of plants and animals, changes in soil properties, erosion rates and sedimentation rates in lakes and estuaries, are all examples. Fire influenced biological productivity by favouring plants of early successional phases, whose growth rates are likely to be rapid. The use of fire at landscape scale seems to have been very widespread so it appears culturally in many forms, especially in legend and in myth, a strand of culture which does not die out with the diminution in the number of hunters and gatherers on the Earth.

The second hybrid system is that of plant and animal populations within the yearly orbit of a group of people. This is difficult to define because any sharp boundaries of territories might be relaxed in times of need. There seem to be many instances, however, of the management of animal and plant populations which include a determination to provide for the future by not overculling. Here is perhaps the biggest identifiable cultural thread in the entire pattern, for many groups recognise this undertaking in their ethics. Beyond the 'conservationists', however, there are people for whom environmental restraint seems foreign. Some may well have disappeared on that account, in ways not traceable in the archaeological record; others may simply have been too few in number to diminish a common animal significantly.[63] The North American bison, for example, survived the transition from being hunted on foot with the bow and spear (often with the aid of fire) to being hunted from horseback with rifles; it succumbed (very nearly totally) only to the incursions of an industrial economy whose use of the animals' tissues was probably less important than the symbolic satisfactions of killing a creature so closely identified with the native population. Moves towards domestication have been inferred from sedentary groups because, in some places, people had no need for an annual cycle of movement. The Indians of the Pacific north-west (*c.*6000 BC) and the Natufian culture of Israel and Syria (from 9600 BC) are examples. The overall degree of modification of the Earth's surface through time and space is rather patchy: there are many regional differences. If fire was not used, then the timespan of alteration of an ecosystem was short: in warm, wet lowlands it could have been measured in months. Elsewhere, a yearly cycle would probably find a band of people returning to a familiar site and finding biotic evidence of their last visit both in the settlement area and the local landscape. Further, we are reliant on patchy evidence from post-contact Europeans and Americans, not all of whom had the professed objectivity of the anthropologist, together with the uneven yield of even the most meticulous archaeologist who has actually written up the results of the dig: provisional views must be accepted.

The details of thought and practice must have been highly variable. Yet one feature stands out. Even if only one species was truly domesticated, then many environments were truly tamed to some extent. Hence, much of the information obtained from gatherer-hunters at 'contact time' was not from the 'wilderness', and these were not 'empty lands' but lands which were altered by the humans living in them, and of which they felt they were a part. Agriculture does not necessarily represent the beginning of either the apparent taming of the land or of species, though it differs in many respects from its precursor economies. Yet, of course, some 70 per cent of the planet's surface was altered practically not at all by the gatherer-hunters. Whether the burning practices of this era were carbon-neutral in the atmosphere is a fascinating question but unlikely to be determined because the early Holocene was a period of rapid change in climate and land cover in which the carbon fluxes will have swamped any human-produced effects.

Foundations of the foragers' environmental history

One underpinning of the gatherer-hunters' tenure of the Earth was a low population density. An often-quoted figure for the average is 0.025 persons per square kilometre in the Lower Palaeolithic and 0.115 per square kilometre just before the spread of agriculture: a typical hunter-gatherer needed 26 square kilometres of land for subsistence. The estimated absolute number of people on the eve of agriculture was between 0.8 and 9.0 million. Though not axiomatic, a low population density is a likely ingredient of a light footprint. These kinds of data have led to the notion that foragers lived below the carrying capacities of their environments, with levels at 20 to 60 per cent of the maximum economic yield being quoted. The mechanisms postulated have been those of conscious population control (infanticide and abortion especially), the consequences of prolonged lactation, the effects of environmental variability, and the incidence of disease, with zoonoses taking the lead. Small populations can oscillate widely, going from x4 to −4 maxima and minima in about 200 years, and few sources of evidence can document beyond dispute the precise course of such fluctuations.[64] Viewed globally, foragers lived in separate worlds, with only limited contact at the edges of particular cultural groups.

In the intellectual climate of the 1960s and 1970s, the existence of groups with 'affluence without abundance' as Marshall Sahlins so famously put it, was often taken as an example for industrial societies.[65] Such societies failed to take on board the need to keep moving (which would not have been easy for a world population of around 5,000 million) and the desire not to utilise resources to their full extent. Three or four decades later, we can observe that the hunter-gatherers' era was not necessarily one of taking the usufruct without altering the natural world at all. For a start, there was no reason *not* to alter the world because none of it was separate from the human members of it. Though stability of the hybrid ecosystems was subject to regional variation, it seems likely that many of the foraging societies of the Late Pleistocene and early Holocene would have qualified for the label of 'sustainable', using today's criteria: wherever they went, there were still 'the stag, the raven and the long-lived tree'.

NOTES

1. J. Donne, 'An Anatomy of the World', lines 114–16 in *The Complete English Poems*, ed. A. J. Smith, Harmondsworth: Penguin Books, 1971. It was first published in 1611. The typical lifespans of deer and raven are perhaps fifteen years but thirty years have been recorded in both cases, which brings them to the same kind of level as prehistoric gatherer-hunters.
2. General works include R. B. Lee and R. Daly (eds) *The Cambridge Encyclopedia of Hunters and Gatherers*, Cambridge: Cambridge University Press, 1999; C. Panter-Brick et al., (eds) *Hunter-Gatherers: an Interdisciplinary Perspective*, Cambridge: Cambridge University Press, 2001. Sometimes I have used 'gatherer-hunters' as a reminder that gathering of plant materials is often more important for survival than hunting animals, especially large ones.
3. The European Palaeolithic culture which is postulated to have travelled from, for example, the Bay of Biscay to Newfoundland is called the Solutrean. The debate over this possibility is vigorous. Evidence now in play includes palaeo-indian skull shapes, dentition, DNA and the dialects of Algonquin Cree and Euskera (modern Basque). The then current evidence and the restatement of the conservative position that the Americas were populated exclusively from Asia is summarised by L. G. Straus, 'Solutrean settlement of North America? A review of reality', *American Antiquity* **65**, 2000, 220–1.
4. O. Soffer and C. Gamble (eds) *The World in 18,000 BP*. Vol. One: *High Latitudes*, London: Unwin Hyman.
5. There is a number of relevant chapters in O. Soffer and N. D. Praslov (eds) *From Kostenki to Clovis: Upper Palaeolithic–Palaeo-Indian Adaptations*, New York: Plenum Press, 1993.
6. The natural history, with some human history, is excellently interpreted in E. C. Pielou, *After the Ice Age. The Return of Life to Glaciated North America*, Chicago and London: University of Chicago Press, 1991. A more strongly archaeological perspective is in S. Mithen, *After the Ice. A Global Human History 20,000–5000 BC*,

London: Weidenfeld & Nicolson, 2003, chs 23–32. The immigration(s) via an ice-free corridor constitute the conservative position mentioned in Note 3, though modified by the presence of ice-free refugia down the west coast of North America. There are complications involving very 'early' dates for settlements in California, Pennsylvania, Chile and Brazil, as well as the unavailability of some skeletal material for study since a Federal Act of 1990 in the United States requires the permission of local native Americans to remove bones if there is any cultural affiliation. See G. Haynes, *The Early Settlement of North America*, Cambridge: Cambridge University Press, 2002.

7. C. V. Haynes, 'Clovis–Folsom geochronology and climatic change', in Soffer and Praslov op. cit. 1993, 219–36.

8. The tempo in northern Europe is explored by G. R. Coope et al., 'Temperature gradients in northern Europe during the last glacial–Holocene transition (14–9 ^{14}C kyr BP) interpreted from coleopteran assemblages', *Journal of Quaternary Science* **13**, 1998, 419–33.

9. Some commentators have made the case for scavenging of dead corpses being more important than hunting, at any rate in the Lower and Middle Palaeolithic; they argue that the energy balance would be improved by eating well-dead animals and that our far ancestors were uninterested in use-by dates. The fights with bears, wolves, vultures, marabou storks and hyenas might have used up some energy, however.

10. This archaeology is dealt with in the context of contemporary gatherer-hunters in my *Changing the Face of the Earth*, Oxford: Blackwell, 2nd edn 1996, ch. 3.

11. 'Each of these [400 at the present time] breeds owes its existence to artificial selection by man, because every dog, whether it is a Great Dane or a Chihuahua, is the descendant of wolves that were tamed by human hunters in the prehistoric period': J. Clutton-Brock, *A Natural History of Domesticated Animals*, Cambridge and London: Cambridge University Press and the British Museum (Natural History), 1988, p. 34. Hard to believe as you watch a major dog show on television, though less so in the 2002 film, *Best in Show*. See also D. Brewer, T. Clark and A. Phillips, *Dogs in Antiquity. Anubis to Cerebus: the Origins of the Domestic Dog*, Warminster: Aris & Phillips, 2001.

12. P. Savolainen, Y. P. Zhang, J. Luo, J. Lundeberg and T. Leitner, 'Genetic evidence for an East Asian origin of domestic dogs', *Science* **298**, 2002, 201–2.

13. O. C. Stewart, 'Fire as the first great force employed by man', in W. L. Thomas (ed.) *Man's Role in Changing the Face of the Earth*, Chicago: Chicago University Press, 1956, 115–33.

14. The dates and places are constantly changing as the techniques for the recovery of evidence improve. The 800 ky date comes from N. Goren-Inbar, N. Alperson, M. E. Kislev, O. Simchoni, Y. Melamed, A. Ben-Nun and E. Werker, 'Evidence of hominin control of fire at Gesher Benot Ya'aqov, Israel', *Science* **304**, 2004, 725–7.

15. J. Goudsblom, 'People, fire and environment', in J. E. van Hinte (ed.) *One Million Years of Anthropogenic Global Environmental Change*, Amsterdam: Koninklijke Nederlandse Akademie van Wetenschappen, 1997, 17–27.

16. J. Goudsblom, *Fire and Civilization*, London: Penguin Books 1994.

17. J. H. Tallis, 'Forest and moorland in the South Pennine uplands in the mid-Flandrian period. III. The spread of moorland – local, regional and national', *Journal of Ecology* **79**, 1991, 401–15.

18. J. H. Tallis and V. R. Switsur, 'Forest and moorland in the South Pennine uplands in the mid-Flandrian period. II. The hillslope forests', *Journal of Ecology* **78**, 1990, 857–83.

19. C. J. Caseldine and D. J. Maguire, 'Late glacial/early Flandrian vegetation change on northern Dartmoor, south-west England', *Journal of Biogeography* **13**, 1986, 255–64.

20. P. D. Moore, 'The origin of blanket mires, revisited', in F. M. Chambers (ed.) *Climate Change and Human Impact on the Landscape*, London: Chapman & Hall, 1993, 217–36; A. U. Mallik, C. H. Gimingham and A. A. Rahman, 'Ecological effects of heather burning. I. Water infiltration, moisture retention and porosity of surface soil', *Journal of Ecology* **72**, 1984, 767–76.

21. S. L. R. Mason, 'Fire and Mesolithic subsistence – managing oaks for acorns in northwest Europe?', *Palaeogeography, Palaeoclimatology, Palaeoecology* **164**, 2000, 139–50.

22 Quoted by S. Pyne, *Burning Bush. A Fire History of Australia*, New York: Holt, 1991, p. 85.

23. D. M. J. S. Bowman and W. J. Panton, 'Decline of *Callitris intratropica* . . . in the Northern Territories: implications for pre- and post-European colonization fire regimes', *Journal of Biogeography* **20**, 1993, 373–81; D. M. J. S. Bowman, 'The impact of Aboriginal landscape burning on the Australian biota', *New Phytologist* **140**, 1998, 385–410; D. M. Yibaruk et al., 'Fire ecology and Aboriginal land management in central Arnhem land, northern Australia: a tradition of ecosystem management', *Journal of Biogeography* **28**, 2001, 325–43; R. A. Bradstock, J. E. Williams and A. M. Gill (eds) *Flammable Australia: the Fire Regimes and Biodiversity of a Continent*, Cambridge: Cambridge University Press, 2002; B. Gott, 'Aboriginal fire management in south-eastern Australia: aims and frequency', *Journal of Biogeography* **32**, 2005, 1203–8.

24. D. R. Piperno, 'Phytolith and charcoal records from deep lake cores in the American tropics', in D. M. Pearsall and D. R. Piperno (eds) *Current Research in Phytolith Analysis: Applications in Archaeology and Paleoecology*, MASCA Research Papers in *Science and Archaeology* vol. **10**, 1993, 58–71.

25. J. B. Kauffman and C. Uhl, 'Interactions of anthropogenic activities, fire, and rain forests in the Amazon basin', in J. G. Goldammer (ed.) *Fire in the Tropical Biota*, Berlin: Springer-Verlag Ecological Studies vol. 84, 1990, 117–34.

26. A. C. Roosevelt et al., 'Paleoindian cave dwellers in the Amazon: the peopling of the Americas', *Science* **272** (#5260), 1996, 373–84. The main thrust of the paper is towards the dating of human occupance of South America rather than details of the environmental impact of gatherer-hunters, but there are a lot of data in it.

27. Summarised in R. C. Bailey and T. N. Headland, 'The tropical rain forest – is it a productive environment for human foragers?' *Human Ecology* **19**, 1991, 261–85. Much of *Human Ecology* **19** (2), 1991 is devoted to this topic.

28. B. Winterhalder and F. Lu, 'A forager–resource population ecology model and implications for indigenous conservation', *Conservation Biology* **11**, 1997, 1354–64.

29. Quoted by P. W. Stahl, 'Holocene biodiversity: an archaeological perspective from the Americas', *Annual Review of Anthropology* **25**, 1996, 105–26. Population estimate from W. M. Denevan, 'The pristine myth: the landscape of the Americas in 1492', *Annals of the Association of American Geographers* **82**, 1992, 369–85. See also J. S. Athens and J. V. Ward, 'The late Quaternary of the western Amazon: climate, vegetation and humans', *Antiquity* **73**, 1999, 287–302.

30. Some of the data are from archaeology, some from early European accounts of contact and some from ethnographers working in near-recent times. All need care in order not to see only what it is desired to see. The exceptions to living on the land surface are a few fishing groups whose dwellings were on stilts over the sea or fresh waters.

31. P. S. Martin and R. G. Klein (eds) *Quaternary Extinctions: a Prehistoric Revolution*, Tucson: University of Arizona Press, 1984. Three examples of the many papers are M. W. Beck, 'On discerning the cause of late Pleistocene megafaunal extinctions, *Palaeobiology* 22, 1996, 91–103; J. Alroy, 'A multispecies overkill simulation of the end-Pleistocene megafuanal mass extinction', *Science* 292 (5523), 2001, 1893–6 and A. D. Barnosky et al., 'Assessing the causes of Late Pleistocene extinctions on the continents', *Science* 306, 2004, 70–5. For Australia see among many, D. Choquenot, and D. M. J. S. Bowman 'Marsupial megafauna, Aborigines and the overkill hypothesis: application of predator–prey models to the question of Pleistocene extinction in Australia', *Global Ecology and Biogeography* 7, 1998, 167–80; R. G. Roberts et al., 'New ages for the last Australian megafauna: continent-wide extinction about 46,000 years ago', *Science* 292 (5523), 2001, 1888–92. Also antipodean: A. Anderson, *Prodigious Birds. Moas and Moa-hunting in Prehistoric New Zealand*, Cambridge, Cambridge University Press, 1989. Overlap with a similar phenomenon in a horticultural group is in D. W. Steadman et al., 'Rapid prehistoric extinction of iguanas and birds in Polynesia', *Proceedings of the National Academy of Sciences* [of the USA] 99, 2002, 3673–7.

32. B. W. Brook and D. M. J. S. Bowman, 'The uncertain blitzkrieg of Pleistocene megafauna', *Journal of Biogeography* 31, 2004, 517–23.

33. S. Cachel, 'Dietary shifts and the European Upper Palaeolithic transition', *Current Anthropology* 38, 1997, 579–603.

34. G. C. Frison, *Prehistoric Hunters of the High Plains*, New York: Academic Press, 1978, 2nd edn, 1981. The volume edited by L. B. Davies and B. O. K. Reeves, *Hunters of the Recent Past*, London: Unwin Hyman, 1990, (One World Archaeology vol. 15) is mostly devoted to the archaeology of the High Plains.

35. There is a summary in W. J. Burroughs, *Climate Change in Prehistory*, Cambridge: Cambridge University Press, 2005, pp. 272–3 which emphasises connections to external forces such as climate and sea-level changes. There are examples from the North American Huron, where post-contact evidence shows that absence of skeletal evidence in burial grounds does not necessarily mean that warfare did not take place, for the fatally wounded may not make it back home.

36. R. K. Nelson, *Make Prayers to the Raven. A Koyukon View of the Northern Forest*, Chicago: University of Chicago Press, 1983. A totally splendid book.

37 C. Martin, *Keepers of the Game*. Berkeley, Los Angeles and London, University of California Press, 1978; and S. Krech, *The Ecological Indian. Myth and History*, New York and London: W. W. Norton, 1999 are both concerned with Indian behaviour in North America, especially under European influence. Recent papers that discuss the impact of unaffected indigenous populations on western animal populations include C. E. Kay, 'Aboriginal overkill – the role of native Americans in structuring western ecosystems', *Human Nature* 5, 1994, 359–98; C. E. Kay, 'Viewpoint: ungulate herbivory, willows and political ecology in Yellowstone', *Journal of Range Management* 50, 1997, 139–45; C. E. Kay et al., 'Historical wildlife observations in the Canadian Rockies: implications for ecological integrity', *Canadian Field Naturalist* 114, 2000, 561–83; B. S. Low, 'Behavioral ecology of conservation in traditional societies, *Human Nature* 7, 1996, 353–79; P. S. Martin and C. R. Szuter, 'War zones and game sinks in Lewis and Clark's west', *Conservation Biology* 13, 1999, 36–45; R. L. Lyman and S. Wolverton 'The late-Pleistocene–Early Historic game sink in the Northwestern United States', *Conservation Biology* 16, 2002, 73–85.

38. R. K. Nelson, *Hunters of the Northern Forest*, Chicago: University of Chicago Press, 1973.

39. N. Bird-David, 'The giving environment: another perspective on the economic system of gatherer-hunters', *Current Anthropology* **31**, 1990, 189–96.

40. There is a good narrative sequence in the maps of ch. 64 of A. Sherratt (ed.) *The Cambridge Encyclopedia of Archaeology*, Cambridge: Cambridge University Press, 1980.

41. Some San bushmen provided rain-making services to farming communities in and near the Kalahari: P. Jolly, 'Symbiotic interaction between black farming communities and the south-eastern San', *Current Anthropology* **37**, 1966, 277–305.

42. M. A. P. Renouf, 'Prehistory of Newfoundland hunter-gatherers: extinctions or adaptations?', *World Archaeology* **30**, 1999, 403–20.

43. J. H. Bodley, 'Hunter-gatherers and the colonial encounter', in R. B. Lee and R. Daly (eds) *The Cambridge Encyclopedia of Hunters and Gatherers*, Cambridge: Cambridge University Press, 1999, 465–72.

44. M. Gadgil and R. Guha, *This Fissured Land. An Ecological History of India*, Delhi: Oxford University Press, 1993, ch. 7.

45. B. S. Trigger, 'Hunting-gathering peoples and nation-states', in Lee and Daly op. cit. 1999, 473–9; R. H. Layton, 'Hunter-gatherers, their neighbours and the Nation State', in C. Panter-Brick et al. op. cit. 2001, 292–321.

46. See *The Guardian* 5 March, 2004, p. 19.

47. R. K. Nelson op. cit. 1983, p. 14.

48. T. Ingold, *The Appropriation of Nature. Essays on Human Ecology and Social Relations*, Manchester: Manchester University Press, 1986; *idem*, 'From trust to domination: an alternative history of human–animal relations', in A. Manning and J. Serpell (eds) *Animals and Human Society. Changing Perspectives*, New York and London: Routledge, 1994, 1–22.

49. M. Eliade, *The Sacred and the Profane: the Nature of Religion*, New York: Harcourt Brace Jovanovitch, 1959, quoted by M. Oelschlager, *The Idea of Wilderness*, New Haven and London: Yale University Press, 1991, p. 20. See also H. L. Harrod, *The Animals Came Dancing. Native American Sacred Ecology and Animal Kinship*, Tucson AZ: University of Arizona Press, 2000.

50. E. N. Anderson, *Ecologies of the Heart. Emotion, Belief and the Environment*, Oxford and New York: Oxford University Press, 1996.

51. The saga is told in R. Kaiser, 'Chief Seattle's speech(es): American origins and European reception', in B. Swann and A. Krupat (eds) *Recovering the Word: Essays on Native American Literature*, Berkeley and Los Angeles: University of California Press, 1987, 497–536; There was a synopsis from the Washington State Library on www.synaptic.bc.ca/ejournal/wslibrry.htm accessed on 25 March 2004.

52. Dog puppies are much better at locating hidden food than wolf puppies raised by humans, with the inference that domestication has selected for a set of abilities that enhance their communication with humans. See B. Hare, M. Brown, C. Williamson and M. Tomasello, 'The domestication of social cognition in dogs', *Science* **298**, 2002, 1634–6.

53. K. Helskog and B. Olsen (eds) *Perceiving Rock Art: Social and Political Perspectives*, Oslo: Instituttet for Sammenlignende Kulturforskning, 1995: the Alta Conference on Rock Art.

54. M. W. Conkey, 'To find ourselves: art and social geography of prehistoric hunter-gatherers', in C. Schire (ed.) *Past and Present in Hunter-Gatherer Studies*, Orlando FL: Academic Press, 1984, 253–76; *idem*, 'Hunting for images, gathering up meanings: art for life in hunting-gathering societies', in C. Panter-Brick et al. op. cit. 2001, 267–91; H. Morphy, 'Traditional and modern visual art among hunting

and gathering peoples', in R. B. Lee and R. Daly (eds) *The Cambridge Encyclopedia of Hunters and Gatherers*, Cambridge: Cambridge University Press, 1999, 441–8.

55. P. G. Bahn, *The Cambridge Illustrated History of Prehistoric Art*, Cambridge: Cambridge University Press, 1998.

56. H. Valladas et al., 'Palaeolithic paintings: evolution of prehistoric cave art', *Nature* **413** (#6855), 2001, 479.

57. M. A. Rappenglück, 'Palaeolithic timekeepers looking at the Golden Gate of the ecliptic; the lunar cycle and the Pleiades in the cave of La-Tête-du-Lion (Ardèche, France) – 21,000 BP', *Earth, Moon and Planets* **85–6**, 2001, 391–404; the paper has a wide-ranging set of references to similar phenomena in other cultures.

58 I can claim to offset the time-conscious with twelve pieces of imagery with 'natural' themes: thirteen if you include a photograph of a dog. The calendrical structuring of spiritual life in the west is an altogether more fragmentary affair, especially since the Christian year seems to have been closely tied to the cycles of pre-industrial agriculture.

59. J. Blodgett, *Kenojuak*, Toronto: Firefly Books, 1985. Her work appears in a wider context in collections such as W. T. Larmour, *Inunnit. The Art of the Canadian Eskimo*, Ottawa: Information Canada, 1967 and on the Cape Dorset website.

60. N. C. Doubleday, 'Sustaining Arctic visions, values and ecosystems: writing Inuit identity, reading Inuit art in Cape Dorset, Nunavut', in G. Humphrys and M. Williams (eds) *Presenting and Representing Environments*, Dordrecht: Springer, 2005, 167–80.

61. A. Barnard (ed.) *Hunter-Gatherers in History, Archaeology and Anthropology*, Oxford and New York: Berg, 2004.

62. Note that they are largely dependent on recently fixed solar energy, though wood can be relatively old. The importation of industrial energies into hunting and gathering societies has not always meant their complete absorption into the more intensive economies. Iron tools, the rifle, the snowmobile and the outboard motor have left some Innuit groups still able to function in a largely hunting mode. Not that agriculture was ever an option in the high Arctic, of course, though herding of caribou might have been, as with the Saami and similar peoples of Eurasia.

63. See, for example, the nuanced analyses of hunter-gatherer behaviour in B. S. Low, 'Behavioral ecology of conservation in traditional societies', *Human Nature* **7**, 1996, 353–79; B. Winterhalder and F. Lu, 'A forager–resource population ecology model and implications for indigenous conservation', *Conservation Biology* **11**, 1997, 1354–64.

64. See the essays in R. S. O. Harding and G. Teleki (eds) *Omnivorous Primates*, New York: Columbia University Press, 1981.

65. M. Sahlins, *Stone Age Economics*, London and New York: Tavistock Publications.

CHAPTER THREE

Pre-industrial agriculture

FIGURE 3.1 *Felipe Guaman Poma de Ayala:* Nueva corónica y buen gobierno (c.1615–16).
Page 32 of GKS 2232 4to, The Royal Library, Copenhagen.

The emperor of the Inca singing with a llama. The chosen llama was a musical guide at court performances. The chronicler, Guaman Pomo de Ayala was also responsible for the first map of Peru (in the sixteenth century) and was probably an administrator in the Spanish government of that country; he was though a native Andean. There are 398 full-page drawings which form an integral part of his 1200-page *Nueva corónica y buen gobierno* of *c*.1615–16.

The idea of the chief of the Inca singing along with (and apparently taking the note from) a llama is somehow symbolic of the relationships in domestication. While talking to plants is regarded in most post-forager cultures as eccentric, talking to animals is not. So from the earliest communication between humans and domesticates (presumably centred on command), there has evolved a diverse set of practices beyond the, so to speak, instrumental. Not many cultures, however, cultivate this kind of intimacy: a few people sing along with their dogs, and play music to cows but to be led is unsual. It does signify, however, a closeness of the human and non-human which was common in pre-industrial agriculture but which began to disappear as the technologies of fossil fuels appeared. In colonial regimes, however, these closenesses might be victims of the social disruptions that occurred. Guaman Poma sought to convince the king of Spain to halt the destruction of Andean society in which the native Andeans were being exploited in the countryside and driven to death in the mines. To escape a dire fate, they were fleeing to the cities. Large-scale miscegenation, armed violence, the exploitation of native labour, and the spread of epidemic disease were, in Guaman Poma's view, inimical to the survival of Andean cultures.

A digital version of the entire manuscript, with commentary, is to be found on http://www.kb.dk/elib/mss/poma/index-en.htm, which is run by the Royal Library of Denmark. The above illustration is on http://img/kb.dk/ha/manus/POMA/poma550/POM0320v.jpg accessed on 3 March 2006.

'No god like one's stomach'

During the time of the hunters and gatherers, the biological evolution of the human species more or less stopped. Thereafter, our adult bodies changed in lifespan when immune systems became more effective or when some of us got too fat, but the characteristics that mark out *Homo sapiens* in the writings of physical anthropologists seem to be at rest. The millennia during which the energy of the sun fuelled human life through the channels of domesticated plants and animals produced a great variety of subsistence types and a positive coruscation of cultural diversity. These did not shift our basic anatomy and physiology though, perhaps unlike hunters and gatherers, most later humans have hoped to make a sacrifice every day to the god of the stomach.[1]

The transfer from dependence upon food collection to food production, from the usufruct of the wild to the reproduction of the tamed, has misty beginnings with no sense of only one place nor any one time. We know of early agriculture only when it becomes archaeologically visible. Pouring into one small pot the hectolitres of academic discussion, we reckon that by about 10,000 BP there were some areas of the world that had shifted irrevocably to the new life: the hill lands of south-west Asia, together with south and east Asia where much land, now below sea-level, had added to the space for human colonisation.

THE CULTURAL ECOLOGY OF AGRICULTURE

The irreducible fact is that a species which had been totally food-collecting became largely food-producing. Even if large areas remained outside the agri-cultural ecumene, the population growth inside makes them seem central to any attempt at an overview. In 10,000 BC the world population was perhaps 4 million but in AD 1750 (an arbitrary but not unrealistic date for the onset of industrialisation) it may have been 720 million and solar-based agriculture was the economic foundation for that growth. The spread of agriculture was not spatially uniform: domesticated plants and animals are not tolerant of many environments, despite skilled breeding and worldwide introductions. There are simply places that were too cold, too dry or too steep to grow crops. Some areas might support domesticated animals, but others remained largely wild or still subject to the hunters' ways of landscape management. If there are more people in less space, then we have the ingredients of a more intensive land cover system, where 'intensive' means a higher ratio of human-directed energy compared with natural flows;[2] overall there was much more human food in the world under agriculture than there was for hunters and gathers.

It follows that the likelihood of environmental manipulation is greater. This apparent ecological simplicity is bound up with immense changes in culture, reaching into depths of human cognition of the world with very far-reaching consequences.

Evolution and dispersal

The history of the earliest agriculture is bound up with two processes: the inten-sification of plant-use by gatherer-hunters and their adoption of permanent settlements. The first centre of such developments to be documented archaeo-logically was in the Levant of south-west Asia, with the site of Abu Hureyra in Syria a key excavation. The core finding comes from the onset of the Younger Dryas period in 11,000–10,500 BC when climatic conditions determined that there was less food. The settlers had cared for wild rye and so intensified their use of it and transformed it into a domesticated species. With the onset of the Holocene at about 9600 BC there was an abundance of plant foods, and rye seems to have been abandoned in the face of other choices, including wheat and barley. The continued hunting of the gazelle was in decline by 8800 BC in favour of sheep-herding. What this site (and a number of others in the hill lands around the Fertile Crescent) demonstrates is that there were phases of interac-tion: the harvest of wild plants, the cultivation of pre-domesticated varieties and then the onset of full domestication.[3] The earliest Holocene saw a dramatic rise in the concentration of carbon dioxide in the atmosphere as well as a reduction in they year-to-year variability of the climate, both of which favoured cultivation. Animals seem to have been domesticated later than plants, with an emphasis on those species which formerly had been hunted, such as sheep and goats, as well as easily tamed taxa like cattle and pigs. The reasons for the transition to domestication have been argued over for decades, but the currently dominant ideas suggest that climatic change and population growth may be the key elements in the shift. There were ecological consequences by about 8800 BC which resulted in some settlements adopting nomadic pastoralism and others budding off migratory groups to other regions. Institutionalised warfare now becomes desirable so as to acquire possessions and territory. The environmental consequences almost certainly included the processes described in the *Epic of Gilgamesh* (*c*.2700 BC) where Gilgamesh defies the gods and cuts down the forests, so they tell him that Sumeria will be plagued with fire and/or drought, and indeed in Sumeria, the earth is turned white, which sounds like a reference to salinification of the soils.[4]

Research in many fields has extended the list of certain centres of comparable importance though none is earlier than the south-west Asian centres unless the north China focus (based on millet) turns out to be equally ancient (Table 3.1). The centres were independent of each other, as shown by the different species domesticated in the earliest phases of each; the differences in time suggest regional variations in the combination of environmental and cultural conditions that led to the adoption of agriculture. The details of the expansion of agriculture from its initial focuses to other regions during the millennia before AD 1750 are given in many accounts.[5] There were many changes during that time, with developments such as irrigation and terracing allowing its spread into areas hitherto occupied by foragers or herders. Food production was characteristic of much of the world by that date: apart from

TABLE 3.1 Dates of transition from intensive hunting and gathering to agriculture (uncontroversial centres only; dates in years BP)

Centre of domestication	Intensive foraging	Agriculture
South-west Asia	15,000	11,500
North China	11,600	>9000
South China	12,000	8000
Sub-Saharan Africa	9000	4500
South Central Andes	7000	5250
Central Mexico	7000	5750
Eastern United States	6000	5250

Deduct 1950 for approximate BC dates, though ^{14}C years do not correspond directly with calendar years.

Source: Extracted from P. J. Richerson, R. Boyd and R. L. Bettinger, 'Was agriculture impossible during the Pleistocene but mandatory during the Holocene? A climate change hypothesis', *American Antiquity* **66**, 2001, 387–411.

the areas incapable of supporting the growth of domesticates, only Australia, parts of western North America, including California, and the southern cone of South America had yet to receive the environmental manipulations associated with this type of economy. The evidence for the dispersion of agriculture is not now confined to archaeology: genetics (especially the recovery of DNA from organic remains) and linguistics are adding to the detail (and the complexity) of the narratives.[6] Stable isotope analysis has provided the remarkable finding that around the North Sea at least the transition to agriculture was sharp and involved even the coastal Mesolithic (forager) populations moving from marine foods to terrestrial domesticates. The menhirs of Atlantic Europe may even be demonstrations of the dominance of the new culture.[7]

The tensions between environmental and cultural explanations of how this immense and far-reaching change came about are highly involving.[8] One of the earliest 'why' discussions was by Gordon Childe, who thought that the climate of the late Pleistocene was becoming so dry that gatherer-hunters congregated in oases where their density forced upon them new ways of subsistence.[9] Variations on this idea emphasised the possibility that the desiccation caused the migration of plants and animals into the Fertile Crescent zone where they were pre-programmed to be domesticates. A more cultural approach simply argued for experimentation in survival under more difficult conditions. It seems reasonable to argue that agriculture was unlikely during the Pleistocene when climates everywhere seem to have been very variable and the global carbon dioxide level low. Its terminal phase had high-amplitude changes on scales which might have been decadal or even less. The Holocene seems to have been (so far) a more stable interglacial than its predecessors, so the agriculturalist's vulnerability to weather has been reduced. Hence climate appears to have been a key factor in the earliest appearances of the new way of life.[10] Warmer, wetter weather and less variability meant, says one writer, 'it was *force majeure*: in effect there was no alternative'.[11]

Other reasons for domestication are also held to be important, either singly or in combination. There has been the claim, for example, that the domestication of cattle was purely cultural: that the horns of the wild beasts were lunate and so a connection with a moon goddess (whose monthly cycles clearly linked her to earthly phenomena) was obtained by the taming of the animals. There is also the possibility that status might be gained by taming wild beasts rather than by killing them, though that sounds a bit unlikely for the young men even if not for the children. Some plants may have been brought in as high-value ceremonial objects (corn fermentable to alcohol perhaps) and were then transformed to be staples. A stronger case is made for the notion that population growth forced a crisis of subsistence: that the wild areas could no longer support gatherer-hunters at their accustomed levels of off-take and so an intensification was needed which led to farming. This conflicts with the view that the foragers limited their numbers to a level below the absolute carrying capacity of the land, but even low rates of sustained growth might have subverted any such intention. Darwinian evolution as a thought-frame can be applied in suggesting that some biota were genetically pre-programmed to co-evolve with human societies; advocates of this set of ideas have to avoid teleology as well as the idea of 1:1 obligate mutualism. Human responses to diseases are also present in this complex of thinking: rising sea-levels and higher temperatures in the early Holocene facilitated the spread of tropical diseases, such as some forms of malaria, to Palaeolithic and Mesolithic populations with no initial resistance; and a great number of other bacteria, protozoans, helminths and fungi must also have made similar journeys, especially into irrigated zones. Schistosomiasis remains to this day a disease of irrigated lands in the near tropics.[12]

The earliest cultural-ecological complexes of domestication are becoming clear. Although difficult to document, there must have been a stage of informal husbandry when useful biota were brought to settlements (even to seasonally shifting 'camps') and the usefulness of a sustained supply was discussed. Grasses with large seeds in the temperate zones, and fast-growing root crops in the tropics are examples. Ways of keeping up the supply might include informal irrigation by channelling a stream's sand and gravel deposits or inserting plant slips into a midden heap. Conscious attempts to modify the genetics were probably absent, though some selection might actually have occurred if wild grass seeds with non-shattering heads were singled out for planting, for instance. Much early agriculture in south-west Asia seems to have transpired in hilly lands from the middle Euphrates through the Jordan valley into southern Jordan with a distinct dry season bearing a vegetation dominated by open woodland and steppe (hence the large-seeded grasses ancestral to, for example, wheat) and populated with wild sheep and goats. This contrasts with the likely origins of domesticated rice in China where it formed part of a wetland complex that was exploited for its wild animals and fish as well as the grass seeds.[13]

In environmental terms, the innovation never stayed put in the environments in which it first developed. In one group of expansions, the successful amplification was into water control. Its most spectacular results have been in the great river valleys such as the Euphrates, Nile, Yangtze and Indus, where elaborate sytems of irrigation stored water from peak flows against a drier season when crops such as rice and wheat might then benefit from a hot sun under which to ripen. Raised-bed agriculture provided a long-term subsistence base for the Maya in the swamps of Meso-America and for others to the south. In both cases a set of humanised environments became characteristic of the landscape. Another group of enlargements was into forests, where systems of shifting cultivation became established. The clearing of trees and the burning of woody material together formed a seedbed in which some varieties of rice (as in Asia) or root crops (as in the Americas and Africa) flourished until lower fertility or weed persistence forced the people to abandon the clearing and create a new one. Though associated mostly with tropical and semi-tropical lands, this system was present in European prehistory and in the coniferous forests of Finland until the nineteenth century. The last major augmentation of domesticate adoption has been the use of herded animals to crop plants inedible to humans in lands that are very dry, such as the arid fringes of savannas, steppes and semi-deserts, or very cold, like high mountain slopes and plateaux or the tundra. This is collectively called pastoralism and the animals herded range from the reindeer and yak through to the more familiar sheep and goats.[14] In the course of time, many have been in contact with a different system and so trade developed: pastoralists and irrigation farmers exchanging meat for grain, for example. It probably originated later than settled dry-land agriculture and on its fringes.

Environmental relationships

The circumstances of the earliest stages of agriculture are different from those of the point where reversibility (in terms of subsistence, environment and culture) was no longer possible. As with large-scale fire management by hunters, there came a point when current practices had to be maintained: there was no going back. Agriculture has to survive pressures of three types: extra-human, such as climatic change or freak weather; inter-human, such as other groups of humans who raid, destroy, engage in arson or deprive of water; and intra-human, where the cultivating group's organisation of planting, cultivation, storage or distribution is poor. Reversions to earlier ways occurred but they are inconspicuous beside the wider scene of the entrenchment of a domestication-based way of life.

The first generalisation of environmental significance is that humans create new genotypes. There was experience with the dog and this was extended to other animal species as well as to a fairly wide range of plants. The key to genetic modification is the replacement of natural selection by cultural selection, through the medium of cultural identification of the desirable outcomes and the controllability of the reproduction of the chosen species. Most of this selection must have happened with the application of traditional knowledge of a diffused kind, though it is not difficult to imagine the rise of specialists in the breeding of dogs, horses and cattle: the Apis bulls of ancient Egypt were bred for their special distribution of black and white in the coat and a black spot on the tongue, and were buried with great ceremony at Saqqara when they died. Sheep breeds appear in Egypt by about 2000 BC and several Roman authors mention the different types of horse that were bred, sometimes in specialised stud farms. Below the species level, many breeds, races and strains were produced, though the definition of those terms is not a settled affair. Selection also comes about accidentally. This was probably very important in the initial stages of wild cereal husbandry, when the selection of non-shattering seed-heads may just have happened rather than being planned. There are numerous examples of animal young being succoured by human communities and thus surviving to reproduce but many will have gone back into the wild to breed. Staying around a settlement to breed is necessary for the firstly accidental and then subsequently deliberate effects of human intervention: women in Papua-New Guinea are recorded as suckling puppies and piglets along with their own offspring.[15] At all events, domesticates both early and later need to be hardy because they have to survive the relevant selection of new conditions of soil, temperature, humidity and infection by parasites and pests. The must be adaptable to living together as fields of wheat or herds of goats, be able to reproduce without too much human aid, and not require inputs of tending beyond that which the human community can supply. In the case of animals, they need to be non-territorial, respond to a herd leader and be slow to flight when alarmed.[16] They need to be energy-positive unless they belong to very high-status groups or individuals for whom this is not a direct concern.

As well as new genotypes, farming creates new ecosystems. Like breeding, some creations are deliberate and some unintentional, though perhaps more accidental consequences are found, or at any rate survive to be chronicled. As with genotypes, many of these are deliberate and some accidental. A first category is the permanent field in which a crop or crops are grown year on year (or more frequently) with or without a fallow period. This alters most of the characteristics of the soil and water regime of the field and adjacent area and, in turn, may require further selection among the genes of the selected crop plant. A variant of this system uses a shifting field which is abandoned when its fertility falls: the tropical versions in forest and savanna are best known but there was a version in Atlantic Europe, for example, that lasted until the nineteenth century. This used an outfield which was taken in from heath or moor and planted with a crop but not intensively fertilised by contrast with the infield which was permanently cultivated. Nutrients (especially nitrogen) had to be recycled as manure from livestock and brought in from non-agricultural sources such as turf, seaweed, shell sand and soil. This usually gave a reasonable margin of energy gain over expenditure, albeit a system that was transformed by the introduction of the potato, usually after 1800.[17] If fertility levels were maintained in agrarian systems with permanent fields, then it was often possible to produce a surplus. This freed people from growing their own food and so made possible many other kinds of land uses, as well as having social consequences. Thus, societies had gardens for pleasure, hunting parks, managed woodland, and land devoted to high-status but inessential crops, for dyes, exotic fruits, drugs, and above all, cities. Though most obvious in, for example, eighteenth-century Europe, the antecedents can be found in most of the ancient riverine civilisations, such as Egypt, China and India.

The second class of relationships is the accidental. Many of these are apparent from sediments accumulated since the earliest days of cultivation. Higher rates of run-off entrain sediments and so floods are more frequent and erosion on slopes enhanced, with concomitant deposition in valleys and at the coast.[18] On-site, a whole class of plants and animals is classified as unwanted and labelled as 'weeds' and 'pests'. On ranges used by pastoralists, grazing above a certain density and frequency means that some plants are unable to regenerate and their place is taken by species unpalatable to the herded beasts. Species with thorns and spines are especially favoured in such places but, in very dry sites, no higher plants may survive and so a move towards desertification can be seen. An agricultural landscape is thus often one of a very large degree of makeover by human influences, the more so if it has been thus used for centuries, during which many environmental hazards have been faced and probably overcome.[19]

Generalised outcomes of these types are, of course, the result of many millions of individual and group decisions over many millennia. It seems clear that, just as farming created new genotypes and new ecosystems, it also generated new patterns of thought and behaviour. We cannot experience the psychological climates that engendered early agriculture, though we might accept the probability that there was an element of compulsion involved: think of the weeding, the storage problems and the organisation of water control, the disposal of wastes and the gathering of firewood, together with the likely accumulation of rats and fleas. A fresh suite of mind-sets had to be adopted. There had to be an ability to develop tools for the more intensive use of plant resources: the digging-stick, plough, and sickle, for example. Foragers' propensities became enhanced: the use of plants and animals as means to allow individuals to acquire social prestige and power, and the development of a form of social relations with biota are instances. Thus, once agriculture had gained momentum and irreversibility, there were more food and more people, more concentrations of both, more specialisation in food production and consumption and in people's occupations, more organisations allocating resources over greater distances, and crucially a differentiation of power among groups of people. Agriculture may even have spread into some regions because it offered alcohol, opium and wine.[20] Tim Ingold has argued that agriculture is a process of submitting to a reproductive dynamic in the natural world rather than converting nature into an instrument for human purposes,[21] and that the idea that production is action upon nature is therefore essentially recent. Yet, given the outcomes in terms of environmental change, where might the difference be?

Many kinds of evidence are used to infer the nature of past experience. In Neolithic Britain, for example, the interpretation of pottery remains has led archaeologists to postulate the development through time of difference and discontinuity in social life; in the later Neolithic the pottery remains suggest that life itself became more complex and fragmented. Hence, what was, initially, a series of interlinked and synchronised temporal cycles broke down after c.3000 BC and more effort was spent on people differentiating themselves from other groups and leading to the single grave of an ancestor from whom descent could be claimed.[22] When early texts, such as those emanating from Mesopotamia and ancient Israel, are added to the evidence, then, for example, the drawing of boundaries now becomes vital: between the farmer and the pastoralist, and within the cultivated area. The last becomes 'home' and the rest is wild, with nature being the place of threats such as wild predators and destroyers such as the locust. Nature does not give produce but it is won by sweat and toil, and the home settlement becomes the all-important site of transformation of the wild into the domestic.[23]

Humans have no innate agricultural calendar and so specialists who mediated between the cosmos and the farmer about when to sow, to water and to reap, and to whom to distribute the crop, acquired special importance: these were probably the first priests. Written sources from the lands of the Hebrew Bible, for example, go further and point to the importance of such castes in controlling scarce foods, such as meat, by insisting that consumption be viewed as sacrifices to gods to whose goodwill only the priests had access. This may even have been taken to the point of dealing with scarcity of meat (pork and beef come to mind) by forbidding its consumption, making it taboo.[24] There seems to be a gap developing here between the existential and the conceptual, and this must have widened with the invention of syllabic writing in Sumer, about 3300 BC. Separations are encouraged, too, by the fact that an early use of cuneiform script was to compile lists: something is either there or not there, like pass lists after examinations.[25] In alphabetic scripts, the written words are separate from, and less than, the life-world but, at the same time, they confer authority on those who can interpret them. When the Greeks from Attica added a focus on the word, then the world-slicing techniques of philosophy were launched. Philosophy bakes no bread, it is said, but once bread is baked by others then philosophers have time to teach, write and think.[26] With western monotheism, the philosophers and theologians posited that human problems were now only soluble by the supernatural, thus drawing a dichotomy which led to ideas of science and knowledge being the only way to heal that division and recreate the original Eden. But any such progress of the type advocated by Francis Bacon (1561–1626), for instance, was hindered by the continuing separation of fact and value, a fragmentation given immense prestige by Descartes (1596–1650) with his *cogito, ergo sum*.[27]

If the shifts in human social attitudes that result from pre-industrial agriculture have to be prioritised then perhaps the most important of all was pronatalism. From the control that existed in forager societies, there was a shift to staffing the farming with children and the army with aggressive young men. The environmental consequences were always notable and persistent.

Fire and the farmer

The control of fire represents a continuity between foragers and agriculturalists. In the earliest types of agriculture, sowing, planting and cultivating are highly correlated with the practice of burning vegetation, which also correlates to dependence on seeds and nuts. There is also a global climatic effect because large-scale fires in shrublands and forests are commonest and most devastating when an ENSO event brings dry conditions to one of the regions that it affects, such as the western Pacific during El Niño years.[28] The interaction of climate and human activity is probably behind the finding that methane emissions to the atmosphere from biomass burning were high in the years AD 0–1000 but reduced by 40 per cent in the next 700 years. One interpretation stresses the influence of wetter climates and population reductions in the Americas in bringing about a reduction in methane after about AD 1500.[29]

At all events, fire remained a feature of the landscapes of pastoralism. Rangelands were often burned to remove dead vegetation of no use to grazing stock and to encourage grasses at the expense of thorny shrubs. Often these lands had been occupied by hunter-gatherers and so the fire continued their practices. In the case of mountain transhumance, fire could be used to keep back woodland in times of warmer climate, as it was on lowland areas which provided wild foods for domestic animals, such as heaths in England. 'Slash-and-burn' cultivation was a widespread type of cultivation in which forest, scrub and savanna were first burned and then a patch of land cultivated. It was practised in Neolithic central Europe, by Melanesians and Maoris, in the taiga of Siberia, and in Finnish forests, such was the versatility of the system, provided there was space into which to move when the plot was no longer prolific in its yield. Food was not the only product: during the European Middle Ages, land around Montpellier (France) was burned to promote the Kermes Oak (*Quercus coccifera*) whose beetle *Kermococcus vermilio* yielded a red dye (cochineal) for wool.[30] If such systems became stabilised, however, with the plots converted to permanent cultivation, then the use of fire diminished. The same is true of 'imperial' fire, where colonists used fire a great deal on their entry into conquered lands (in eighth century Iceland it was used as a marker of possession of a man's land)[31] but then suppressed it whenever possible, especially in woodlands and savanna, seeing it as a deleterious influence. In Australia, the indigenous use of fire prepared the land for European pastoralism rather well, and fire produced the 'green pick' upon which cattle thrived. As the land filled up with colonists, control became more difficult and required the transcendence of property boundaries and a degree of neighbourly agreement that progressively became harder to achieve. Hence fire suppression took over as the dominant policy, with all its potential for fuel accumulation and wildfires.[32] A final landscape use of fire is in warfare, where it can frighten horses, remove the sight-lines of weapons, smoke out enemies and remove cover. 'Firepower' is still the measure of military might, though not always of effectiveness. As cities grew, their demands for fuel wood increased the pressure on local woodlands, scrub and savanna, for it is generally not economical to transport it very far: Medieval London received wood from a zone roughly 120 by 40 kilometres in area, even at a time such as the year 1320 when the price index for firewood (with 1261–1270 = 100) was 277, compared with 166 for wheat.[33]

It is even possible that burning, together with deforestation and agriculture, changed the global climate, populating the atmosphere with many aerosols. Thus, the end of the Little Ice Age (LIA) in the nineteenth century is helped along by the cessation of many practices that had resulted in the firing of biomass.[34]

The role of fire in ritual should not be ignored. It has an important place in Hinduism and Zoroastrianism, for example, and was at the centre of sacrificial rites in ancient Israel. Potters and smiths were fire masters and both produced high value objects, especially weapons: the ability to produce and to work iron was probably never a purely material process. Fire was a sign of both ineffable power and divine anger: the whole is encapsulated in the myth of Prometheus who stole fire from the gods to give to humankind, but who then suffered nightly evisceration of his liver as a sign of Zeus' anger. But by then he had given humans 'all the arts'. So the adjective has come to mean large-scale and potentially risky human endeavour which brings about changes of great magnitude, though many users ignore the second half of the story.

Management and impact

Fire was only one technique harnessed by humans: metals, draught animals, wind- and water-mills, the plough, explosives and paper-making do not exhaust any list of inventions that had environmental consequences. Similarly, it is not possible to list all the environments and all the makeovers for which our species was responsible even before industrialisation. But we can give enough examples to convey a story of massive change over a long period of time, constantly bearing in mind that the predominant source of energy for human communities was the sun, mostly mediated via photosynthesis but also through wind and falling water. The environmental impact of humans was nevertheless increasingly channelled through technology, and the possession of a hard metal such as iron is the key to many alterations and impacts: the term 'iron age' deserves a wider circulation outside the discipline of archaeology.

The main categories of land cover which require attention are firstly those land-based systems that produced mostly comestibles but also other organic products such as wool and cotton, namely rain-fed permanent cultivation, irrigated cultivation and pastoralism, together with their intensifications, such as colonial ranching. These are aided by accessory systems such as salt production and the construction of ships and harbours. A second main group includes land cover where the totality may be important, as with forestry, hunting grounds and gardens; then comes the inorganic materials in the instance of minerals and the construction of the mostly non-living urban centres, with the largest cities having a higher inorganic content. Water management provides the link to a third totally organic unit, which is the use of the waters of the planet for fishing and whaling.

Land-based systems for organic production

Rain-fed agriculture cannot be ousted from a leading position in the environmental history of the pre-industrial world, even though irrigation agriculture (discussed below) became immensely important. Since some rain-fed systems may benefit from deliberate storage, and most need water control in the form of drainage, the water content of a soil is on a par with its fertility. This relates in part to the physical nature of the soil's top horizons and also to its content of plant nutrients. The management of these features starts with ensuring that the soil is physically present: that it is not washed or blown away. Thus, the soil cover may be important in preventing heavy rainfall from moving the soil into the run-off. Simple mounding was widespread in Africa, with units as large as 2 metres across and 1 metre high. Here, too (in Zimbabwe and Zambia for example) tree roots and live trees were left in cleared areas so as to retain soil even when there is no crop. Leafy litter used as a mulch, in places as far apart as Tanzania and some Pacific Islands, performs the same function. The movement of soil downslope can be checked by the construction of terraces, of which there are many different types. The origin of terracing is not precisely known but the first examples are probably from south-east and south-west Asia before AD 1. During pre-industrial times, the practice was common in many environments other than sub-arctic and the extreme fringes of the cool temperate zone, though the flights of terraces were never so elaborate as in irrigated regions. South America probably displayed the widest applications, in mountainous, dry and marshland environments.

Everywhere, keeping the soil high in nutrients produced a great variety of practices: ploughing, fertilising (including manuring) and rotations are simple terms for multifarious actions to keep nutrients in soils, to add to the store, and to recycle whatever was not consumed by humans or stock. Pre-industrial techniques encompassed composting, mulching, and the addition of materials such as calcareous marls, shells and mud from river beds. China provides good examples of the use of all of these. Above all, nitrogen levels must be kept up and so fallowing, crop rotation, fixation by leguminous intercropping, and the addition of animal excrement (including that of humans) are employed, as they were in the crop rotations of medieval Europe. Keeping down competition for the chosen crop is central and so weeding and pest control have a very long history: children may not be able to plough but they can scare birds. The improvement of yield took a number of forms which included the selection of varieties most suited to a place (and a culture), and the import of new crops which flourished in a different place. Wet rice, for example, must have been a radical change for people hitherto growing rice in temporary forest clearings. Some seeds turned out to be revolutionary in their capacity for high yields and thus potential for trade. The use of the grape for wine is a good example, transforming many slopes between the Caucasus and the Mediterranean by about 2000 BC and thereafter being spread into numerous environments[35]. Lastly, the management of the crop after harvest is environmentally influenced because deterioration by fungi and bacteria, and competition from, for example, rodents depend on the local conditions.[36]

The impact of an agriculture uninfluenced by fossil fuels is by no means minor because, in many environments, it has been continuous for perhaps 8,000 years. When an area is taken in for cultivation, its ecology changes markedly. Most unaltered ecosystems have mechanisms for a 'tight' recycling of nutrients between the soil and the biota but cultivation opens that system to spatial translocation, so that soil particles and nutrients may move downslope, then further down a river basin and possibly into the sea. Upslope fertility will be lost even though it can be caught by terracing or diverted into flood plains lower down the watercourse. Catastrophic loss at times of flood is common. In case after case, the coming of agriculture and its development under increasing population pressures are followed by higher rates of soil translocation, some of which finds its way into lakes or into the sea: little retelling of this story is needed. What is often less well relayed is the capacity of land to regenerate under a reversion to semi-natural vegetation or under a management regime which minimises soil loss, as happened in many pre-industrial cultures. Some examples of soil loss, as in the famous case of Attica as described by Plato ('only the bones of a sick body, all the fat and soft of the earth having fallen away') has been re-read as the result of earthquakes and landslides rather then poor watershed management.[37] In central Mexico in the sixteenth century, high population densities tested some environmental thresholds but the Spanish conquest took away such threats by reducing the native population; when the conquistadores expanded livestock production they introduced transhumance, which accelerated changes in plant species' composition. In places this in turn led to degradation from overgrazing.[38] Impact may extend to changes in animal populations as some species are attracted to human presence and crops whereas others are frightened away or killed. Domestic animals are often used as carriers of nutrients from forests or open grasslands into enclosed agricultural lands and in the woods and pastures they alter the species composition by their selective grazing and browsing. An agricultural society may bring about chemical concentrations (though not on the scale of industrial communities) which have environmental impacts: the crushing of mineral ores will dump possibly toxic fines into streams and townsmen will not often bathe in the river downstream from the tanneries.[39] In times and places of aridity, it is perhaps possible that extra dust, from bare areas of marginal cultivation, and herding would contribute to an atmospheric burden that was by nature global.

Technique is not the whole story in the environmental relationships of pre-industrial agriculture.[40] Inside a social group, arrangements may alter environment relationships: the roles of individuals may vary with small distances of space and time (he/she may be a slave, a squatter, a peasant or a hired labourer) and thus exert different pressures on a local ecosystem.[41] Indirectly, population growth is affected by the efficacy of patterns adopted to ensure the survival of newly born children: if they work, then the environment is subject to further immediate pressures. At times of acute food shortage, famine foods may be subject to rigorous cultural evaluation: a product such as milk might be unacceptable even if life-saving because its nature or its acquisition is subject to powerful prohibitions. At another extreme, groups (perhaps commonest in Melanesia) which emphasise the necessity of social ties through the lavish exchange of gifts may create environmental change in the acquisition of the materials given away. Sometimes they are the same materials as are used in economic transactions but which are transformed by the social and ceremonial context.[42] In shifting agriculture, the decision to move on may be brought about by the incidence of fleas or the occurrence of a death just as much as by the falling fertility of the plot or the rampant success of weeds. War, too, may alter land cover indirectly: in the Balkans various episodes of conquest moved out one population to be replaced by another: for example, farmers were replaced by transhumant herders whose impact on the forests of the higher slopes was profound.[43] In earlier times, the incursion of the Visigoths into Italy in the fourth century AD removed the imperial protection of water management systems and allowed the return of wetlands, the flooding of plains and even landslides as at Piacenza in Emilia.[44] Likewise, the 'scorching' of the earth of northern England (after an uprising put down by Norman troops) in 1069 led to the spread of scrub and rough grassland, both detectable by pollen analysis.

Environmental relationships may be different through space when one economy may be possible but is not adopted for cultural reasons. The non-use of milk in a swath of central Africa and most of east and south-east Asia has been explained on ecological and pragmatic grounds but the categorisation of milk as unclean (cattle may be kept but the milk not taken) is probably more powerful. The introduction of dairy produce into parts of south-east Asia as ceremonial items in Indian-origin religion during the early centuries AD lasted only about 1,000 years, for instance[45]. Hinduism, argue two Indian writers[46], is well suited to differentiated land conditions since different castes can occupy dissimilar niches and adapt to the particular ecologies of hunting and gathering, plough agriculture, and shifting cultivation. In the Andes, a self-sufficient Inca community (the Ayllu) ideally had access to pastures, potato lands, irrigated maize fields and land for cotton and coca. It was also a spiritual entity that worshipped the natural features of the unit, especially the places where the founding ancestors had emerged from the earth in order to provide irrigation water. Work on irrigation canals then became restricted to men of a certain spiritual and legal status. After the Conquest, the Spanish retained much of the social structure tied to irrigation rather than replace it with a European style of legal ownership of rights[47].

Transformation of relationships may be internally generated over time. In Iceland in the Middle Ages, the word *náttúra* took on the meanings not only of the objective conditions of human life but also the ideas of a natural 'peculiarity' or 'quality' as well as a supernatural being or power. The sea was part of this 'natural' world and to fish was to engage in an appropriation of the wild which exposed people to dangerous forces beyond their control, such as jumping whales and polar bears. If starvation threatened then fishing might be taken up but at some times of famine even the easier route of hunting small animals was occasionally eschewed in favour of the choice of enduring hunger and staying in bed. Farming, by contrast, was an 'inside' activity which was domesticated, settled and controllable. Hence, when people had to retreat from upland grazing (*atréttur*) during periods of climatic downturn, such as the Little Ice Age, they retreated into the settlement to protect the agricultural core rather than adopting new shepherding practices that would be better adapted to changing conditions[48]. The introduction of a new religion can produce rapid change. Within a 45-kilometre radius in Senegal, there have been contemporaneous wet-rice communities and drained-land areas devoted to groundnuts. The introduction of Islam made some, but not all, of the Jola people give up palm wine, pigs and ducks and adopt the trade-oriented production of groundnuts; so, in the same regional conditions, contrasting economies have transformed environments in highly contrasting ways.[49] Culture may determine that the landscape of one period may be regarded as iconic and almost everything else is a degradation: hedged enclosures of early modern England and the combination of corn, vineyards and olives of Renaissance Italy are two examples.[50]

The position is summarised by the way in which technology and symbolic practices, such as ritual, are often regarded as equally necessary and effective. To modern western eyes, technology is essential but ritual is effective only in the symbolic realm. In pre-industrial societies they are usually inseparable: in the case of the Mnong Gar of Vietnam, the technology of burning the forest had to be right in order to achieve a total destructions of the trees but it was also essential to call on the whole human and spirit community in a co-operative effort.[51] Christian imagery opposed the wolf and the lamb and so wolves became, as it were, demonised and killed as symbols as well as predators on deer and sheep.[52] Likewise, meteorology might be moral in the sense that bad weather was sent from heaven as a punishment for human failings and transgressions: medieval Europe, pre-industrial China and twenty-first century Louisiana are instances. The reception of the Lisbon earthquake of 1755 shook science, religion, and the literary world alike, and may in fact have dented the optimism of the Enlightenment.[53]

The management and impact of water control for farming cover a great variety of practices. At one extreme was the creation of virtual aquariums in the form of flooded fields, dry for only limited periods; at the other is the creation of ridges of soil dry enough to grow crops in what was essentially a permanent swamp. Thus, irrigation as commonly conceived is but one type (albeit a very important one) of regime within a wide spectrum of patterns. Some brought water, some carried it away and some did both. The classic historical locale for irrigation farming is Ancient Egypt. The Nile floods were diverted into a system of canals and ponds whch held water back primarily to irrigate the cereal crop. To produce the fields and watercourses, of course, the existing vegetation of the valley had to be cleared (there are tomb paintings of scrub clearance) and the animal communities extirpated: hippos and crocodiles are poor companions to cereal-growing. Simple flooding plus ponding allowed the irrigation of some 8,000 square kilometres of land, but the advent of water-lifting devices such as the *shaduf* from *c.*1570 BC and the animal-powered *saqiya* from *c.*320 BC each added about 15 per cent to the irrigable total. Irrigation methods spread around the Mediterranean with Roman and Muslim conquests: in eastern Spain, the Moors intensified and adapted the pre-existing Roman systems. Here, their first main stimulus seems to have been a royal whim to display exotic plants in palace gardens but then they were responsible (by about AD 960) for the introduction of many new cultivated plants: lemon, apricot, rice, cotton, bananas, cauliflower, watermelon, eggplant, henna, safflower and jasmine do not exhaust the list. Spinach and sugar were added before 1038.[54]

South and Central America display management techniques at their most diverse. Canals were dug and maintained for agriculture in the mountain areas and for the royal baths in Mexico but much food production came from a large range of irrigation methods that used pits dug in dry river beds, sub-surface irrigation using gallery flow and floodwater retention. The latter encompasses one continental speciality, namely the raised field lowland system in which the surplus water is confined to canals and crops are grown between on raised ridges. The canals might be dug right through swamp soils to bedrock and the displaced soil used for the planting beds; or the swamp soil might be pared off and ridged leaving canals to form between. Called *chinampa* in central America, both types were common in Classic Maya times AD 300–1000. About half of the area around Lake Titicaca in AD 1000–1450 was part of a canal system that both drained and irrigated; its high altitude, however, made it vulnerable to any changes in climate towards cooler and wetter conditions.[55] A similar diversity of water control for the production of taro in Polynesia was found: raised beds and pit cultivation drained water and impounded it respectively, on islands with different rainfall regimes.[56]

In a classic description of wet rice cultivation in Indonesia, Geertz suggested that terraced fields with a controlled water supply were, ecologically speaking, aquariums. Although dried out seasonally after harvest when ploughing would incorporate the stubble back into the soil, the water was the key element because it not only allowed the plants to grow but functioned as a medium for nitrogen-fixing blue-green algae and the water-fern *Azolla filiculoides* as well as a source of animal protein such as fish and shrimps. Like all such systems, it was labour-intensive and depended upon complex social systems of water control. It could always be intensified in times of increased demand and its survival suggests an adaptability and resilience of a very high order.[57] One source lists twenty-six management practices common in South Asia wet rice cultivation[58] and the key element was keeping up the nitrogen levels, something that was recognised as crucial (though the term nitrogen was not, of course, used) along with water temperature by the Chinese of the Han Dynasty (206 BC–AD 220); leakage of nitrogen to urban centres was in part restored by the use of composted human excreta. The potential for wet rice to promote environmental change extends outwards from the realigned drainage systems on controlled slopes to a manipulated landscape focused upon the flooded fields. These were surrounded by forests that were the homes of shifting cultivators but which might in time be converted to terraced fields and become home to the dam, the sluice gate and the treadle pump.[59]

Lest we are too easily lost in wonder, the literature on irrigation systems is peppered with references to failure. Many irrigation systems are vulnerable to drought: in 1413 the Valencia region suffered famine because there was not enough water to power the flour mills.[60] In Meso-America there was periodic famine (for example, AD 1446–54) caused by stress on productivity or by the demands of unequal distribution of harvests, by the lack of royal munificence at difficult times and by locusts, flooding, drought, frost and hurricanes. In the Nile valley, the bureaucracy of the Old Kingdom seemed unequal to coping with large-scale disruptions when the annual flood was meagre, though the relationship to political instability at the same time is uncertain. The Nile broke down again in the fifteenth century when about half the population was eliminated by the Black Death: the central control of irrigation was unable to contain local conflicts and so manage the system with a smaller labour force. Hence agrarian output fell by about 70 per cent. Scholars disagree on the causes of the breakdown of the Lake Titicaca systems around 1480: climatic cooling is one possibility but this is disputed by proponents of evidence of political instability at that time. The repeated references to political factors in breakdown suggest that any cultural discussions must include the structure of political life.[61] The cultural and technological achievement of the societies of some of the major river valleys, such as Egypt, Mesopotamia and the Indus, led the scholar Karl Wittfogel (1896–1988) to term them 'hydraulic civilizations'.[62] He believed that only centralised authoritarian government, including many civil servants, technical experts and unfree labourers, could run them successfully. The control of water and the control of people went together. Inevitably, counter-examples have been provided of powerful regimes that did not develop irrigation, but the notion that the surpluses created by successful irrigation helped to consolidate the power of those at the top of, as it were, pyramidal societies has not died. It is clear that in the major Old World river valleys, those with power on Earth received their authority from those on High and that the relationship had to be good. Thus, all such societies ran on a fuel of technology, tax, royal power, magic and pro-pitiation of the gods of Earth, sky and water.[63] The existence of central pol-itical control is also found in Polynesia but here it is associated with the dry-land agriculture that is most prone to crumble under drought, hurricanes and rapid population growth. Chiefs in such systems are the most likely to embark on the aggressive acquisition of territory in order to keep their power. So a drought that produced crop failure might also trigger a predatory war of territorial expansion.[64] The role of religion is brought out by the suggestions that wet rice in Sri Lanka was introduced together with Buddhism.[65] The founding texts prescribe begging for the monks and so perhaps highly pro-ductive farming made dependence easier.

There are alternatives. In Bali, the productive system was an engineered landscape of rice fields, irrigation channels and tunnels, threshing soci-eties and markets. The symbolic network is even more complex because it embraces the production system, the wild flowers and transcendental con-cepts as well. At the apex of it was the Temple of Crater Lake. This institution had no control over any irrigation system and no coercive powers but it could create new systems, control water flow and decide disputes. In effect it regu-lates the flow of holy water from hilltop to the sea.[66] There are choices within the same environments, too. In Al-Andalus (Islamic southern Spain), Muslim culture produced two different types of water distribution systems: a Syrian type and a Yemenite type. (At heart, the one distributed water proportionately and the other sold it.) The difference persisted after the *Reconquista* in the thir-teenth century and its remains persist. A development like the *qanat*, adapted to very arid areas of the Near East and Mediterranean was taken to Spain in the eighth century along with the Arab conquests; from there to the Canaries, Peru and Chile with the hispanic conquests and eventually to Mexico's Guadalajara region during its colonial period (1521–1821). The culture changed but the technology remained virtually the same.[67] In China's Miju River basin, the period from the fifteenth to the nineteenth centuries wit-nessed a shift from a community-based system of water management to a tightly bureaucratised one, as land was developed and native vegetation cleared.[68] Long-lived success leading to surpluses and the reinforcement of elites undoubtedly happened, and later in the chapter there is some discussion of stratification in pre-industrial societies and the environmental effects of those social separations.

The ecological basis of all forms of pastoralism is the transformation of cel-lulose, inedible by humans, into meat, milk, blood, horn, hide and dung. The vegetation in these regions was usually low in stature And the regions were often cold, dry, mountainous, high altitude, high latitude or swampy or, indeed, some combination unsuited to domesticated crop production. The wide variety of environments led to the selection of many different species of domesticates and some remained semi-wild in the sense that there was little genetic selection: the reindeer and the camel are examples. The list of pastoralism's animals is very long: cattle, sheep and goats are the most numerous, but horses, camels, reindeer, mules and donkeys, llama and yak are all used.[69] Though able to eat a broad spectrum of plants, many exert constraints on their management by needing fresh water daily or at other intervals that depend on the lushness of the vegetation. Management is helped by species which are gregarious and have a herd leader. The chosen beasts are not always native species: the domesticated humped cattle of Africa (the zebu strain) seem to have been imported into the Horn of Africa as part of Indian Ocean trade and to have diffused from the east rather than from any northern imports from Egypt and the Middle East.[70] The history of pastoralism also includes changes through time. In south-west Asia between 6000 and 3500 BC ovicaprids were dominant, with some cattle; from 3500 to 1900 BC there were more equids and a high level of mobility, and after 1500 BC the camel became a staple in arid regions.

In contrast to the rice terraces of China or Bali, the management and impact of pastoral agriculture seem small. Yet over large areas and through many millennia, the production of food and other materials from domesticated animals has been a significant environmental force. The stereotype is derived from nomadic pastoralism, with a group of culturally distinct and self-contained people following an annual cycle around remote parts. Most pastoralists had some knowledge of plant production, might have engaged in it and almost certainly traded with its practitioners. Further, pastoralism took other forms: there were sedentary groups as well as those who seasonally moved to good pastures, often at high elevations (transhumance). The origin of pastoralism in the Old World is generally thought to post-date settled agriculture and to have started on the fringes of the farming cores of south-west Asia, though one school is more convinced by the case for the west and central Sahara from about 4500 BC. We might add examples like the rabbit in medieval Britain, which was cosseted in carefully managed warrens but where in the Lakenheath Warren of the Breckland (Norfolk) in the seventeenth century the wind carried loosened soil 4 miles (6.4 kilometres) to affect 4,000 square metres of land, including 800 square metres of arable, and houses collapsed under the weight of blown sand.

The management of pastoralism mostly started with animal breeding. It was desirable to breed individuals with the lowest possible need for water but with good resistence to heat and disease. Inevitably, the major impact was on the vegetation grazed by herds on their daily, seasonally or yearly movements. Different animals have varying eating habits (cattle pull off leaves rather than nibble them like sheep, for instance) and plants are variable in their resistance to repeated mastication. Hence, lush edible grasses can be replaced by short wiry species, thorny shrubs can colonise grassland and, in extreme conditions, plants may be lost altogether and open soil remain. This is then prone to erosion. In a transhumance system in Argentina, grazed slopes with grasses were found to erode whereas those with trees stayed intact. The unanswered historical question is whether this system was modified by the Hispanic conquest or whether this has been a feature since the system evolved.[71] That such patterns are not constant is shown by the prehistoric period in the Zagros Mountains where, in 6500 BC, keeping animals was essentially a supplement to a village-based crop agriculture. As the village population expanded and animals had to be moved over larger distances, then the organisation of labour had to be modified to move larger herds over longer distances: nomadic pastoralism was the result.[72]

Relations between humans and domestic animals can be very close, and it is not surprising that many features of human culture impress themselves on to the animals and on to the ways of keeping them. The size and strength of animals can be impressive and, in farming societies such as those of ancient Egypt and Mesopotamia, there was an unbroken nexus between the power of the bull, fertility, and kingship: wild bulls must have been 'the largest, strongest and most successfully libidinous animals familiar to the founding peoples of most ancient civilizations'.[73] Those characteristics were taken into the pastoralism groups that lived at irrigation's margins: the bull has cults to this day around the Mediterranean. The first letter in western alphabetic systems is a bovine head: in cuneiform it is ∀ (an inverted triangle with two dots over it), in Greek on its side with curved horns α and in Latin on its back as A. The connections between cattle and kingship in many African societies are close: the prosperity of the kingdom, the fertility of animals, humanity and crops all depended on the repeated performance of ritual activities which linked kingship to cattle through the idioms of pastoralism. Where life depended upon cattle, then women tended to be excluded from pastoral activities but when the effects of capitalism, colonial policies, and Islam are taken away, then women may have had great influence in political and economic realms. All pastoral groups, however, demonstrate practical knowledge of their animals, and with cattle especially there are often symbolic systems of naming, identification and classification which add up to a resilient adaptation to the environmental conditions.[74]

In many pastoralist societies, the meat of the main animal is rarely eaten: the sustained yield products are more valuable and so smaller species may be kept for flesh. Many groups, however, do accord their focal taxon the privilege of being sacrificed ceremonially. This is especially true of the role of the scapegoat in which the failings of people are laid upon an animal which is sacrificed. In the case of the Inca, llamas deemed to bear human sins were eviscerated and the innards washed out of the city down the river. In addition, llamas evoked songs and poetry, and a red llama joined the king in song at certain ceremonies.[75] Nobody knows how these close and complex relationships turned some nomadic pastoralists into aggressive entities. Between AD 1000 and 1400, those of central Asia shifted from running protection rackets on caravans, to unleashing attacks on centres of civilisation to their east and to their west. Competition with agriculturalists for land with both potential uses may have been the environmental linkage but seems unlikely to be the whole explanation: fidelity to an image of the imperatives of power derived from the possession of animals (in this case the horse) might well be another likely ingredient.

Hints have been given that none of these food-producing systems remained static. During pre-industrial times, one of the most widespread changes was one of the forms of intensification in which more crop per unit area per unit time was demanded, either because of factors such as population growth, of withdrawal into a smaller area because of environmental change, colonial demands, or entry to a cash-crop economy. In the case of pastoralism there is the convenient term of 'ranching' but, in dry-land and wet-land agriculture, intensification occurred before the waves of European-style industrialism reached them. One example of intensification of dry-land agriculture that needed no fossil fuel input is the enclosure movement in lowland England and the coeval increases in productivity that have attracted the label of 'revolutionary'. In parts of England (though by no means all), the end of the Middle Ages saw a pattern in which common fields were the basis of food production. They were 'common' in the sense of being communally managed and often 'open' in the sense that there were no fences or ditches between holdings. Between 1500 and 1800 most of these areas were transformed into separate properties under private ownership, usually with hedged and ditched fields. This movement meant more control over a central aspect of productivity, namely that of soil fertility. The key element is nitrogen (though the farmers of the time were unaware of the science, they knew about the pragmatics) and there had been many ways of keeping its levels up. Fallowing was a key practice because soil nitrogen could then accumulate by bacterial activity; folding animals was another, because cattle and sheep tended to eat more during the day but excrete equally by day and night so manuring the field in which they rested; and planting leguminous crops such as peas and beans fixed nitrogen, though these crops soaked up other elements, such as potassium, and grew best on limed soils with a controlled acidity. The adoption of new crops was usually made easier by private ownership and so the turnip, swede and mangel became available from the 1620s onwards though, initially, they were not popular outside East Anglia. Their leafy growth helped slow down nitrogen loss from leaching and kept down weeds. In the seventeenth century clover was added to the suite of available plants: it can fix 55–660 kg/ha/yr of nitrogen (N_2) against the 30–160 kg/ha/yr of peas. Both could be grazed and so fallow was replaced by manure production.[76] From the end of the eighteenth century, the rising populations were then supported calorifically by the potato though its greatest contribution came in the succeeding decades. New varieties of corn and animals and the concentration of manure by stall-feeding cattle added to the rises in outputs that turned farming from something largely directed at self-sufficiency into a series of commercial operations aimed at a much wider set of consumers.[77] In environmental terms, the new patterns reduced soil erosion, especially on lighter and thinner soils. Animals such as the skylark and corncrake diminished in number though edge species such as the fox, whitethroat and blackbird increased, as did the rabbit at the expense of the hare.[78] On the edge of the city, the common land of a community could be enclosed for gardens, 'for show and pleasure' in a double insult to the poor.[79]

A different kind of product and a different spatial scale are represented by the early history of sugar production, which is essentially an addictive luxury, never having been a cheap source of energy[80]. It was introduced into the Mediterranean under Islamic influence but then Hispanic merchants and adventurers were responsible for a trans-oceanic breakout. By 1452 there was a water-driven sugar mill on Madeira, and the island was exporting to England by 1456. Madeira devoted itself to a monoculture and the *levadas* which ring the slopes represent a re-shaping of the island's slopes and watercourses (there are some 700 kilometres of conduits and tunnels) to irrigate the cane. Gran Canaria had its first mill by 1484 but in both cases the need for wood to boil the cane began to limit production: it took 100 cubic metres of fuel wood to crystallise 1 tonne of sugar. Westward shifts then included Brazil where the forests supplied the fuel from the 1620s. Sugar transformed the economy and ecology of West Indian islands, such as Barbados, where a variety of export crops were replaced after 1660 by a sugar monoculture. The consequences of a very rapid removal of forests (they had all gone by 1665) were dire in the sense that the crop quickly ran into economic trouble because there was insufficient wood and the soils were depleted of essential minerals. In order to combat the latter, cattle were imported in large numbers but their heroic efforts at dunging were too late. Environmentally, the disappearance of the forest took with it two species of tall trees plus shrubs and ground plants. The woodpigeon had gone by 1654 together with so many birds from forest-canopy habitats that there was a noticeable absence of bird song on Barbados. Such animal species as were left were increasingly referred to as 'pests'. The vacant ecological niches meant that two-thirds of the flora now came from outside, including acacias, guava, grasses and tamarind. The soil lost its fertility and was itself washed away, with cane planting in trenches exacerbating the rate of downslope movement. Watercourses draining forested watersheds might have concentrations of 280 milligrams per litre (mg/l) of sediment but bare slopes would raise that to 26,000 mg/l.[81] The European (and especially English) sweet tooth exacted a heavy price environmentally as well as in more directly human terms.[82]

In neither early modern England nor the West Indies were these changes brought about by a shift in access to energy sources nor by external environmental factors. The notion of culture embraces institutional arrangements and so the favourable attitude of the Portuguese and Spanish royalties towards trade and exploration was helpful to the colonisation and sugar growing in the Atlantic islands and Brazil; the acceptability to all Europeans of slavery meant that immense supplies of cheap labour could be obtained for work in the subtropics. But the improvements in sea-going vessels (and especially the Portuguese *caravela*) acted as catalysts in opening up the widest oceans to navigation.[83] In the case of enclosure, there needed to be agreement between landlord and tenants, and within the tenantry, for the procedures to be implemented which often fell out in favour of the better off and the dispossession of the poorest who were most likely to depend upon common rights. This was worst in the case of the enclosure of heath and moor where many controls on, for example, small-scale cultivation and on grazing had not been enforced and a marginal existence had been possible. The broader-scale enclosure by Act of Parliament characteristic of the nineteenth century was an even worse time for such poorer people. Both examples show, however, that cultural attitudes are often paramount but that a technological shift will produce an apparently sudden change in the conception of what is possible. Acquisitiveness then kicks in with some force.[84]

Accessory changes

Pre-industrial communities were not simply isolated patches of environmental change at intervals across the world. They exchanged ideas, people, species and goods. Lists of examples would be endless: the worldwide travels of the Chinese and the Portuguese in the fifteenth century, Bronze Age trade between the British Isles and the Black Sea, the Silk Road between Europe and China, the Muslim introduction of sugar into the Mediterranean, slave trades across and out of Africa, are all examples. All kinds of scales and all levels of environmental linkages were found.

One example of a commodity that was bulky but still precious enough to be traded over long distances was salt. It was extracted from underground brine wherever that was possible and its value was such that in the sixteenth century 48 kilometres of wooden pipe for the brine were built from Halstatt to Ischl (Austria) and a canal constructed from Lüneburg to Lübeck in north Germany. Another major pre-industrial source was sea water, evaporated by the power of the sun in warmer climes such as the Bay of Biscay and the Mediterranean and by burning peat, wood or coal beneath clay or lead pans of brine in cooler regions.[85] Such was the demand for one of the chief preservers of foodstuffs that the Roman Empire imported large quantities from the The Wash on the east coast of England, and in medieval times the same area was largely controlled by a few monasteries which needed the salt badly enough to transport it to southern England.[86] Since the ratio of discarded silty material from the brine to extracted salt was very high, there were large quantities of waste material. Some of this was added to the coastline to form a raised ridge about 1 kilometre wide and 25 kilometres long parallel to the coast (called The Tofts); other disjecta can still be seen forming parallel mounds some 20 metres long and 3 metres high. If the demand for salt fell, then the ponds could be coverted to grazing land or, as in the Mediterranean, to oyster beds. For high-volume materials, water transport was vital, and there is another environmental consequence in the modification of waterways by dredging both inland and at the coast. Rivers, harbours and estuaries silt up by movement of suspended matter along coasts and down the courses of rivers as well as filling up canals. To keep them clear, the cheapest way is scouring: damming up a good head of water and then releasing it so as to exert maximum flushing power. This was practised for example in China in the sixteenth century along the Xuzhou branch of the Yellow River when it was virtually reconstructed under Pan Jixun, a feat which included as one of its minor achievements the dredging of 35,000 metres of riverbed, possibly using the iron 'river-deepening harrow' which was made by the thousand.[87] In the slow-water context of The Netherlands, the Dutch invented machines which loosened the mud at high tide (for example, the *krabbelar* or 'scratcher') so that it might be flushed out on the ebb, perhaps aided by a tidal reservoir whose contents added to the scour. A floating horse-powered bucket dredger (the Amsterdam mud-mill) was in use before 1600.[88]

Whole environments: forests, hunting and gardens

Environmental management in the pre-industrial millennia often extended to environments where the organic whole was as important as any component product. In all three cases there might well be an immediately useful output such as wood, food and herbs but the whole might well reflect the owners' culture in being an exhibition of non-material values, such as pleasure or status. In agrarian economies, forests have been prized mostly for their products rather than their presence. Individual trees can be manipulated to produce, for example, axe handles or tethering posts for animals.[89] On a wider scale, the contest between their usufruct and their potential as agricultural land generally ended in forest removal. (The period of greatest destruction starts in about 1800 and is dealt with as part of the discussion of industrialisation.) In 1700, the Earth held perhaps 6×10^9 hectares of 'primary forest', 2×10^9 hectares of 'secondary forest' and 1×10^9 hectares of 'unexploitable forest'.[90] Before 1800, wooded land on a world scale was seemingly illimitable. Locally, though, the loss of tree cover might cause concern: in England, maritime supremacy was threatened by the lack of oak for shipbuilding, In eighteenth-century Sweden iron production was reduced by central government edict so as to curtail the demand for charcoal, and in Mauritius there was concern that the rainfall of the island was diminished by cutting down trees, an idea extended to British India. We do not know whether the wholesale conversion of forest to *garrigue* for Roman charcoal production caused concern but it created large areas of *saltus* that contrasted with the previous *silva*.[91] Large timber is needed for big buildings though even a farmhouse would eat up 330 trees[92] and a cathedral's roofing, vault frames and scaffolding could require the import of the right size and shape of timber as in the case of the straight poles needed for Ely Cathedral in England which came from Scandinavia. Ships' masts needed straight trunks from, preferably, coniferous trees, so opening trade from Britain with the Baltic and Russia and eventually with North America and the Malabar coast of India. The special shapes needed for larger wooden vessels after the fifteenth century came best from oaks in wood-pasture, where they had space to grow in a more crooked fashion. Money might then be made by opening out woods and selling the timber, grazing beasts on the grassland and the harvesting the larger oaks in due course, all to supply the 20 hectares of mature oak needed to build a seventy-four-gun ship of the line.[93] Smaller trees supplied many other needs including, for example, fuel, implements, fencing, containers. To get a renewable supply of a variety of lengths and widths, coppicing was often used. When cut at the base, many species send forth a multiplicity of shoots which can be harvested on a rotational basis at the desired size. This technique might also be used along with pollarding and shredding to take leafy branches for animal fodder. Another source of loss was fire. Many coniferous forests are fire-adapted but times of particular drought might allow immense fires: some correlation with ENSO events is therefore likely. Warfare, too, consumed woodlands by fire when opponents were smoked out, and instances are known from the Classical Mediterranean to the Seven Years War of the eighteenth century.

The protection of watersheds for 'modern' reasons is shown in seventeenth- and eighteenth-century Japan, where much cutting on steep slopes was forbidden and restoration enforced where there had been clear-felling and shifting agriculture, thus laying the foundations of *midori no rettō*, 'the green archipelago'.[94] In India, King Ashoka protected all four-footed creatures of the forests that were neither useful nor edible but this Buddhist act (*c.*256 BC) was unusual because the preservation of forests was more often as a land resource primarily devoted to hunting. Hunting wild animals by violent means for pleasure was largely an aristocratic pastime and, in some societies, the killing of a large predator was a rite of passage for young men. The management of hunting at its crudest might involve the capturing of an animal to be released in front of the hunters but a more usual practice involved management of habitat so as to foster the chosen species. In many regions of Europe after around AD 1000, red deer (*Cervus elephas*) became the most esteemed quarry and so predators were themselves killed, cover was maintained for laying up and breeding, and there might be closed seasons so that reproduction was as successful as possible. Louis XIV of France (1638–1715) spent three to five days a week chasing deer and is credited with killing 10,000 of them over fifty years.[95] Common people were excluded whenever possible because they might poach the animals (incurring heavy penalties for doing so) or might attempt to convert land to crops or pasture domestic beasts. Thus, large areas of terrain might carry some form of designation in which the hunting interests of royalty were paramount though these might be delegated to lesser groups. In medieval England the king had the rights to the red deer but lesser nobles might be granted the right to chase lesser animals; some bishops had to make do with rabbits. Eventually, many landowners created parks by fencing in order to keep their deer under control, with one-way leaps to encourage deer to enter but not to leave. Such parks might also be reserves of timber and grazing, just as larger areas of royal hunting ground might be sold off to replenish the royal purse. In many places these parks became the scenes of the killing of driven animals and even 'hunts' by women, and were eventually segued into public open space. St James's Park and Hyde Park in London were once royal hunting parks, for instance. In certain limited circumstances, netting might be allowed as noble but mostly was condemned as suitable only for '. . . fat men, old men, idle men and churchmen . . .' Falconry, on the other hand, was always esteemed: the Altai saker (*Falco cherrug altaicus*) was the great pride of Arab falconers, just as the merlin (*F. columbarius*) was for ladies, with the peregrine (*F. peregrinus*) as 'jentyll fawcoun', for a prince.[96] Thus, some species were singled out for special recognition, whether as prey or as aids to the hunters, as were many specialised breeds of dogs.

Gardens, in contrast, have a wider social spread. The word is etymologically akin to 'yard' and betokens an enclosed space. Within are grown plants for utility (such as fruit, vegetables, herbs) and pleasure, (such as trees and flowers) along with expanses of grass and water. Early examples are found in Mesopotamia and Egypt, and there is a continuous history in most agricultural societies of having an equivalent plot of land near the back door of a residence. Only rarely does subsistence depend upon the garden, and it is more likely to yield herbs and spices for flavour and herbs for medicines; the rich are likely to cultivate exotic crops, such as vines and peaches in northern climes, aided eventually by expanses of glass.[97] The energetics of gardens demand an intensive input of labour so that they rank among the most manipulated and managed of all environments. So much is obvious of kitchen and cottage gardens but it is largely true of the multi-hundred-hectare gardens of great houses and palaces throughout Eurasia and the Americas where a park might be turned into a 'landscape garden' by the planned planting of trees, the diversion of watercourses and the removal of the dwellings of the poor. A big enough expanse might also retain some of its wild species to be hunted or to be a reserve of timber and grazing by way of ancillary income. As shooting technology evolved in the later eighteenth century, then birds like the pheasant could be reared en masse to provide park-based sport for even the visually challenged gentleman, just like the red grouse on the uplands of Britain in the nineteenth century.[98] The rich man in his castle and the poor man at his gate are likely to have different ideas about the relative importance of dietary supplements and visual, olfactory and tactile pleasures. The poor might concentrate on vegetables, herbs and perhaps chickens and a pig, whereas the rich have the run of exotic fruits, smooth lawns, belvederes, ponds with brilliantly coloured fishes, arboretums, mazes and *jardins d'amour*. The conflation of the park with the garden in the English landscape garden of the seventeenth and eighteenth centuries shows the social implications very well: it was above all a

symbol of exclusion and separation even if no village had actually to be razed
and rebuilt out of sight in any particular instance.[99] In its apparent informal-
ity, rigid lines nevertheless persist in avenues of trees framing the approach to
the great house, in the classical form of the house itself, and in walled kitchen
gardens. No mazy lines were found in that other nexus of the exploding world
of the Renaissance and early modern times, the botanic garden. Its origins
seem to have been in the herb gardens of the medieval monasteries of Europe,
which exchanged plants over a continent-wide zone.[100] The gardens were then
transformed into a site for curiosity (both proto-scientific and popular) and
later into a kind of market for the exchange of species between continents as
European exploration opened up new trade and new empires. Culturally, it
became a microcosm of what might be achieved by way of improvement of
the natural order (provided that the right people were in charge) perhaps even
to recreating the Garden of Eden. Such possibilities were also explored in
poetry and painting and, in essence, they and gardening can be regarded as a
cultural continuum.[101]

   The cultural context of forests is also complex and historically very deep. It
tends to be recounted in terms of those attitudes which made for clearance
and those which helped protect the trees. There is, naturally, a tension between
the two but one which is resolved in favour of control and consumption if
there is any doubt. The Benedictine orders had no doubt that the reclamation
of all wild places was holy labour, and woodland was included: the Abbot
blessed the woodcutters and their implements because, '. . . a wild spot . . . is,
as it were, in a state of original sin.'[102] There is a take-up of those kinds of ideas
into the Puritan view of the forests of North America, the clearance of which
was an act of redemption and clear sign that a divine purpose was being fulfilled: progress, control and usefulness were the watchwords. It is an interesting speculation as to what extent these outlooks drew upon a deeper fear in
European history of the woods as places of terror, with dangerous mythical
inhabitants whose lineage might well be descended from hunter-gatherers.
The resonances of *Little Red Riding Hood* may well go beyond even the more
springy branches of feminist theory. The protectionists can point to political
writers such as John Evelyn writing in seventeenth-century England not
merely about the likely shortage of oaks for ships but of trees as visible
symbols of human society. The oak became a symbol not only of maritime
power but somehow standing for the nation, like John Bull and roast beef. Add
to that the undercurrent of feeling from the colonies that forests and rainfall
were somehow connected, and the likelihood of keeping woodland rather
than felling it is increased. Stir in the probability of profit and the prospect of
pleasure and landowners began to initiate plantations, and governments to
look towards some kind of 'scientific' forestry: all this before the end of the
eighteenth century. In such circumstances, the approbation of Romantic
poets such as Byron or Wordsworth: 'one impulse from a vernal wood may
teach you more of man . . . Than all the sages can' is soon forthcoming.[103] This

can be seen as a counterbalance to the notion that the clearance of forest and the cultivation of land was a founding myth in western history from Classical times onwards.[104] Trees, woods and forests, then are tied in with our individual psyches and our national ethos, and fully amplify W. H. Auden's more recent poetic assertion that 'A culture is no better than its woods'.[105]

A cultural context for hunting might run to many pages. The male ego is much implicated in the sense that hunting for status seems likely in hunter-gatherer times, that male aristocrats were the chief pursuers in cultures such as that of ancient Mesopotamia and feudal Europe, and that in many places the ability to kill large predators was an important rite of passage especially to high office such as kingship, so one form of control is mirrored in another. The pleasures of the hunt spun off into other activities, as noted by the sixteenth-century Indian king Rudradeva of Kumaon: 'The sentiment of self-importance makes the enjoyment of women all the more pleasant after hunting' and in Europe, courtly love was often part of the experience.[106] There are wider social incorporations as well. In Classical Greece, the hunt was seen as the triumph of the rational and humanist over the bestial and irrational, but the medieval Christian view was somewhat modulated by the many legends of the saving of the white hart, which had become a religious or royal symbol, and that of St Eustace, who was converted by a vision of Christ in the antlers of a stag that he was hunting; thereafter the wild beasts in the arena would not touch him. Eventually, hunting might become a symbol of the tyranny of the aristocracy and poaching acquired a risky respectability among the peasants and any minor gentry who might feel excluded. The rise of science loaded the scales a little further in favour of the wild animals though, in the nineteenth century, Darwinism provided a counter-weight.[107]

Inorganic production: minerals and cities

The attraction of rock as a resource goes back as far as humanity. Once farming became well established, then ores were sought so that first bronze and then iron might enter the tool-kit. Stone's durability attracted the builders of monuments and other important buildings, and shiny forms of rock joined gold and silver as precious items. Salt from underground deposits was always sought since production per man-hour was higher than coastal salterns. 'Mining' has strong industrial associations but it was widely practised in pre-industrial times:

> Man puts his hand to the flinty rock and overturns mountains by the roots.
> He cuts out channels in the rocks, and his eye sees every precious thing.
> He binds up the streams so that they do not trickle. [108]

The Classical economies of Greece and Rome were famous for their wide use of stone and metals, and the flood of silver and gold from the Americas worked as a spur to European colonisation during the fifteenth and sixteenth centuries. The amount of environmental alteration caused by the extraction of rocks of all kinds was very variable but usually nothing was allowed to get in the way of the cheapest route to mining and processing. Where a small local quarry was concerned, then there was no huge impact, but a major quarry was there for all time. Deep mining required timber for shoring tunnels and shafts and would probably need some form of drainage. Hence the impact of mining extended to the surrounding area as effluent water was led into streams and rivers, forests felled to provide mine timber, and heaps of solid wastes piled up near the mine entrance. Wherever smelting occurred, then these processes were intensified and air pollution might well be added, with waste gases adding an especially toxic element detrimental to most forms of life. Pliny the Elder (AD 23–79) noted the use of bladders as respirators by the workers in zinc smelters. As examples of the reach for inorganic materials, we can note Roman mines as deep as 150 metres in Etruria and five to six hundred Bronze Age mines in Austria each using 20 cubic metres of timber per day. Medieval tin mining in England consumed farmland at 120 hectares per year, and a Cistercian monastery took 40,000 cartloads of stone to build; toxic gases affected cattle for an 11-kilometre radius around a lead smelter in the Tyne Valley of northern England. Moreover, there was an incipient trade in fossil fuels: these were to become the foundation of the Industrial era but, even before, say, 1750, coal was in widespread use where its transport (especially by water) was cheap: the River Tyne exported 40,000 tonnes a year in 1700. Thus, London brought in coal by boat (so it was called sea coal to distinguish it from charcoal) to supplement its wood supplies, and a number of medieval and early modern proclamations against its use were made, mostly to no avail, so that corrosion of buildings was noted by 1661. In more minor usages of fossil hydrocarbons, seepages of tar from oil shales in China helped caulk ships, and natural gas was led from surface leakages into bamboo pipes and thence burnt as street lighting as well as used to boil brine in the refining of salt. Seepages of crude oil were also known, and in the eleventh century AD it was thought that they would replace pine wood when the forests were exhausted; the oil was seen, however, as a source of ink, not energy. Marco Polo talked of the black stones that kept a fire going better than wood and gave out great heat. In AD 800–900 Islamic rulers of the Baku region (now Azerbaijan) developed the production of crude oil on a commercial scale for medicine, to burn, and for weapons. They also knew how to distil it for the lighter fractions. Neither culture seized upon the use of hydrocarbons to generate steam on an industrial scale: the earliest uses of steam technology were developed from the seventeenth century onwards, in Europe and their floresence is one of the great environmental manipulations of the Holocene.[109] The cultural context is, as always, complex but two points can be made: why do some substances, such as shiny metals and jewels, attract human attention to the point where everything, including the environment, can be sacrificed in order to possess them? Secondly, in pre-industrial times, it was mostly the slaves who worked in mines.

Slaves are useful in cities, too. The concentration of people brings a concentration of materials, some of which need preparation, as with peeling and chopping vegetables, and others need disposal such as excreta and general rubbish. This is just a microcosm, of course, of the ecology of a city. It is a crystallisation of energy and materials which it stores for varying lengths of time.[110] The pre-industrial city has a high organic content because much of it is usually built in wood and thus has a relatively rapid turnover of its storage components, not least because severe fires are frequent. The city must take in enough energy to feed and perhaps warm its inhabitants, build enough structures to house them and their industries, and dispose of harmful wastes. This ensures the creation of many microhabitats. Under calm conditions, a late medieval city might build a heat island of 4 °C, and the higher temperatures allowed urban vineyards to persist into the Little Ice Age in western Europe. Much of the heat came from wood: processing London's beer and bread requirements in 1300 for a population of perhaps 80,000 would have required about $29-32 \times 10^3$ tons of wood per annum, equivalent to at least 518,000 acres (209,631 hectares) of coppice.[111] When, in the eleventh century, herring was the chief source of animal protein, its preservation by smoking needed a good deal of fuel, a demand somewhat diminished by a shift towards meat in the mid-fourteenth century. Warmth encourages many species, most notably fleas and mites, and it is not noted that rats and mice often die of cold. Towns with poor waste-disposal systems need scavengers, and the pig, feral dogs and birds (in Europe notably the red kite, *Milvus milvus*) all helped. If sewage went into backyard cesspits then the organisms that cause human diseases, such as dysentery and cholera, were encouraged. The reject material from the butchers and the tanners needed a reasonable flow of water to wash them away to be someone else's problem or food for the gulls. It is easy to overstate however, the contribution of urban centres to worldwide environmental change, for until the nineteenth century most of them would now only rate the label of a 'small town': as late as 1700, only fourteen cities exceeded 200,000 inhabitants. A city might well control the land cover of a large surrounding area though in medieval Europe the monastic contribution to land-use patterns was probably every bit as great.

The culture of cities is a topic that has spawned an immense literature though most of it deals with the nineteenth century and after. But one apparent paradox comes down to us. On the one hand there is the phrase *Stadluft macht frei,* in which the towns' escape from many of the bonds of feudal society is celebrated: the town was a place of innovation, both technical and social. In symbolic opposition, many cities in many cultures were walled: partly to keep out enemies but mainly to control who entered and left and to make sure they paid the appropriate tolls.[112] Some at least of the freedoms were illusory or at any rate bought at premium prices.

Water and life

All societies have to have access to water to function. Indeed adult humans must have roughly 2 litres per day if they are to stay alive.[113] The planet has great quantities of it but 97 per cent is in the oceans and 2 per cent in polar and mountain ice; of the rest, 95 per cent is in underground aquifers. The encounter with human societies is therefore mostly in the weather which cannot be controlled, and in surface run-off which is a more likely scene for attempts at manipulation.[114] Intervention in the hydrological cycle must be of great antiquity though no one knows when the earliest attempts took places: small-scale attempts to divert a stream into a stand of wild grasses that might yield harvestable seed heads might have followed the success of the grass stand in attracting grazing animals, but this is highly speculative. Nevertheless, effective water management became crucial in many solar-powered agricultural societies, as discussed above under irrigation. The energetics of water management are much helped by the pull of gravity, and so water from springs and precipitation needs only a little help to be poured, so to speak, into containers on its way from hill to the sea. Getting rid of too much water generally means enlarging the containers and speeding the water on its path or the one chosen for it. Freshwater ecosystems are usually high in biodiversity and wetlands high in biological productivity. Water for direct human use, by contrast, is best if low in biodiversity (especially of bacteria but nobody wants frogs in the fountains, either) but all kinds of content are acceptable downstream provided that swift flushing takes place. Moving water against gravity led to some early technologies being used in irrigation and also to the development of the windmill which, like the watermill, might be used to grind corn or mineral ores but might also lift water from areas subject to flooding or overall sogginess, as with the medieval reclaimed lands around the North Sea in England, The Netherlands, and Frisia.

The management and impact of water management therefore centre upon three phases: supply, storage and removal. The supply may be to people, to crops and to industry, and the removal from farmland, industry and urban concentrations. Storage is less necessary in climates with year-round rainfall and constant levels of demand, but this leaves a large number of societies wanting to even out the supply. The requirements are also different upstream- and downstream: supplies need if possible to be pure (though high mineral content may help farmers) but the fluid needed to disperse toxic substances need not be all that pure. Until the arrival of steam it was only by water that bulk, low-value goods could be economically moved any distance at all. The river lock and the canal are environment-altering pieces of technology that catered to this demand. Apart from irrigation agriculture, the outstanding example of water management must be its supply to cities. The Romans are famous for their construction of aqueducts into cities and garrisons, and for their ability to remove sewage (in such diverse places as the *cloaca maxima* of Rome itself and the multi-seat flushing latrine at Housesteads Fort on Hadrian's Wall in northern England), to the point where it took some time for the equivalent achievements of medieval cities and monasteries to be recognised. Among many examples, the Consiglio Generale of Siena in the early fourteenth century had been responsible for the construction of 25 kilometres of underground channels (*bottini*) to bring water to public fountains and also dedicated at least one to industrial supply via the guild of the Arte della Lana. In London at about the same time, the Great Conduit in 'Cheapeside' was a 'Cesterne of leade castellated with stone' and its supply was contested between the ordinary people and the brewers, a competition seemingly ended only by the Black Death.[115] All such supplies needed upstream diversions of streams or springs.

Another direct impact on freshwater life has been the demand for fish as food. Preserved sea fish and locally caught fish such as eels, carp and pike were highly desirable, especially if meat was sacrce or prohbited. So weirs and traps were elaborated and ponds dug, from China to Ireland. In the 1600s the demand for carp (*Cyprinus caprio*) turned the eastern edge of the Paris basin from woods and wetlands into 'mist-shrouded expanses of water and reeds'. Both habitat manipulation and their classification as 'pests' meant that beaver and otter were subject to population decline: beaver were classified as 'fish' in Roman Catholic Europe and thus were substitutes for meat on fast-days. Management in France failed to prevent over-fishing, provoking a full-scale royal ordinance in 1289.[116]

Springs and watercourses invite propitiation: metal-using prehistoric cultures threw precious objects into water, the Romans at Bath in England venerated the goddess of the hot springs. The centrality of the Nile's water in Egypt was shown in rituals that related water to the hope and promise of life beyond the grave. During Roman rule (200 BC–AD 200 was the peak time) there were processions in honour of Isis in both northern Egypt and central Italy in which a pitcher of water was carried by a cult official; the water made visible the god Osiris.[117] The religious rite of baptism is an obvious successor and fits well into a land where water was scarce. Water plenty that needs control also spawns a hierarchical mind-set: the example of Bali has already been mentioned (p. 70), and the covenants of governments of the Netherlands meant that from medieval times water engineering was part of the national psyche, and its practitioners were allotted high rank. In thirteenth-century England, all royal ponds south of York were co-ordinated by William, the King's Fisherman. Consider then that the pitcher of water and the sea-bank are entirely isomeric: both of them contain and make control of the liquid possible.

There go the ships . . .

'. . . and there is that Leviathan: whom thou hast made to take his pastime therein', runs part of Psalm 104, summarising two important parts of human relationships with the sea.[118] Even if not potable, seawater can bear vessels, and yield salt, fish, whales and other forms of marine life. All require technology, aim to exert control, and have some impact upon the natural systems. At the margins of the oceans, the extraction of salt resulted in the transformation of upper salt marshes into dry land. In addition, many fish traps were set up in intertidal zones to catch fish left by the falling waters and, of course, harbours were constructed and piers led out to protect harbour entrances. Fish populations may not have been affected much by the traps and nets but from modern experience we know how the movement of sediments along a coast can be affected by training walls. In pre-industrial times these effects were lower in magnitude but not absent. Whaling was carried on from medieval times and the oil much sought for lighting; yet again, the populations were probably little affected in the long term except in very popular hunting grounds, such as the Bay of Biscay. Fishing was not normally managed in any modern sense, nor were the animals bred for particular qualities, and so the whole process has been much like hunting and gathering without the conservation elements discussed in chapter 2. Many a medieval commentator echoed a view that there was masterless wealth. Yet, in the thirteenth century the commercial herring fishery of the southern Baltic disappeared when there was enhanced run-off from land clearance in the Odra and Wisła basins; the breakdown of herring stocks in the North Sea and off Skåne around 1400 may have come from an interaction of heavy fishing and climatic change. As often in later eras, one response was to search for stocks further away: off Iceland and Newfoundland, for example.[119] A European movement between AD 500 and 1500 from subsistence to artisanal and then to commercial fishing saw impacts on progressively more pelagic species. As early as 1376 an English document inveighs against the use of a beam trawl, which dragged up from the bottom of the sea 'all the bait that used to be the food of great fish'. The herring was the heart of massive trade, as improvements in boats and preservation methods became common; allied to changes in weather and climate, the herring fisheries of the southern North Sea broke down after 1360 and those of Scania between 1410 and 1420.[120]

Apart from the harbours, sailing ships are environmental factors only in a secondary sense, if we accept the power of the oceans to process the wastes ejected from the vessels. The development of the technology of the ocean-going vessel is critical and, though the Portuguese carrack leads the top list, the contribution of Islam was great: in the tenth century, several dozen kinds of boats (all carvel-built) were catalogued and most had the lateen sail later added to the square rig by the Portuguese. In Egypt, forests for the production of construction timber were cultivated, to grow *Acacia nilotica* (in Arabic, *sant*) but much more wood had to come from the western Mediterranean provinces such as Al-Andalus in southern Spain. Dockyards with a military function were the *dar al'sen'a*, which (via the Arsenale of Venice) give us today the power and might of a great British football club.

Human cultures and the sea are prone to romantic interpretations. The early modern period and the nineteenth century alike purveyed the image of the British as an island race with salt in their blood, which is no doubt why the navy had often to press-gang men into its service. Small ships no doubt provided an escape from conformist pressures, and all voyages and explorations were an outlet for the authoritarian personality once he [*sic*] had become the skipper, and in English at least 'the ship of state' figure of speech conveniently overlooked the less savoury aspects of service at sea.[121] Thus, fear is probably bound up inexorably with navigation: the vastness of the seas, their power to overwhelm, and, even for the stay-at-home, the risks of investment in maritime trade. The anonymous Anglo-Saxon poet who wrote 'The Seafarer' probably both echoed and prefigured a widely held view when he sang that,

> He knows not,
> Who lives most easily on land, how I
> Have spent my winter on the ice-cold sea,
> Wretched and anxious, in the paths of exile
> Lacking dear friends, hung around by icicles,
> While hail flew past in showers.[122]

So in pre-industrial times, the seas were open to partial control at a few places on the coasts, but worldwide, very little. Beyond the sight of land, there appeared to be practically no human influence upon the ecosystem processes produced by evolution. That was an illusion even in medieval Europe yet still an attitude to be perpetuated right into the middle of the twentieth century, when the ships still went but the leviathans (and many other species) became fewer and fewer.

Diminution and disappearance

In one sense at least, this pre-industrial world has not disappeared. Though there cannot be any terrestrial ecosystems that have not received the impacts of industrialisation, many parts of the world function largely on flows of solar power and not on fossil fuels. Many forest areas, unmanaged grasslands, mountains and icefields receive only a fraction of their energy inputs via technology from coal, oil or natural gas, and so belong to this era rather than that of fossil-fuel-powered industrialisation. Areas of crop production have been moved towards dependence on oil and its products (usually termed 'modernisation') but there are still places where its penetration is shallow and second-hand, so that human and animal power are dominant. Metals are ubiquitous, but then iron was very widespread in most cultures before the eighteenth century, albeit in much smaller amounts than in later centuries. Nevertheless, the incidence of pre-industrial agricultural systems ('traditional farming') is declining, often as local subsistence is abandoned in favour of participation in wider markets by producing cash crops.

The 'old ways' have a certain romantic appeal, especially from a distance. Where they resulted in a landscape with high aesthetic appeal and therefore likely also to be one which attracted well-heeled tourists, then the pressures to maintain such technologies and social structures have been high. Farmers are paid subsidies to perform in a manner reminiscent of the past, for example, by eschewing certain types of machinery, fertilisers and pesticides or by using high summer pastures otherwise abandoned. So far the most obvious manifestation has been in 'organic' farming which usually accepts machinery but not bagged chemicals so that its energy inputs are lower than intensive 'modern' farming, but not at the levels of pre-industrial 'subsistence' farming.[123] All, however, operate in a matrix of an industrialised world. Such places, though, are appealing to well-off travellers: an advertisement for the 'real' Spain or Thailand is likely to lead to a landscape with relatively low levels of modernisation of the rural economy, and very likely some poverty as well.

A last manifestation of the former ways is seen in groups with an ideological rejection of today's practices. Probably the best known of these are the Amish cultures of the north-eastern United States, where the stricter adherents travel in horse-drawn vehicles and eschew inventions such as the zip fastener. Their farming methods, too, rely more on manure than on bags of nitrogenous fertilisers.[124] In consequence, they are the focus of a lively tourist trade and are gradually marrying out; disappearance or Disneyfication seem likely fates. Occasionally to be seen in western countries, too, are communities of the type left over from the alternative societies of the 1960s and 1970s. The majority of those self-sufficient communes have disappeared but a few remain in favoured climates and live by producing their own food, with a little surplus for sale or barter. It is difficult not to suspect that they receive cash subsidies but most rely on generating their own electricity, if they use it. Again, the matrix is important: if the axe slips then the local Accident and Emergency Department is no doubt a welcome sight.

Overall, therefore, we can discern islands of lower energy-intensity systems within today's world. These have more affinity with the pre-industrial rural world and its environments (much less so with its cities) than those of today. Some commentators regard them as seedbeds for a post-industrial future, others as mere relics of the past, preserved for those on holiday or for childrens' history lessons. Such places (often in island form) are the common basis for imaginary places where conditions are usually better than the prevailing reality.[125]

SEWING THE WORLD TOGETHER

The Silk Road can be used an an image for the transfer of materials and ideas which inevitably will lead to economic and ecological change. It connected

China and Europe overland and was active from Roman times until the fourteenth century AD. News of Chinese technology and materials (rhubarb as well as silk) came westwards, as did pathogens such as bubonic plague; Buddhism and Islam went eastwards. It was open when empires created stable conditions such as the *pax romana* and the *pax mongolica*, themselves sources of coalescence of economies and ecologies as well as the mixing of human genetics.[126] The romantic view of this route makes it a symbol for the many channels of trade and exchanges of materials and thinking that stitched together parts of the world during the agricultural era. Less romantic and more effective were the volcanic eruptions which disrupted climatic conditions over wide areas for a few years at times when vulnerability would have been high: AD 536 was one such (though that might have been a comet), 1258 and 1783 were others. Thick 'dry fogs' were one widespread consequence and cold winters another. The 1258 eruption in the tropics was certainly recorded in the tree rings of Mongolia.[127]

There were different scales at which human societies coalesced during the era of solar-powered agriculture. Obviously, it happened at local levels as the population grew and people came into contract with their neighbours, with whom they traded, fought and married. Larger-scale coalescences involved migrations, when a culture was taken long distances, as with the Austronesians who voyaged from Taiwan to Easter Island/Rapanui one way and Madagascar the other. Nation states evolved from smaller kingdoms and principalities and some aspired to become empires within which there was a dominant culture, as with Rome. Empires might be contiguous pieces of terrain or separated patches linked by sea or land communications and thus not always very different from trade routes organised by one powerful elite: the transition of India from a company fiefdom to a jewel in an imperial crown was relatively simple. Trade often started as the transfer of precious goods and remained thus while the carrying capacity of ships remained small and included the carriage of slaves as one of the most ubiquitous of cargoes. Three-fifths of the people who crossed the Atlantic before 1500 were African and only they made sugar and tobacco successful export crops. Simple exploration led by curiosity was a relatively rare determinant of pre-modern expeditions. The immense voyages of China's circumnavigations of 1421–23 seem to have left only a few memorial stones and perhaps some DNA.[128] The acquisition of materials or the conquest of peoples were usually part of the leaders' instructions even if they were cloaked with religious motives, as when Islam and Christianity both assumed that each of their faiths was destined to become the sole religion of humanity. There was, however, no global consciousness before the coming of cheap print, faster communications and better education, all following industrialisation: the earlier era was a series of separate elisions rather like the blobs in a lava lamp. Some bubbles were large, however: in the time of the maximum extent of the Roman, Mauryan and Han Empires, a precious object might be traded from Galloway in Scotland to Seoul in Korea. In the thirteenth century, there was a world system in which its eight 'circuits' allowed trade (in expensive items at least) from England to Indonesia. Chase-Dunn and Grimes identified twelve such entities (which they called 'political/military interaction networks') and, by AD 1500, nine of them could be identified largely as a single network, with only Indonesia, the Far East and Japan standing outside until the nineteenth century.[129] After 1500, largely regional trade became worldwide, though with some areas of greater intensity than others: few merchantmen voluntarily put into Tierra del Fuego. Yet by 1500 the older practices of simple bringing-together of merchants had been amplified and overlain by the transfer of technology such as the gun, and diseases which brought about exterminations of peoples in the Americas, Caribbean and Pacific.[130]

The results in environmental terms are well known. The highest profile is accorded to the transfer of species between continents, particularly where a large-scale export crop was emplaced. Sugar is perhaps the best example but there are several others, such as tobacco and maize. Tea and coffee were expanded well beyond their original scales and ranges of production as well. The range of species introduced to Europe, for example, is very wide. The main crops of the Neolithic came from south-west Asia (that is, wheat, barley, pulses, sheep, goats and cattle), and the Mediterranean zone acquired sorghum from Africa as well as rice, cotton and sesame from south Asia. In Classical times, the empires took in fruit trees such as apricots, peaches and walnuts, pears and oranges from Asia. Muslim expansion took with it sugar, the citrus family, rice and cotton. A selection of these crops was then taken to the Americas (sugar and cereals being the most important but also sheep, cattle and the horse) accompanied by many diseases of humans which proved efficient at killing off 60 to 90 per cent of native American populations.[131] Exotic animals populated aristocratic menageries: George III of Britain had a zebra, for example, and there was a royal zoo at the Tower of London between 1235 and 1835.[132] The ecosystemic effect in terms of coalescence was the spread of large-scale cultivation which involved the disappearance of native vegetation and, fauna and its replacement with a relatively uniform set of exotics, as described for sugar on p. 74.

There was transfer of ideas, too. Without doubt, the people of the Hebrew Bible put into place the idea of time as linear (though not infinite) rather than as annually renewed, as had been the case with many foraging groups, and this basic notion became virtually universal (excepting notably in Hinduism) with the spread of European and Mediterranean cultures. Where an empire holds sway, then it can enforce (usually with varying levels of success) its hegemonic modes of thought which may well extend to the natural world. In theory, for example, Muslim and Christian worlds should have pursued the idea of the stewardship of nature rather than its conquest, though hard-and-fast evidence is hard to find.[133] In a classic paper, Lynn White argued that Christian thought (as seen in iconography) favoured technology and that therefore it was at the basis of the 'ecological crisis' of the twentieth century;

many trees have been felled in refuting and restating that controversy, which resonates still.[134] Empires of all kinds (Venetian, Dutch, British and French especially) encouraged trade and territorial expansion, and the city's place as a site of technological innovation was enhanced with every vessel that arrived.[135] Their scientific elites influenced governments with ideas which sound very modern and environmentalist and were often a direct response to the destructive environmental policies of colonial rule.[136] Lastly, there are deployments of thought and practice that are capable of bringing about greater coalescence but also deeper division. The development of measurement (including that of time) in the late Middle Ages and Renaissance of western Europe turned that culture into one that put quantification and visual apprehension at the heart of modes of thought which were then taken on everywhere else.[137] Map-making can be seen as one expression of this and one which, while allowing navigation and exploration, also permitted the drawing of accurately placed boundary lines: Gerard Mercator's (1512–94) sixteenth-century contribution deserves special mention.[138] Ambiguity of coalescence and division includes the transfer of species in the sense that, although the world is the more uniform in its crop distribution, the production methods induce a separation of masters and servants (indeed, slaves in some places) and a separation of profitable crop and unnecessary nature. At an even deeper level, much of western culture in general is underlain by the thinking of Classical Greece, which included environmental matters: the names of Aristotle and Theophrastus are usually included in any discussion of environmental ideas.[139]

ALL COHERENCE GONE?

One of the abiding themes of the world's agricultural millennia must be the separations that became possible. Certainly by the early seventeenth century, when John Donne wrote of his world 'tis all in pieces/all coherence gone' rather than of its 'joint tenants' and when J. S. Bach separated the fundamental harmonics of nature from those of a musical instrument in 'The Well-Tempered Clavier' in 1720–40, the creative arts seemed to have latched on to a resonating theme in human history. As populations grew and diversified culturally, and as access to resources became differentiated within and between societies, the developments typified by the discussion of coalescence came into tension with those of stratification and fragmentation. Archaeological research has pointed out the way in which the fragmentation of objects – with parts being buried with the dead – both differentiated and cemented a social identity in Neolithic Europe. This suggests that there were social practices which effected both tendencies at the same time.[140] One major avatar of fragmentation in western societies has been the growth of individualism where the rise of the private sphere in Europe between 1000 and 1800 anticipates the many trends of the twentieth and twenty-first centuries. This emergence had periods of acceleration, as in the wake of the Black Death when there was less feudal control and the family unit became dominant. Much of this was cemented by the rise of the 'artist' (as distinct from the artisan) in the Renaissance and the affirmation of the standing of the individual human in the Reformation and the Enlightenment.[141]

There are many archaeological examples deriving from 'prehistoric' farming cultures, when it appears that the division of land resources between groups becomes important and that markers need to be visible symbols of such division, whether they be burial mounds on conspicuous sites or ditches dug across the landscape. Few social historians fail to point out the way in which separations of role and withdrawals of behaviour come along with successful agriculture. It appears to hasten the dominance of men, for example, and sharpens the focus on the rich as they sequester resources which then appear as conspicuous displays of power. The grain surpluses of the Nile valley are thus transmogrified into the pyramids and the other galaxies of treasure designed to procure eternal life. Such societies often felt threatened and so engaged in warfare with their neighbours which, in turn, required a pro-natalist policy to sustain the warrior element in the population. Smaller-scale actions have also attracted notice: the withdrawal of the lord and lady from the common hall of the medieval manor to their private room ('the solar') behind the dais, and the early modern development of the corridor in large houses are further examples. Increased wealth meant an ability to own and organise land for pleasure as in hunting parks and landscape gardens, and to exclude the lesser folk from them by means of restrictive (and indeed often draconian) laws.

The environmental consequences of these social crystallisations were many and varied. Agriculture provoked a cultural evaluation of non-crop species, with many of them being regarded as 'other' and therefore liable to be extirpated. 'Predators', 'weeds' and 'pests' became sharply differentiated in a way not native to foragers. Inevitably, some became extinct. The best-known examples are perhaps the ground-dwelling birds of Polynesia, whose species diversity plummeted under the land clearance and the introduced rats of colonising people. Small populations of easily killed animals succumbed relatively quickly, as the history of the dodo, another flightless bird, shows. Our focus is often on the visible loss of tree cover and its fauna (as in the Caribbean) or the disappearance of a noted species (as with the retreat of the elephants in China) but we need also to remember the multiplicity of fungi, bacteria and other micro-organisms, and the complex communities in which they lived that were also farmed out of existence. Farming may leave fragments of earlier habitats in which a species is isolated so that the breeding population is too small and too genetically uniform to survive for long, just as it may improve the chances of species which flourish in edge habitats. In Europe the common fox (*Vulpes vulpes*) is one of the latter category. The highest profile of all such retreats has been given to forest cover though, as

Michael Williams shows, the industrial era has been the time of most clearance. The calculation that a hand-powered pitsaw can convert trees to planks at about 100 to 200 board feet per day, whereas a water-powered saw of 1621 can raise that figure to 2,000 to 3,000 board feet a day, is indicative of the advance of technology, though it is eclipsed by the steam-powered band saw of 1876 at over 20,000 board feet.[142] The result in pre-industrial times is summarised by Williams as,

> . . . the pre-agricultural closed forest probably once covered 46–28 million km^2, and more open woodland 15–23 million km^2, and these have been reduced by 7.01 million km^2 and 2.13 million km^2, respectively. Other evidence based on historical reconstructions of clearing supports the general magnitude of change as being between approximately 8.05 million km^2 and 7.44 million km^2. . . . [there was a] global reduction of between 7 and 8 million km^2 of closed forest, and between 2 and 3 million km^2 of open woodland and shrubland . . .[143]

The other great change was in wetland habitats, where there were worldwide shifts in land cover and land use on a variety of spatial and temporal scales. Attention has been centred on massive schemes in Russia, China, Holland, England and the United States in times before steam power was available to pump up water and so drainage relied on gravity or wind-mills and to a lesser extent horse mills.[144] Many habitat changes were enormous (much of the central Netherlands was stripped of peat bog for agriculture, and also mined for the peat)[145] but so were the smaller projects, which dried out ponds and odd bits of reed swamp which might well have been a refuge for migrating waterfowl. The same is true of patches of semi-wild vegetation left from earlier land-use regimes, such as lowland heaths in Europe, home to birds such as the nightjar (*Caprimulgus europaeus*) but which are cheaply convertible to cash crops and grassland without the use of industrial power.

Such lists are potentially endless and ought not to exclude the seas where, for instance, the silt run-off from the land can affect the breeding success of fish or even the food webs of adults; it is even possible that over-fishing can be achieved in pre-industrial times if a heavy effort coincides with alterations in water quality or with climatic shifts bringing about alterations in water temperatures. But they all stem from basic human states of mind and actions. The most important is that there were no overriding cultural reasons why nature should not be altered by breeding of species, by extirpation of pests or by 'improving' the land cover. The second was the ability to bring about the changes, increasingly through the use of technology as a pathway through which to direct human and solar energies. In the linking across the double helix there was an acceptance of the results, both social and environmental.

REPRESENTING THIS WORLD

There is no shortage of self-representation of human and nature during the agricultural era. There is writing in many scripts, pictures in several media, and sculpture. Much of this is direct portrayal: of the way to reclaim marsh in China, or of the desired landscape of Renaissance Italy. There are also indirect messages, as when medieval Christian iconography showed that it was the elect that had the technology: it was not the Devil's work at all. There is also the tantalising message of the image-rich walls at Çatalhöyük (in south-central Turkey) at 9500 BP: men hunting, vultures and headless people are depicted. One interpretation is that hunters became sedentary and then began to domesticate, and that making pictures of wild nature was a way of taming it in the mind (basically, overcoming fears) before taming the actual organisms. On this hypothesis, the shifts in the mind came first and were followed by the biological changes.[146]

Out of the material that populates thousands of libraries, museums and galleries (and not a few private and sometimes illegal collections), perhaps some selection can be made. The first item might be the portrayal of landscape as a major feature in a picture rather than simply appearing as a backdrop. Art historians are apt to disagree about such matters but let us plump for Giorgione's *La Tempesta* (The Storm) of the late fourteenth century. There are human figures that might signify both fertility and destruction and there is, indeed, a stormy sky. We can, though, read into it a moral ecology of birth, fertility, destruction and change. They are all interwoven and so it is truly ecological, with the natural world having a significant part.[147] The second item is the appearance of the kind of empirically based and systematic writing that shows the emergence of the underpinning concepts of modern science. Competition for the Founding Father [*sic*] award is fierce and who is to say whether Newton, Galileo, Bacon or Descartes ought to claim the podium? Possibly it was Newton (1642–1727) because he showed that the known universe could be represented by equations and was therefore knowable; maths and physics could become instruments of human happiness. But given that we want an environmentally oriented hero, let us spotlight Stephen Hall who was elected to the Royal Society of London in 1717. He laid the basis for work on the human impact on air quality and vegetation change, not least by influencing colonial authorities in France and Britain.[148] The rise of the natural sciences as the primary way to represent the world is, however, a nineteenth-century story, as is Adam Smith's view of the Industrial Revolution as a salvation for humankind. We must not forget that writing is an essential medium for scientific developments because, among other things, it enforces autonomy (and allows permanence) for all manner of thoughts where definition and accuracy are essential, as it has done since (at least) the time of Plato.[149]

Such discussions make us realise that what we have now is a narrow band of material, from the people at or near the top of any hierarchy: a history written by the victors, so to speak. Many groups are little represented: women come first to mind, but the colonised and enslaved are also under-mentioned. The native people of most regions were stigmatised as backward and 'savage' and only recently has Traditional Ecological Knowledge been revalued. Their environmental attitudes are often found in folk tales and folk songs, as with the God of the Stomach in Note 1. Thus, the image of the llama singing along with the Inca king shows at once the importance of a domesticated and hence controlled animal but also one which had something to tell even a king.

OUTCOMES

As cultures became differentiated, so describing and interpreting their history becomes more complex. To end this chapter, three themes will be condensed from the mass of ideas and data available. An empirical picture from standard historical information will be given first, looking at the actual extent to which, by about AD 1750, the world was physically changed. This is followed by a discussion of the technology by which this was achieved but which discusses the conceptual impact of technology as much as its actual physics. It acts as a bridge to thinking about changes in the human imagination during the agricultural era which had relevance for environmental change. The obvious point at which to stop is the emergence of a written philosophy of the environment. The cliché has it that philosophy bakes no bread, but once agriculture is established, philosophy springs forth in abundance.[150]

The world on the cusp of industrialisation

There are two main ways of assessing the state of the world in about 1750. The first is look at current indicators of human-driven environmental change, such as population, carbon dioxide or sulphate emissions and see if they were showing upward trends by 1750; the second is to inspect maps of land use and land cover for that date as a prelude to tracking change in later centuries. In the case of the first, the picture is relatively simple: population has clearly started to take off towards its global explosion after about 1850. It was not, however, launched upon the steepest part of the curve. None of the other main indications, within the limits of accuracy of their reconstruction, shows any major rises in the eighteenth century with the exception of nitrous oxide (N_2O) which is probably correlated with the area under agriculture, the introduction of legumes in crop rotation and the use of animal manure.[151] A map of the land cover of the Earth in 1700 would show some very definite areas of cultivated land and pasture land, in places where we would expect: Europe, south and east Asia and Meso-America, with a fringe in eastern North America and a scattered presence in the southernmost parts of Africa. But many maps underestimate the human presence: many savannas, for example, were the sites of shifting cultivation, a practice also found in boreal forests.

Many maps show tropical forests as if they were virgin lands whereas it is likely that some at least were quite intensively occupied (see pp. 33–4). Add in those lands where fire was still an important tool for hunting and gathering populations, as in Australia, and the world's land surfaces show ample traces of enduring human presence, the more so since the stability of agriculture was high in places such as the Nile valley, the wet rice lands of southern China or the formerly wooded areas of north-western Europe. The seas are generally left out of any discussion but land-use changes may well have added to artisanal fishing in impacts upon some fish populations; the early dynamics of whale populations are speculative but the shift of whaling centres from the Bay of Biscay to the Atlantic and then to the Arctic might be interpreted as moving on from depleted populations to untapped resources.[152] Certainly, elephant seals and penguins were cropped in large numbers during the eighteenth century in the remote Falkland Islands.[153] Nor do most accounts include mining which changed land surfaces and contributed wastes to land, water and air, and the islands of Polynesia and Micronesia are mostly too small to appear on world maps but had long histories of profound human-induced change.[154]

Technologies of a solar-powered era

Many technologies were introduced in this era. Some hunter-gatherers' devices were greatly 'modernised': the hollowed-out log eventually becomes the ocean-going galleon, and fire was deployed in the service of shifting agriculture. But the list of new ways of altering the natural world is very long: metals, the plough, gunpowder, mills, water-lifting, surveying and the classification of plants are just some of the developments. Most of them increased human access to energy sources: by tapping draught animals, wind or water, for example. Some improved the energy intensity of crops by introducing plants with a greater photosynthetic area than their predecessors: maize is one instance. Animals with a greater efficiency of conversion of plant to animal tissue were also worth taking around the world: cattle are representatives. More interesting than mere chronologies of inventions and their dissemination is the effect of technology on attitudes to the natural world. Writing, for example, is more than a set of graphic techniques, for it always enforced a separation of one entity from another in a way not necessary in oral communication; this led to the analytical thinking encouraged by the list, the formula and the table.[155] This activity, it has been argued, is a great way of separating the objects of the world into those which belong to humans and those which are definitely 'other'. Technology, as both a social force and a social product, can be seen in the example of the mechanical clock which made possible the tighter co-ordination of more complex communities; it could also be a mechanical representation of the universe or an automaton for entertainment, but especially a celebration of human mastery over nature.[156] Though John Donne's world may have seemed lost, the Renaissance was also busy formulating a more worldwide and less regionally constrained view of the cosmos. The organism with an autonomous mind and a map was now dominant, aided by the clock, as we have seen, and also by good navigation, more precise astronomical observation, selective breeding of crops and the beginnings of the chemical analysis of substances.[157] Here was conceived the environmental world of what J. F. Richards calls 'the unending frontier'.[158] Not that such calculating rationality was universal: the great Lisbon earthquake of 1755 evoked a great deal of sensational journalism, attitudinal theology and plain panic as well as helping to formulate the scientific turn in the philosophy of Immanuel Kant.[159] Another natural phenomenon, the resurgence of the Little Ice Age created tensions after about 1560 which, it has been argued,[160] resulted in the blaming of witches for climatic anomalies and their consequent persecution.

The emergence of philosophies

The combination of tools, ideas, trade and hegemony seems to have resulted in the emergence of actual philosophies of environment. Some of these were local and pragmatic but others achieved widespread adoption.[161] In China, for example, Taoism's creed of going with the flow of nature, following the model of water, was a major philosophical system. Its emphasis on quietism, however, did not seem to stop the making over of much of China in the early millennia AD.[162] In India, the Vedic scriptures were called forest books (*Aranyakas*) and celebrated the diversity of life in the forest and hence in human society as well. In Europe, especially, hierarchies of esteem developed, with heaven and its inhabitants at the top, humans a good second, and then the various attractive animals and so on down, with the non-human world being desacralised. In parallel comes a widespread human tendency to make dualisms and so the notion of the environment being part of 'the other' allows it to be changed whenever a human demand is discerned. This requires that any religious sanctions on the alteration of the non-human world be lifted, a process fraught with consequences for the whole world once industrialisation drew all places the more together. Sanctions of another kind were always needed in cities in order to control redistributive functions, and stratification was the usual result. One cultural effect of widespread control was that Asian pre-industrial cities were cleaner than their European equivalents because human excrement was normally sold for manure in Asia but regarded as a discardable waste in much of Europe. At one more level, this era saw the emergence of the nation state, which became a powerful agent of environmental change, when, for example, it initiated large-scale projects like land drainage as in the fenland of eastern England in the seventeenth century. The state might also allocate 'marginal' lands to individuals or companies to raise profits for the crown and thus hasten the conversion of wilder ecosystems into more immediately productive terrains.[163]

Agriculture was by and large a success. A few groups went back into hunting and gathering because of conquest or expulsion but mostly its adoption was an irreversible process in which one Malthusian trap was avoided even if another eventually took its place. Looking back, it has an air of inevitability even though failures are well chronicled.[164] Some of the development of agriculture took place in what has been labelled the 'Medieval Warm Epoch' (*c.*1000–1200) with temperatures of 0.5–1.0 °C above 1970 means, but the phase may not have been global. In places like the western United States, drought was widespread. More pervasive was the downturn between 1550 and 1850 (that is, the Little Ice Age) which straddles the agricultural era and the early industrial eras,[165] and in which there were consequences for agriculture and settlement for a particular year or sequence of years even if not over those centuries as a whole. Yet, although populations were sometimes devastated by disease and famine, a global total of perhaps 50 million in 1000 BC had grown to 800 million by AD 1750. The energetic underpinning was the ability of hunter-gatherers to maintain densities of 0.01–1.0 persons per square kilometre whereas, with shifting cultivation 10–80 people per square kilometre was possible and sedentary farming garnered solar power to underpin 100–1000 people.[166] The outcomes of the agricultural era are therefore both physical (and were emphasised by the many millennia that it has occupied) and mental, disseminated above all by the spread of writing in its various forms. Both these themes must be taken forward and explored again in their industrially powered transformations when many stomachs were filled by agriculture but many more were born to growl.

NOTES

1. Part of a Yoruba poem called 'Hunger' in R. Finnegan (ed.) *The Penguin Book of Oral Poetry*, Harmondsworth, Penguin Books, 1982, p. 166:

   > There is no god like one's stomach:
   > We must sacrifice to it every day.

2. The treatment of energy in agriculture and pre-industrial resource use is outstanding in V. Smil, *Energy in World History*, Boulder CO: Westview Press, 1994, chs 3 and 4.
3. A. M. T. Moore, G. C. Hillman and A. J. Legge, *Village on the Euphrates*, Oxford: Oxford University Press, 2001.
4. T. Watkins, 'From foragers to complex societies in Southwest Asia', in C. Scarre (ed.) *The Human Past*, London: Thames & Hudson, 2005, 200–33 is clear, up to date and very well illustrated, as are the subsequent chapters on the equivalent process in East Asia (ch. 7) and the Americas (ch. 9).
5. See especially D. R. Harris, 'The origins and spread of agriculture and pastoralism in Eurasia: an overview', in D. R. Harris (ed.) *The Origins and Spread of Agriculture and Pastoralism in Eurasia*, London: UCL Press, 1996, 552–73; *idem*, 'Domesticatory relationships of people, plants and animals', in R. F. Ellen and K. Fukuyi (eds) *Redefining Nature. Ecology, Culture and Domestication*, Oxford and Washington DC: Berg, 1996, 437–63. An explanatory account of the earliest

stages of domestication is in D. R. Harris, *Settling Down and Breaking Ground: Rethinking the Neolithic Revolution*, Amsterdam: Stichting Nederlands Museum voor Anthropologie en Praehistorie, 1990.

6. See for example M. Jones, *The Molecule Hunt*, London: Penguin Books, 2001; L. L. Cavalli-Sforza, *Genes, Peoples and Languages*, London: Allen Lane, 2000; for linguistics the innovative tome is C. Renfrew, *Archaeology and Language: the Puzzle of Indo-European origins*, London: Cape, 1987 and Pimlico, 1998; also P. Bellwood and C. Renfrew (eds) *Examining the Farming/Language Dispersal Hypothesis*, Cambridge: McDonald Institute for Archaeological Research, 2002.

7. M. P. Richards, R. J. Schulting and R. E. M. Hayes, 'Sharp shift in diet at onset of Neolithic', *Nature* **425**, 2003, 366; A. Whittle, 'Very like a whale: menhirs, motifs and myths in the Mesolithic–Neolithic transition of Northwestern Europe', *Cambridge Archaeological Journal* **10**, 2000, 243–59.

8. In Table 1 of A. B. Gebauer and T. D. Price, 'Foragers to farmers: an introduction', in A. B. Gebauer and T. D. Price (eds) *Last Hunters, First Farmers: New Perspectives on the Transition to Agriculture*, Santa Fe NM: School of American Research Press, 1992, p. 2. The authors list thirty-eight 'suggested causes' from the literature on the transition to agriculture.

9. V. G. Childe, *The Most Ancient East: the Oriental Prelude to European Prehistory*, London: Kegan Paul, 1928; *idem, Man Makes Himself*, London: Watts, 1951.

10. P. J. Richerson, R. Boyd and R. L. Bettinger, 'Was agriculture impossible during the Pleistocene but mandatory during the Holocene? A climate change hypothesis', *American Antiquity* **66**, 2001, 387–411. It is interesting that climate has come round again, since after Childe it was often pushed into the background; it is connected in part, of course, with the renewed curiosity about climatic history brought about by 'greenhouse' anxieties.

11. W. J. Burroughs, *Climate Change in Prehistory*, Cambridge: Cambridge University Press, 2005, p. 263.

12. This paragraph derives from a large number of sources. They include, E. Isaac, *Geography of Domestication*, Englewood Cliffs NJ: Prentice-Hall, 1970; J. Cauvin, *The Birth of the Gods and the Origins of Agriculture*, Cambridge: Cambridge University Press, 2000, trans. T. Watkins (first published in French 1994, 2nd edn 1996); M. A. Blumler, 'Ecology, evolutionary theory and agricultural origins', in D. R. Harris (ed.) op. cit. 1996, 25–50; D. Rindos, *The Origins of Agriculture: an Evolutionary Perspective*, Orlando FL, Academic Press, 1984; L. Groube, 'The impact of diseases upon the emergence of agriculture' in D. R. Harris op. cit. 1996, 101–29; M. N. Cohen, *The Food Crisis in Prehistory*, New Haven: Yale University Press, 1978. Most authors refer to Carol O. Sauer even if only to point out that there is now a great deal more empirical evidence, but they usually acknowledge his seminal work in, for example, *Seeds, Spades, Hearths and Herds: The Domestication of Animals and Foodstuffs*, Cambridge MA: MIT Press, 1969, 2nd edn.

13. New finds and new dates invalidate any account as soon as it written. But both regions have domesticates in the early Holocene.

14. They are often central to transhumance, where there is a once-yearly movement from a permanent base to summer pastures. Pastoralism usually involves more movements so as not to overgraze the forage in areas of slow plant growth.

15. J. Clutton-Brock, *A Natural History of Domesticated Animals*, Cambridge: Cambridge University Press/London: British Museum (Natural History), 1987.

16. *Idem*, 'The unnatural world: behavioural aspects of humans and animals in the process of domestication', in A. Manning and J. Serpell (eds) *Animals and Human Society*, London and New York: Routledge, 1994, 23–35.

17. A detailed budget is given for highland Scottish examples in R. A. Dodgshon and E. G. Olsson, 'Productivity and nutrient use in eighteenth-century Scottish Highland townships', *Geografiska Annaler* **ser B 70**, 1988, 39–51.

18. Care is needed in extrapolating erosion rates from present-day measurements to historic times. See S. W. Trimble and P. Crosson, 'U.S. soil erosion rates – myth and reality', *Science* **289**, 2000, 248–50.

19. R. C. Sidle et al., 'Interaction of natural hazards and society in Austral-Asia: evidence in past and recent records', *Quaternary International* 118–19, 2004, 181–203.

20. J. Goodman, P. E. Lovejoy and A. Sherratt (eds) *Consuming Habits: Drugs in History and Anthropology*, London and New York: Routledge, 1995.

21. T. Ingold, 'Growing plants and raising animals: an anthropological perspective on domestication', in D. R. Harris (ed.) *The Origins and Spread of Agriculture and Pastoralism in Eurasia*, London: UCL Press, 1996, 12–24.

22. J. Thomas, *Understanding the Neolithic*, London and New York: Routledge, 1999. This book is a revised second edition of *Rethinking the Neolithic*, 1991. It takes its evidence from southern Britain only.

23. M. Oelschlaeger, *The Idea of Wilderness*, New Haven and London: Yale University Press, 1991, ch. 2; K. Anderson, 'A walk on the wild side: a critical geography of domestication', *Progress in Human Geography* **21**, 1997, 463–85.

24. The idea is derived from M. Harris, *Good to Eat: Riddles of Food and Culture*, New York: Simon & Schuster, 1985, and developed by J. Goudsblom, 'Ecological regimes and the rise of organized religions', in J. Goudsblom, E. Jones and S. Mennell, *The Course of Human History*, Armonk NY and London: M. E. Sharpe, 1996, 31–47.

25. J. R. Goody, *The Domestication of the Savage Mind*, Cambridge: Cambridge University Press, 1977; *idem, The Power of the Written Tradition*, Washington DC and London: Smithsonian Institution Press, 2000.

26. Something might be said about philosophy and sliced bread, but perhaps better not. See the extended discussion of writing in D. Abram, *The Spell of the Sensuous*, New York: Vintage Books, 1996.

27. M. Oelschlager op. cit. pp. 65–72.

28. An easily accessed source of chronology for ENSO events is in the tables provided on http://sharpgary.org

29. D. F. Ferretti et al., 'Unexpected changes to the global methane budget over the past 2000 years', *Science* **309**, 2005, 1714–17.

30. L. V. Trabaud, N. L. Christensen and A. M. Gill, 'Historical biogeography of fire in temperate and Mediterranean ecosystems', in P. J. Crutzen and J. G. Goldammer (eds) *The Ecological, Atmospheric, and Climatic Importance of Vegetation Fires*, Chichester: Wiley, 1993, 277–95.

31. J. L. Byock, *Medieval Iceland. Society, Sagas and Power*, Enfield Lock (UK): Hisarlik Press, 1993, p. 55, first published by the University of California Press, 1988.

32. S. J. Pyne, *Burning Bush. A Fire History of Australia*, New York: Henry Holt, 1991, ch. 12.

33. J. Goudsblom, *Fire and Civilization*, London: Penguin Books, 1994; J. A. Galloway, D. Keene and M. Murphy, 'Fuelling the city: production and distribution of firewood and fuel in London's region, 1290–1400', *Economic History Review* **49**, 1996, 447–72; D. H. Fischer, *The Great Wave. Price Revolutions and the Rhythm of History*, New York: Oxford University Press, 1996.

34. A. Robock and H-F. Graf, 'Effects of pre-industrial human activities on climate', *Chemosphere* **29**, 1994, 1087–97.

35. Wine-making was dependent on the development of sound containers of wood or pottery. Once developed, though, there was probably little technological change in wine-making between 1500 BC and the 1950s; H. Hobhouse, *Seeds of Wealth*, London: Macmillan 2003.

36. Most books on world agriculture contain historical material which relates to the subject of this paragraph. Useful concentrations are in G. A. Klee (ed.) *World Systems of Traditional Management*, London: Edward Arnold, 1980; W. Denevan, *Cultivated Landscapes of Native Amazonia and the Andes*, Oxford: Oxford University Press, 2001; C. L. Crumley (ed.) *Historical Ecology*, Santa Fe: School of American Research Press, 1994; J. D. Hughes, *Ecology in Ancient Civilizations*, Albuquerque: University of New Mexico Press, 1975. Terracing has spawned a couple of classic writings: J. E. Spencer and G. A. Hale, 'The origin, nature and distribution of agricultural terracing', *Pacific Viewpoint* 2, 1961, 1–40 and R. A. Donkin, *Agricultural Terracing in the Aboriginal New World*, Tucson: University of Arizona Press Viking Fund Publications in Anthropology no. 56.

37. O. Rackham, 'Ecology and pseudo-ecology: the example of ancient Greece', in G. Shipley and J. Salmon (eds) *Human Landscapes in Classical Antiquity. Environment and Culture*, London and New York: Routledge, 1996, 16–43.

38. A. Sluyter, 'From archive to map to pastoral landscape. A spatial perspective on the livestock ecology of sixteenth-century New Spain', *Environmental History* 3, 1998, 508–28. E. Melville, *A Plague of Sheep: Environmental Consequences of the Conquest of Mexico*, Cambridge: Cambridge University Press, 1994.

39. There is a worldwide overview with many detailed examples by A. M. Mannion, *Agriculture and Environmental Change*, Chichester: Wiley, 1995.

40. See the complexities explored in D. Herlihy, 'Ecological conditions and demographic change', in R. L. DeMolen (ed.) *One Thousand Years. Western Europe in the Middle Ages*, Boston MA: Houghton Mifflin, 1974, 3–43.

41. R. Halperin and J. Dow (eds) *Peasant Livelihood*, New York: St Martin's Press, 1977.

42. P. Sillitoe, *An Introduction to the Anthropology of Melanesia*, Cambridge: Cambridge University Press, 1998.

43. B. Fürst Bjeliš, 'Triplex confinium – an ecohistorical draft', in D. Roksandić (ed.) *Microhistory of the Triplex Confinium*, Budapest: Central European University Institute on Southeastern Europe, 1998, 147–55; there is a much wider context in J. R. McNeill, 'Woods and warfare in world history', *Environmental History* 9, 2004, 388–410.

44. N. Christie, 'Barren fields? landscape and settlement in late Roman and post-Roman Italy', in G. Shipley and J. Salmon op. cit. 1996, 254–83.

45. F. J. Simoons, 'The determinants of dairying and milk use in the Old World: ecological, physiological and cultural', in J. R. K. Robinson (ed.) *Food, Ecology and Culture*, New York, London and Paris: Gordon and Breach, 1980, 83–91; P. Wheatley, 'A note on the extension of milking practices into Southeast Asia during the first millennium A.D.', *Anthropos* 60, 1965, 577–90.

46. M. Gadgil and R. Guha, *This Fissured Land. An Ecological History of India*, Delhi: Oxford University Press, 1993.

47. J. E. Shermondy, 'Water and power: the role of irrigation districts in the transition from Inca to Spanish Cuzco', in W. P. Mitchell and D. Guillet (eds) *Irrigation at High Altitudes: the Social Control of Water Control Systems in the Andes*, American Anthropological Association, 1994. (Copy gives no place of publication.)

48. K. Hastrup, *Nature and Policy in Iceland 1400–1800*, Oxford: Clarendon Press, 1990; I. A. Simson, A. J. Dugmore, A. Thomson and O. Vésteinsson, 'Crossing

the thresholds: human ecology and historical patterns of landscape degradation', *Catena* **42**, 2001, 175–92.

49. O. F. Linares, *Power, Prayer and Production. The Jola of Casamance, Senegal,* Cambridge: Cambridge University Press, 1992. The introduction of Islam came under French colonial influence but in a largely pre-industrial economic context.

50. For Italy, see E. Sereni, *History of the Italian Agricultural Landscape,* Princeton: Princeton University Press, 1997, trans. R. B. Lichfield, originally published in 1961 as *Storia del paesaggio agrario italiano.*

51. G. Condominas, 'Ritual technology in Mnong Gap swidden agriculture', in I. Nørlund, S. Cederroth and I. Gerdin (eds) *Rice Societies,* London: Curzon Press, 1986, 28–41. The study was in the late 1940s but the group (Montagnard in French) seemed little affected by colonial influences.

52. A. Pluskowski, *Wolves and the Wilderness in the Middle Ages,* Woodbridge: Boydell Press, 2006.

53. J.-P. Poirier, 'The 1755 Lisbon disaster, the earthquake that shook Europe', *European Review* 14, 2006, 169–80.

54. K. W. Butzer, Early Hydraulic Civilizations in Egypt, Chicago and London: University of Chicago Press, 1976; K. W. Butzer, J. P. Mateu, E. K. Butzer and P. Kraus, 'Irrigation agrosystems in Eastern Spain: Roman or Islamic origins?' *Annals of the Association of American Geographers* **75**, 1985, 479–509.

55. W. M. Denevan, *Cultural Landscapes of Native Amazonia and the Andes,* Oxford: Oxford University Press, 2001; T. M. Whitmore and B. L. Turner, *Cultural Landscapes of Middle America on the Eve of Conquest,* Oxford: Oxford University Press, 2001.

56. J. V. Kirch, *The Wet and the Dry. Irrigation and Agricultural Intensification in Polynesia,* Chicago and London: University of Chicago Press, 1994.

57. C. Geertz, *Agricultural Involution: the Process of Ecological Change in Indonesia,* Berkeley and Los Angeles: University of California Press, 1963.

58. B. J. Murton, 'South Asia', in G. A. Klee op. cit. 1980, 67–99.

59. M. Elvin, *The Pattern of the Chinese Past,* London: Eyre Methuen, 1973.

60. T. F. Glick, *Irrigation and Society in Medieval Valencia,* Cambridge MA: Belknap Press of Harvard University, 1970.

61. S. J. Borsch, 'Environment and populations: the collapse of large irrigation systems considered', *Comparative Studies in Society and History* **46**, 2004, 451–68.

62. K. A. Wittfogel, *Oriental Despotism: A Comparative Study of Total Power,* New Haven: Yale University Press, 1957.

63. There are several papers on these themes in B. Menu (ed.) *Les problèmes institutionels de l'eau en Egypte anciennne et dans l'Antiquité mediterranéene,* Cairo: Institut Français d'Archaeologie Orientale, 1994.

64. J. V. Kirch op. cit. 1994, ch. 13.

65. J. Shaw, 'Sanchi and its archaeological landscape: Buddhist monasteries, settlement and irrigation works in central India', *Antiquity* **74**, 2000, 775–6. (The suggestion about Sri Lanka occurs in the discussion section of the paper.)

66. J. S. Lansing, *Priests and Programmers. Technologies of Power in the Engineered Landscape of Bali,* Princeton: Princeton University Press, 1991; J. S. Lansing, and J. N. Kremer, 'Emergent properties of Balinese water temple networks: coadaptation on a rugged fitness landscape', *American Anthropologist* 95, 1993, 97–114.

67. C. S. Beekman, P. C. Weigand and J. J. Pint, 'Old World irrigation technology in a New World context: qanats in Spanish colonial western Mexico', *Antiquity* **73**, 1999, 440–6.

68. K. W. Butzer, 'Irrigation, raised fields and state management: Wittfogel redux? *Antiquity* **70**, 1996, 200–4; K. Elvin, *The Retreat of the Elephants: an Environmental History of China*, New Haven: Yale University Press, 2004.

69. J. Clutton-Brock (ed.) *The Walking Larder. Patterns of Domestication, Pastoralism, and Predation*, London: Allen & Unwin, 1989; J. Galaty and D. Johnson (eds) *The World of Pastoralism: Herding Systems in Comparative Perspective*, New York: Guilford Press/London: Belhaven, 1990; A. B. Smith, *Pastoralism in Africa. Origins and Development Ecology*, London: Hurst, 1992.

70. O. Hanotte, D. G. Bradley, J. W. Ochieng, Y. Verjee, E. W. Hill and J. E. O. Rege, 'African pastoralism: genetic imprints of origins and migrations', *Science* **296**, 2002, 336–9.

71. M. F. Molinillo, 'Is traditional pastoralism the cause of erosive processes in mountain environments? The case of the Cumbres Calchaquies in Argentina', *Mountain Research and Development* **13**, 1993, 189–202.

72. K. Abdi, 'The early development of pastoralism in the central Zagros mountains', *Journal of World Prehistory* **17**, 2003, 395–448.

73. C. W. Schwabe, 'Animals in the ancient world', in A. Manning and J. Serpell, *Animals and Human Society. Changing Perspectives*, London and New York: Routledge, 1994, 36–58 at p. 37.

74. P. Bonte, '"To increase cows, God created the king": the function of cattle in intralacustrine societies', in J. Galaty and P. Bonte (eds) *Herders, Warriors and Traders. Pastoralism in Africa.* Boulder CO: Westview Press, 1991, 62–86; D. L. Hodgson, 'Gender, Culture and the myth of the patriarchal pastoralist', in D. L. Hodgson (ed.) *Rethinking Pastoralism in Africa*, Oxford: James Currey, 2000, 1–28; J. C. Galaty, 'Cattle and cognition: aspects of Maasi practical reasoning', in J. Clutton-Brock (ed.) *The Walking Larder. Patterns of Domestication, Pastoralism, and Predation*, London: Allen & Unwin, 1989, 215–230.

75. G. Brotherston, 'Andean pastoralism and Inca ideology', in J. Clutton-Brock op. cit. 1989, 240–55. The drawing of the king and llama singing together in the epigraph to this chapter forms the frontispiece to Clutton-Brock's volume.

76. M. Overton, *Agricultural Revolution in England. The Transformation of the Agrarian Economy 1500–1850*, Cambridge: Cambridge University Press, 1996.

77. The story is much more complex than this. Orientation towards urban markets was often found in medieval times but massive changes preceded much industrialisation. In parallel with improving lowland agriculture, there was massive enclosure and transformation of heath, moor and fen but there is no space to discuss that here.

78. J. Chapman and S. Seeliger, *Enclosure, Environment and Landscape in Southern England*, Stroud: Tempus, 2001.

79. The quotation is from near London in 1515 and given in J. R. Siemon, 'Landlord not King: Agrarian Change and Interarticulation', in R. Burt and J. M. Archer (eds) *Enclosure Acts. Sexuality, Property, and Culture in Early Modern England*, Ithaca and London: Cornell University Press, 1994, 17–33 at p. 30.

80. H. Hobhouse, *Seeds of Change. Six Plants that Transformed Mankind.* London: Sidgwick and Jackson, 1985, revised edition Macmillan 1999.

81. D. Watts, *The West Indies: Patterns of Development, Culture and Environmental Change since 1492*, Cambridge: Cambridge University Press, 1987; J. R. McNeill, 'Agriculture, forests and ecological history: Brazil, 1550–1984', *Environmental Review* **10**, 1986, 122–33; J. H. Galloway, *The Sugar Cane Industry: An Historical Geography from its Origins to 1914*, Cambridge: Cambridge University Press, 1989.

82. An eighteenth-century observer wrote, '. . . these two vegetables [coffee and sugar] have brought wretchedness and misery upon America and Africa. The former has been depopulated, that the Europeans may have land to plant them in; and the latter is stripped of its inhabitants, for hands to cultivate them.' Quoted by M. Williams, *Deforesting the Earth*, Chicago and London: University of Chicago Press, 2003, p. 159.

83. This is of course a Eurocentric view. The Chinese may have circumnavigated the globe (with vessels that would have dwarfed the caravel) and explored the southern oceans in the fifteenth century, although they left relatively little trace of their journeys compared with Europeans' metamorphoses. See G. Menzies, *1421. The Year China Discovered the World*, London: Bantam Press, 2003.

84. L. Jardine, *Worldly Goods. A New History of the Renaissance*, London: Macmillan, 1996. John Clare (1793–1864) is the great poet of nineteenth-century enclosure and its deprivation of livings for the poor and space for nature.

85. R. P. Multhauf, *Neptune's Gift. A History of Common Salt*, Baltimore and London: Johns Hopkins University Press, 1978.

86. S. Rippon, *The Transformation of Coastal Wetlands: Exploitation and Management of Wetland Landscapes in North West Europe during the Roman and Medieval Periods*, Oxford: Oxford University Press for the British Academy, 2001.

87. M. Elvin and Su Ninghu, 'The influence of the Yellow River on Hangzhou Bay since AD 1000', in M. Elvin and Liu Ts'ui-jung (eds) *Sediments of Time. Environment and Society in Chinese History*, Cambridge: Cambridge University Press, 1998, 344–407; J. Needham, with Wang Ling and Lu Gwei-Djen, *Science and Civilisation in China*, Cambridge: Cambridge University Press, vol. 4, Part III, 335–9.

88. J. Van Veen, *Dredge, Drain, Reclaim. The Art of a Nation*. The Hague: Martinus Nijhoff, 1962, 5th edn.

89. See, for example, L. Östland, O. Zackrisson and G. Hörnberg, 'Trees on the border between nature and culture. Culturally modified trees in Boreal Sweden', *Environmental History* 7, 2002, 48–68.

90. M. G. Wolman and F. G. A. Fournier, 'Introduction to land transformation in space and time', in M. G. Wolman and F. G. A. Fournier (eds) *Land Transformation in Agriculture*, Chichester: Wiley, 1987, for ICSU: SCOPE **32**, 3–43. Their 'primary forest' would no doubt include such areas as Amazonia, the ecology of which is now coming under revision, and many areas of forest where fire history was also its human history. Their estimates, however, give some kind of baseline.

91. C. Delano-Smith, 'Where was the wilderness in Roman times?', in G. Shipley and J. Salmon op. cit. 1996, 154–79.

92. O. Rackham, *The History of the Countryside*, London: Dent, 1987, p. 87.

93. These data come from M. Williams, *Deforesting the Earth*, Chicago and London: University of Chicago Press, 2003, which is an immense store of data and historical knowledge. An abridged edition is available from the same publisher, 2006, in paperback.

94. C. Totman, *The Green Archipelago. Forestry in Preindustrial Japan*, Berkeley, Los Angeles and London: University of California Press, 1989.

95. M. Cartmill, *A View to a Death in the Morning. Hunting and Nature through History*, London and Cambridge MA: Harvard University Press, 1993, p. 66.

96. J. Cummins, *The Hound and the Hawk. The Art of Medieval Hunting*, London: Phoenix Press, 2001, first published 1988. The saker (*Falco cherrug*) was like the gyr falcons of northern Europe.

97. There is an immense literature and most of it concentrates on the gardens of the wealthy as distinct from the less rich: see, for example, J. S. Berrall, *The Garden. An Illustrated History*, Harmondsworth: Penguin Books, 1966. More critical material can be seen in J. D. Hunt (ed.) *Garden History: Issues, Approaches, Methods*, Washington DC: Dumbarton Oaks Research Library, 1992.

98. T. Williamson, *Polite Landscapes. Gardens and Society in Eighteenth Century England*, Stroud: Sutton Publishing, 1995. For estimates of the quantity of parkland in England see, for example, H. C. Prince, 'England *circa* 1800', in H. C. Darby (ed.) *A New Historical Geography of England after 1600*, Cambridge, Cambridge University Press, 1976, 89–164; and for descriptive work that takes the story through to recent times, S. Lasdun, *The English Park. Royal, Private and Public*, London: André Deutsch, 1991.

99. T. Williamson, *Polite Landscapes. Gardens and Society in Eighteenth-Century England*, Stroud: Sutton, 1995.

100. There is a paradox in the sense that, for example, Cistercian monasteries were supposed to be set apart from the world but, in fact, exchanged materials and ideas between each other frequently in the course of disciplinary visitations within the order. No doubt their lay brethren also brought in new ideas when admitted to the order.

101. J. Prest, *The Garden of Eden. The Botanic Garden and the Re-Creation of Paradise*, New Haven and London: Yale University Press, 1981 and for connections in other arts, J. D. Hunt, *The Figure in the Landscape: Poetry, Painting and Gardening during the Eighteenth Century*, Baltimore: Johns Hopkins University, 1976, pb edn 1989; it is almost exclusively about England.

102. Quoted in M. Willams op. cit. 2003, p. 112.

103. William Wordsworth (1770–1850), 'The Tables Turned'. It *might* have been true then but is now rather nonsensical.

104. R. Waswo, *The Founding Legend of Western Civilization. From Virgil to Vietnam*. Hanover NH and London: University Press of New England, 1997. Waswo points out that the chain-dragging bulldozer used to clear forests during the Vietnam War was called the Rome Plow.

105. 'Woods' (Bucolics, 2), in *Collected Shorter Poems 1927–1957*, London and Boston MA: Faber & Faber, 1966, pp. 257–8.

106. M. Cartmill, *A View to a Death in the Morning. Hunting and Nature through History*, Cambridge MA and London: Harvard University Press, 1993, p. 233, quoting from India, M. M. H. Shastri (ed.) *Syainka Sāstram: the Art of Hunting in Ancient India*, Delhi: Eastern Book Linkers, 1982; J. Cummins op. cit. 2001 describes some of the erotic associations of hunting.

107. M. Cartmill op. cit. 1993.

108. Holy Bible: The Book of Job, ch. 28 vv. 9–11, RSV. Probably about the fourth century BC.

109. Most of this comes from various earlier works of mine, especially the two books on environmental history and *Changing the Face of the Earth*, 2nd edn. See also J. Needham (with Wang Ling) *Science and Civilisation in China*, Cambridge: Cambridge University Press, 1959, vol. 3; A. Y. al-Hassan and D. R. Hill, *Islamic Technology*, Paris and Cambridge: UNESCO/Cambridge University Press, 1986. Marco Polo is quoted by J. L. Abu-Lughod, *Before European Hegemony: the World System A.D. 1250–1350*, New York: Oxford University Press, 1989, p. 349.

110. I. Douglas, *The Urban Environment*, London: Edward Arnold, 1983.

111. J. A. Galloway et al. op. cit. 1996, Appendix I.

112. The famous occasion on which J. S. Bach improvised a three-part fugue for Frederick the Great upon the composer's arrival in Potsdam in 1747

immediately followed the normal presentation to the King of a list of arrivals that day at the town gate. 'Gentlemen', Frederick is supposed to have said, 'Old Bach is here'. (J. Gaines, *Evening in the Palace of Reason*, London: Fourth Estate, 2005.)

113. Water is an unusual molecule found naturally in solid, liquid and gaseous forms. 'Liquid' may be incorporated in, for example, plant tissues.

114. There have been attempts at affecting the weather before industrial-era techniques such as cloud-seeding from aircraft. Cannons were fired into thunderstorms in France to try to pre-empt hail falling on the grapes just before harvest. As with cloud-seeding, nobody really knows whether or not it has worked.

115. R. J. Magnusson, *Water Technology in the Middle Ages*, Baltimore and London: Johns Hopkins University Press, 2001. More hygienic disposal of sewage did not come to London until the industrial era.

116. R. C. Hoffman, 'Economic development and aquatic ecosystems in medieval Europe', *The American Historical Review* **101**, 1996, 631–9; idem, 'Carps, cods, connections. New fisheries in the medieval European economy and environment', in M. J. Henninger-Voss (ed.) *Animals in Human Histories*, Rochester NY: University of Rochester Press, 2002, 3–55.

117. R. A. Wild, *Water in the Cultic Worship of Isis and Sarapis*, Leiden: E. J. Brill, 1981.

118. The leviathan of The Bible is a mythical beast but in modern Hebrew simply means 'whale'.

119. R. C. Hoffmann, 'Frontier foods for late Medieval consumers. Culture, economy, ecology', *Environment and History* **7**, 2001, 131–67.

120. R. C. Hoffmann, 'A brief history of aquatic resource use in medieval Europe', *Helgoland Marine Research* **59**, 2005, 22–30. The material on the beam trawl is from the Parliamentary Rolls temp. Edward III, quoted at http://sharpgary.org/FisheryTimeline.html, accessed on 10 July 2006.

121. It was first used by King Creon in Sophocles' *Antigone*.

122. In R. Hamer (ed.) *A Choice of Anglo-Saxon Verse*, London and Boston MA: Faber & Faber, 1970, p. 187.

123. Much 'subsistence' farming was, in fact, enmeshed in local and regional networks of trade and exchange.

124. '[Emerson] tells me he does not like Haynes as well as I do. I tell him he makes better manure than most men': Henry David Thoreau in 1852, quoted in D. Foster, *Thoreau's Country. Journey through a Transformed Landscape*, Cambridge MA and London: Harvard University Press, 1999, p. 40.

125. A. Manguel and G. Guadalupi, *The Dictionary of Imaginary Places*, London: Bloomsbury, 1999.

126. J. R. McNeill, 'Europe's place in the global history of biological exchange', *Landscape Research* **28**, 2003, 33–9. The Silk Road was in fact a set of routes rather than a single road. Central Asian mtDNA seems to be intermediate between European and eastern Asian sequences: D. Comas et al., 'Trading genes along the silk road: mtDNA sequences and the origin of central Asian populations', *American Journal of Human Genetics* **63**, 1998, 1824–38. There is a broad but detailed account of the silk roads which includes the complementary sea routes in J. H. Bentley, *Old World Encounters. Cross-Cultural Contacts and Exchanges in Pre-Modern Times*, New York and Oxford: Oxford University Press, 1993.

127. J. P. Grattan and F. B. Pyatt, 'Volcanic eruptions, dry fogs and the European palaeoenvironmental record: localised phenomena or hemispheric impacts?' *Global and Planetary Change* **21**, 1999, 173–9; R. B. Stothers, 'Climatic and demographic consequences of the massive volcanic eruption of 1258', *Climatic Change* **48**, 2000, 361–74.

128. The Portuguese saying, 'E se mais mundo nouvera, là chegara' ('if it had been bigger, we would still have gone round it') sounds a little like boasting. This paragraph derives mostly from chapters in A. G. Hopkins (ed.) *Globalization in World History*, London: Pimlico 2002; and R. Robertson, *The Three Waves of Globalization*, London: Zed Books, 2003. An holistic account of Europe per se is in D. Levine, *At the Dawn of Modernity. Biology, Culture and Material Life in Europe after the Year 1000*, Berkeley, Los Angeles and London: University of California Press, 2001, and the Old World is the focus of J. H. Bentley op. cit. 1993. See also J. L. Abu-Lughod, *Before European Hegemony: the World System A.D. 1250–1350*, New York: Oxford University Press, 1989; A. Gundar Frank and B. K. Gills (eds) *The World System: Five Hundred Years or Five Thousand?*, London and New York: Routledge, 1993; see also G. Menzies, *1421. The Year China discovered the World*, London: Bantam Books, 2002.

129. C. Chase-Dunn and P. Grimes, 'World system analysis', *Annual Review of Sociology* **21**, 1995, 387–417, especially Figure 1; C. Chase-Dunn and T. D. Hall, 'The historical evolution of world systems', *Sociological Inquiry* 64, 1994, 257–80.

130. J. H. Bentley op. cit. 1993.

131. J. R. McNeill op. cit. 2003. The classic text is A. W. Crosby, *Ecological Imperialism. The Biological Expansion of Europe, 900–1900*, Cambridge: Cambridge University Press, 1986/Canto Books, 1993. Note that other areas were recipients as well: the Japanese word for bread is *pan*, presumably from the Portuguese missionaries of the sixteenth century; the 'traditional' Japanese lantern is said by some to have been copied from the stern lanterns of European galleons.

132. D. Hahn, *The Tower Menagerie: Being the Amazing True Story of the Royal Collection of Wild Beasts*, London: Simon & Schuster, 2003.

133. The 'Franciscan tradition' is sometimes quoted but in actual practice seems not to have had much effect until much later, as Keith Thomas (*Man and the Natural World. Changing Attitudes in England 1500–1800*, London: Allen Lane, 1983) points out for England. The legend of St Francis preaching to the birds seems summative: he preached to them rather than listened. But see Stanley Spencer's reversed-boots version of it: *St Francis and the Birds*, 1935 (London: Tate Britain: can be seen online).

134. L. Whyte, 'The historical roots of our ecologic crisis', *Science* **155**, 1967, 1203–7.

135. C. Chant and D. Goodman, *Pre-Industrial Cities and Technology*, London and New York: Routledge, 1998.

136. R. H. Grove, *Green Imperialism. Colonial Expansion, Tropical island Edens and the Origins of Environmentalism, 1600–1860*, Cambridge: Cambridge University Press, 1995.

137. A. W. Crosby, *The Measure of Reality. Quantification and Western Society, 1250–1600*. Cambridge: Cambridge University Press, 1997. In the case of time, there is the unification caused by the adoption of uniform times but also the division of time into periods of specific activities.

138. N. Crane, *Mercator. The Man who Mapped the Planet*, London: Weidenfeld & Nicolson, 2002.

139. Greek ideas are put alongside those of the Romans, Jews and early Christians in R. French, *Ancient Natural History*, London and New York: Routledge, 1994.

140. J. Chapman, *Fragmentation in Archaeology. People, Places and Broken Objects in the Prehistory of South-eastern Europe*, London and New York: Routledge, 2000.

141. P. Spierenburg, *The Broken Spell. A Cultural and Anthropological History of Preindustrial Europe*, New Brunswick: Rutgers University Press, 1991, first published in Dutch; D. Levine, *At the Dawn of Modernity. Biology, Culture and*

Material Life in Europe after the Year 1000. Berkeley, Los Angeles and London: University of California Press, 2000.

142. M. Williams op. cit. 2003, p. 247.

143. *idem*, 'Dark ages and dark areas', *Journal of Historical Geography* **26**, 2000, 28–46, at p. 42.

144. M. Williams (ed.) *Wetlands: a Threatened Landscape*, Oxford: Blackwell, 1990.

145. P. J. E. M. Van Dam, 'Sinking peat bogs: environmental change in Holland, 1350–1550', *Environmental History* **6**, 2001, 31–45.

146. M.Balter , 'The seeds of civilization', *Smithsonian Magazine*, May 2005, npg; see also M. Cauvin op. cit. 2000. Further detail of this interesting debate can be found in I. Hodder (ed.) *On the Surface: Çatalhöyük 1993–95*, Cambridge: McDonald Institute, 1996; D. Lewis-Williams, 'Constructing a cosmos. Architecture, power and domestication at Çatalhöyük, *Journal of Social Archaeology* **4**, 2004, 28–59; M. Verhoeven, 'Beyond boundaries: nature, culture and a holistic approach to domestication in the Levant', *Journal of World Prehistory* **18**, 2004, 179–282.

147. Giorgione lived from 1477 or 1478 to 1510, mostly in Venice. See B. Wittkower, 'Georgione and Arcady', in R. W. Wittkower (ed.) *Idea and Image. Studies in the Italian Renaissance*, London: Thames & Hudson, 1978, 161–73.

148. See chapter 4 of R. H. Grove op. cit. 1995.

149. D. Abram op. cit. 1996.

150. In his restatement of this, Oelschlaeger adds theology and has them spring forth 'with a vengeance'. Revenge for what? (M. Oelschlaeger, *The Idea of Wilderness. From Prehistory to the Age of Ecology*, New Haven and London: Yale University Press, 1991, p. 29.)

151. D. Reynaud, T. Blumer, Y. Ono and R. J. Delmas, 'The Late Quaternary history of atmospheric trace gases and aerosols: interactions between climate and bio-geochemical cycles', in K. D. Alverson, R. S. Bradley and T. F. Pedersen (eds) *Paleoclimate, Global Change and the Future*, Berlin: Springer, 2003, 14–31.

152. E. L. Jones, 'The environment and the economy', in B. Purke (ed.) *The New Cambridge Modern History*, vol. **XIII**, Cambridge: Cambridge University Press, 1979, chapter 2. For one historic reconstruction see L. Hacquebord , 'Three Centuries of Whaling and Walrus Hunting in Svalbard and its Impact on the Arctic Ecosystem', *Environment and History* **7**, 2001, 169–85.

153. P. H. Armstrong, 'Human impact on the Falkland Islands' environment', *The Environmentalist* **14**, 1994, 215–31.

154. P. V. Kirch and T. L. Hunt (eds) *Historical Ecology in the Pacific Islands. Prehistoric Environmental and Landscape Change*, New Haven and London: Yale University Press, 1997.

155. J. Goody op. cit. 1977, p. 162.

156. O. Mayr, *Authority, Liberty and Automatic Machinery in Early Modern Europe*, Baltimore and London: Johns Hopkins University Press, 1986.

157. J. Opie, 'Renaissance origins of the environmental crisis', *Environmental Review* **11**, 1987, 2–17; L. Jardine, *Ingenious Pursuits. Building the Scientific Revolution*, London: Little, Brown, 1999.

158. J. F. Richards, *The Unending Frontier. An Environmental History of the Early Modern World*, Berkeley, Los Angeles and London: University of California Press, 2005.

159. There is a symposium on the earthquake's effects on European thought in *European Review* 14 (3) 313–67.

160. W. Behringer, 'Climatic change and witch-hunting: the impact of the Little Ice Age on mentalities', *Climatic Change* **43**, 1999, 335–51.

161. J. Freidman, 'Ecological consciousness and the decline of "civilisations": the ontology, cosmology and ideology of non-equilibrium living systems', *Worldviews* **2**, 1998, 303–15.

162. The subject of a classic paper by Yi Fu Tuan, 'Discrepancies between environmental attitude and behavior: examples from Europe and China', *The Canadian Geographer* **12**, 1968, 176–91.

163. R. A. Butlin, 'The role of the state in the initiation and development of land drainage schemes in England in the seventeenth century', in P. Cereno and M. L. Sturani (eds) *Rural Landscape between State and Local Communities in Europe Past and Present*, Alessandria: Edizioni dell'Orso, 1998, 121–9.

164. Diamond gives eight examples of 'collapse' from agricultural societies (ten if recent times in Haiti and Rwanda are included) and only two from modern industrial economies. The latter two are Australia and China where judgement is perhaps suspended for the moment. (J. Diamond, *Collapse. How Societies Choose to Fail or Succeed*, London: Allen Lane, 2005.)

165. R. S. Bradley, K. R. Briffa, J. Cole, M. K. Hughes and T. J. Osborn, 'The climate of the last millennium', in K. D. Alverson, R. S. Bradley and T. F. Pedersen (eds) *Paleoclimate, Global Change and the Future*, Berlin: Springer, 2003, 105–41; J. M. Grove, *Little Ice Ages. Ancient and Modern*, London and New York: Routledge, 2004, 2 vols, 2nd edn.

166. V. Smil, *General Energetics. Energy in the Biosphere and Civilization.* New York: Wiley Interscience, 1991.

CHAPTER FOUR

An industrious world

FIGURE 4.1 *Derwentcotes Steel Mill.*
Photograph by I. G. Simmons.

A little-known but crucial place in the development of the industrial-isation of the world. The remants, now preserved by English Heritage, of the Derwentcotes Steel Mill near Gateshead in north-east England. Built in the 1720s, this mill's coal-fired furnaces turned iron into steel, in this case especially for cutting tools. Although iron was critical in the development of industrialisation, the much harder steel was in many industries an essential breakthrough. The mill packed iron bars into a furnace with coal, to allow the carbon to be absorbed into the iron and so harden it.

This 2006 photograph might fancifully be thought to depict the workers having their midday break in the sun but in fact shows the site as a lunch-time stopping point for a ramblers' club. There is a connec-tivity in this because it has been the development of industrial methods of production that has released workers into a life-pattern with much more leisure time. Also, it has permitted a revolution in attitudes: in the

eighteenth century, not many working people would have spend precious leisure time on a Sunday going for a 10-mile 'ramble' unless it had
some powerful attraction such as courtship. The coming together of
people who live apart but have common interests has been contrasted
with those who have to live and work together: Ferdinand Tonnies's
(1855–1936) German terms *Gesellschaft* and *Gemeinschaft* were for a
long time the accepted descriptions of the new (company or goal-
oriented society) and the old (community) styles of living and hence of
environmental relationships, especially in terms of travel patterns and
what would now be called ecological footprint. Both terms are perhaps
less useful in post-industrial conditions: indeed, why are these people not
shopping?

The photograph was taken in April 2006. The building is supposed to
be open on Sunday afternoons in the summer but not on this one.

A SECOND IRON AGE

The treatment of industrialisation must give prominence to the development
of fossil fuels as an energy source. But energy for human use has always to be
channelled, whether via plant and animal tissue, inside a vast concrete and
steel sphere or, putatively, inside a magnetic field that is independent of temperature.[1] In the case of the application of coal, oil and natural gas the critical
material which formed the conduit down which the energy was applied
depended upon iron, increasingly in the form of steel. Not for nothing did the
chief driver of Soviet industrialisation take the name 'Stalin.'[2]

This chapter is mostly about the period 1750 to 1950 when the possibilities
of exploiting fossil hydrocarbons expanded first into the western world and
thereafter taken to almost everywhere else.[3] Some use for coal and natural gas
had been found in the previous era, and peat was a vital fuel in the Netherlands
from the seventeenth century onwards, but the scale and the technology for
railways, steamships and high-volume chemical production belong to the
years from the eighteenth century onwards. The inhabitants of the first centres
of industrialisation lived in a society where the per capita use of energy was
much higher than in the previous era. In Britain in 1870, 100 million tons of
coal were used, which produced the same quantity of calories needed to feed
850 million adult males for a year, Britain having about 10 million of them in
that year. Energy carries value socially and environmentally when it is embedded in materials; in terms of this new iron age, we can contrast Britain's pig-
iron production of 17,350 tons in 1740 with 2,701,000 in 1852.[4] The key stage
in the transformation of an energy source such as coal to a metamorphosed
environment such as a mine and its wastes was the knowledge of how to
control steam under pressure, which was a central piece of technology for the
whole planet. The phrase 'steam power' has many resonances beyond the
(admittedly evocative) locomotive whistle.[5]

The cultural ecology of industry

The underlying essential of the industrial era is the human ability to supplement the flows of solar energy with that of fossil fuels which are essentially mineral in form. Like living tissue, which they once were, coal and oil (and the derivative natural gas) are carbon compounds; reflecting their geological heritage, they are distributed worldwide though not uniformly so.[6] But their adoption meant that industrial energy (and, indeed, most other materials) was no longer dependent upon the use of surface area together with imports. Yet we often overlook the fact that the energy subsidy was also applied to agriculture. Producing more wheat per hectare may be ecologically as significant as producing millions of cheap tin trays but the latter has a higher profile since the development of manufacturing is the new element. It implies the production of a high volume of goods and their fabrication in factories on a concentrated site rather than piecemeal in workers' houses dispersed through a rural area. It comes to an apogee in this era in mass production on production lines of the type associated with Henry Ford in Detroit from about 1900. One outcome was the suburban family in the 1920s western country which now had access to about 100 times as much energy (mostly from fossil fuels) as their agrarian predecessors gained from child labour, a hired hand and an ox team. Other relationships were transformed as well: until coal replaced wood for firing bricks, more trees were consumed in building a house in brick than in timber.[7] The social context was important, too. Industry created developments which could often be accommodated in land hitherto held in common and used for sport or grazing. If it was enclosed then landlords could develop mines, houses, water reservoirs and transport links in a more or less unhindered fashion.[8]

Since neither the raw materials nor the social context were favourable everywhere, there were core areas of development of the new ecology and an uneven spread in time and space via, mostly, trade and empire. Moreover, none of the previous energy supplies was lost and new technology gave added utility to such sources as falling water once electricity became known and controllable. The cultural basis of the new economy cannot be overlooked, the more so because the changes wrought were truly Promethean in scale and in diversity of mode ('all the arts'), admitting to no sense of limits in any form or direction. James Watt (1736–1819), a Scottish improver of steam engine technology, wanted to 'find out the weak side of nature and to vanquish her'.[9] The complexity of the idea (and also the mythology) of 'industrial revolution' is beyond the present volume's scope but its historical centrality and its ecological reality are beyond question.[10]

Evolution and dispersal

Even though steam power was critical, the application of water power was a key precursor. New machine technology, especially in the English textile industry of the eighteenth century, made factory-based production more economical provided that the machines could be powered. The harnessing of rivers was the solution, so that places like Cromford in Derbyshire tapped the power of the River Derwent as it came off the uplands of the English Peak District. From about 1750 canals were important in transporting bulky materials, such as coal, and commanded environmental changes such as the diversion and tapping of tributary water supplies. Water engineering for mills was not a new skill but it was applied at new scales and densities so that flights of water-control mechanisms were commonplace: a knowledge that came in especially useful when furnaces had to be supplied with blasts of air. Many early industrial developments in Britain, Europe and eastern North America could tap the skills and materials of the earlier era in water management and in fuel supply where charcoal was the all-important material. To that extent the 'industrial revolution' had many evolutionary traits and apparent quirks: British iron-making was about the only industry not to change from wood to coal by the end of the seventeenth century; it was complete only by 1790. The new machine of star status was the steam pump, developed by pragmatic engineers from the seventeenth century onwards, with Thomas Newcomen's 'atmospheric engine' of 1705, notably developed by James Watt in the 1770s.[11] These engines were especially valuable in allowing coal to be raised from deeper mines than hitherto and, unlike horses and people, they did not tire; one colliery in Warwickshire had employed five hundred horses to lift out water bucket by bucket. Mills on rivers hitherto dependent on climate could now rely on steam power if there was a good coal supply. The stationary steam engine was converted into portable form as the railway locomotive and the steamship, and its efficiency improved many times, with a type of culmination in the liquid-fuel powered internal combustion engine of the late nineteenth century: industrialisation could now go anywhere, so to speak. The new economy also called forth a large chemical industry with improved versions of acids, alkalis and dyes in which organic materials were replaced whenever possible with inorganic inputs and the products became available in more concentrated forms.

The core areas of industrial growth are well known: they evolved where there was coal, iron ore and access to other raw materials such as the limestone used a flux in blast furnaces. If England was the core zone, then many other parts of Britain, Europe as far east as the Urals, the eastern United States and eventually Japan, followed by 1900. Industrial production was spread outwards from these core territories through the media of trade and imperial direction, so that everyone in the world knew about their existence even if only through new iron-edged tools or cheap guns. By 1950, even those countries without the basic industries of chemicals, iron and steel, wanted factory-type production and so might develop electricity supplies from damming rivers or importing oil. Some larger nations, such as India, regarded heavy industry as essential to their post-independence economies.

In cultural terms, it is interesting that very few nations turned their backs on industrialism even when they had a choice. There might be a short sharp debate about 'modernisation' that resulted essentially in economic and political revolution, as with the restoration of the emperor in Japan in 1866–9, but essentially Walt Whitman in his *Leaves of Grass* of 1860 summed up the scope and penetration of the new power source that had been harnessed:

> His daring foot is on land and sea everywhere
> He colonises the Pacific, the archipelagos
> With the steamship, the electric telegraph
> The newspaper, the wholesale engines of war
> With these and the world-spreading factories
> He interlinks all geography, all lands.

To sum up the development of fossil fuel-powered industry until 1950, we might adopt four main but overlapping phases of dominant systems:

1750–1820   Water power, turnpikes, iron castings, textiles
1800–1870   Coal, canals, iron, steam power, mechanical equipment
1850–1940   Railways, steamships, steel, coal, chemicals, telegraph
1920–1950   Electricity, oil, cars, road-building, radio. From 1930 onwards the beginnings of mass consumption.[12]

The ecology of the first one hundred years of industrialism clusters around the processes of extraction (of mineral ores, coal and oil), of processing in factories, and in applying the goods in social and environmental fields. Each stage may need transport and each will have secondary effects: early coalfields not immediately near iron ore evoked canals to transport the coal, for example; the wastes from alkali plants poisoned fish in rivers many kilometres downstream; coal mines had a strong impact on forestry practices because of the demand for timber underground; every trend combined to foster the growth of towns and cities. The environmental consequences were inevitably strongest near the centres of industrial activity; nevertheless long-distance transport by water or railway meant that frontier industries sprang up well isolated from the burgeoning cities the inhabitants of which staffed the new factories: the best example is probably logging of primary forests in lightly populated areas of North America during the nineteenth century. So it is impossible to underestimate the outreach of the industrial economy: where we do not detect it then it is probably because the traces have been overlain by later changes. But look in the river sediments or in the layers of the peat bogs or in the health records of the populations and in the diaries that record the disappearance of familiar species of wild animals.[13]

Environmental relationships

The closer we come to the present, the more material we have and so the narratives become more complex. In the case of 1750–1950, we are dealing with an era in which there was widespread literacy aided by printing, a greater curiosity about the world, and the desire to record the results of that wonder in writing, graphics and in collections of specimens. So our main problem is about how to treat so much information.

One major driver of environmental relationships is also one where estimates have to be accepted: the growth of the human population. Even today, censuses are not always reliable and the first systematic counts in the more organised nations were not until the late eighteenth to early nineteenth centuries: for example, in Scandinavia and Prussia, then the United States in 1790 and Britain in 1801. The data that emerge from demographic research suggest that, in 1750, the world population was 720 million, rising to 900 million in 1800, then to 1625 million in 1900, and 2500 million in 1950. Before 1750 the annual percentage increase was about 0.01 but between 1750 and 1950 it rose from 0.5 per cent per annum. to 1.5 per cent per annum. Making some *bravura* assumptions about populations since 50,000 years ago then perhaps 12.5 per cent of all the people who have ever lived were present between 1750 and 1950.[14] Their impact can be imagined by taking the level of industrialisation of Britain in at 1900 to equal 100. Then the whole world in 1750 was at 127 but at 1360 in the 1920s, with 950 the level of the industrial core of 'western' countries.[15] The summative indicator in terms of historical data is the fall of the death rates in the world. This happened first in the industrialised countries but then elsewhere, longevity increased and birth rates took off as infant mortality declined. Behind these trends we see not simply the effects of the political and scientific exports but also of the worldwide distribution of effective crops such as maize and the potato and the build-up of resistance to infectious diseases as these spread following trade and exploration and allowed people to build up resistances once the initial epidemics had been survived. So an Asian population of 480 million in 1750 became 1386 million in 1950; Africa grew from 95 million to 206 million in the same period and Latin America from 11 million to 162 million.[16]

The onset and development of industrialisation are akin to a tsunami wave: they tend to blot out everything. In this case the technological changes can let us forget that the natural world was not necessarily in a state of equilibrium. Post-glacial changes in sea-level, for example, were not complete since the relative effects of isostatic recovery and ice-melt were still in contention. Some plant and animal species were still migrating under the impetus of their own behaviour and not that of humans, and human populations themselves were acquiring resistance to diseases. The largest-scale disequilibrium in the natural world, however, was a global downturn in climate that is usually labelled the 'Little Ice Age' (LIA). This had been in effect since about 1550 but its worst phase was the first half of the nineteenth century, with harshest winters recorded in the northern hemisphere. The onset was not sudden and there were anomalies (positive and negative) in both time and space but it was probably the coldest interval in the Holocene. Its end was, however, relatively abrupt for, soon after 1900, temperatures started the rise into today's values.[17] We must not infer that the LIA caused industrialisation but no doubt the easy availability of coal was a cushion in some places, specially where centuries of lower temperatures had retarded tree growth.

There were other climatic impacts as well. The ENSO cycle's ability to affect many countries with, variously, floods and droughts seemed to concentrate upon China, India, South Africa, and Brazil, with lesser impacts upon the Sahel, Sudan, Egypt, Ethiopia and Indonesia. Thus, very strong El Niño phenomena were observed in 1790–3, 1828, 1876–8, 1891, 1899–1900 and 1925–6. The years 1876–8 saw a once-in-two-hundred-years drought which was virtually worldwide. In many instances, there was drastic famine followed by epidemics of cholera, malaria and tuberculosis. Colonial governments helped little if at all and independent areas were left to manage as best they could. In the Horn of Africa in 1888–1902 scorched fields were followed by rinderpest, caterpillars, locusts and rats, then social and agrarian collapse because, if there were no cattle, then there was no cultivation even when the rains came.[18] Volcanic eruptions do not have a cyclic pattern but climatic cooling was certainly a result of the eruption of Tambora, Indonesia, in 1815, followed by 'the year without a summer' in 1816.

The world was cool in the early nineteenth century but the reception of technological change was warm. The advent of controlled and efficient steam power stimulated the invention of thousands of machines for almost every conceivable purpose. Queen Victoria's journal gives her reaction to the Great Exhibition of 1851,

> Went to the machinery part . . . which was excessively interesting and instructive . . . What used to be done by hand and take months doing is now accomplished in a few instants by the most beautiful machinery . . . We saw hydraulic machines, pumps, filtering machines of all kinds, machines for purifying sugar – in fact every conceivable invention . . .

Those machines were powered by aeons of stored photosynthesis: a year's fossil fuel use in the twentieth century consumed perhaps 400 times the global net primary productivity (NPP) of one year in that century[19]. Above all, coal was the fuel of the nineteenth century, though being overtaken in usefulness by oil in the twentieth. Hence, the steam locomotive is a cardinal invention and emblematic of the nineteenth century, yet one which, by 1950, was waning in favour of diesel traction. The all-important machine of the twentieth century is the internal combustion engine, made possible by the extraction and refining of oil after successful drilling was achieved in 1859, in

Pennsylvania, its presence having been noted on a map of the 1750s. This engine underlay not only cars, trucks and, perhaps emblematic of its environment-altering power, the bulldozer but also powered flight. To add to these two directly environmentally altering technologies there is a third crucial, but indirect, piece of applied science in the form of the electric telegraph and its successors, such as the telephone, radio and television. All of these speeded up both the transmission of ideas and the pace of action, adding to the sense of mobility characteristic of this industrial era.

As we saw in the discussion of agriculture, there has to be a receptive cultural climate for the adoption of technology. One of the outcomes of that era was the formation of the nation state with an ability to provide wide-ranging cultural contexts. In the epoch of industrialism, there was a state-backed urge towards imperialism which had many environmental consequences. Not the least of these was the export of the new technology to areas previously low in its presence and the resultant impact on agriculture, forestry, wildlife and urban growth. Most of this was devoted to the welfare and profits of the imperial power and in many cases expanded its 'ghost acreage', freeing land from the production of food and fibre at home since it could be produced cheaply in the empire and transported cheaply in large ships. Then, after World War II, the *Zeitgeist* went into reverse and colonies largely became independent, albeit at varying rates. One carry-over from imperial days was the taste for large-scale projects so that having a large dam became a symbol of national status as did the possession of with an army, expensive military aircraft, a national airline and Mercedes saloons for the governing class. The urge to profitable (for some) export of minerals and crops was kept as well, often aided by international development institutions such as the Food and Agriculture Organization (FAO). Before 1950, however, it is unrealistic to say that any nation state had an environmental policy, though outstanding individual ills might be tackled in the richer places or in those with an environmentalist legacy from their colonial years.

The industrial era can be seen as being one in which the human population came to demand three 'products' from its environments. The first is utility in the form of resources: energy as the binding resource that determines the availability of many others, including food, water and minerals. The second is sanitary, in the sense that the environment is expected to receive and preferably process the wastes produced by industrial and urban conglomerations. Thirdly, there is outdoor pleasure, once confined to few social strata but increasingly becoming available to many in the industrial nations if less so beyond them. To provide these social benefits, new genotypes and new ecosystems have been created. The new genotypes have mostly been deliberate creations, as in the case of plant and animal breeding once these had been put on a scientific basis by an increasing knowledge of genetics. But incidental creations include species of plant adapted, for example, to grow on industrial wastes of high toxicity. New ecosystems fall into two types. The deliberately created systems include the extension of agriculture at the expense of forest and the reclamation of wetlands and the conversion of all kinds of terrain to urban and industrial use. As with genotypes, there are creations that were not foreseen, such as the extension of desert vegetation in steppe areas subject to heavy grazing, the diminution of many kinds of marine life as a consequence of heavy fishing, and the death of corals reefs if silted up by run-off from terrestrial land manipulation. The social appraisal involves the acceptability of such changes. Although they are in general regarded as a price worth paying for enhanced access to resources, reactions set in through the nineteenth century which formed the basis for today's conservation and environmental movements. One generalisation which emerges gradually in the period to 1950 is that some human-driven effects can be truly global: that material injected into the upper atmosphere (be it then acceptable like carbon compounds or less so like radioactive particles from bomb testing) can affect almost every part of the global environment.

So the extension and intensification of the ecological footprint of humanity are major features of the industrial age. The relationship is not so much driven by population growth alone but by a 'population × technology' equation. This also needs a factor for the rate of population growth as this was often accompanied by rapid environmental change.

Management and impact

Nothing can diminish the importance of fossil hydrocarbons for the world's ecology after about 1800. The getting of this concentrated supply of energy from under the earth, its harnessing in many ways and its embedding into so many materials, and the consequent creation of waste products were revolutionary. It is no exaggeration, therefore, to called it a 'binding resource' in the sense that access to coal, oil and then natural gas defined so many things: economy, political power, health status and birth rate included. The ability to get at energy supplies has also defined environmental relationships in terms of the ability to manipulate the natural world, to transfuse it with wastes and to encompass it symbolically as never before. If Newton's equations constructed a knowable theory of the universe then modern communications technology has made another virtual world but one with emotional appeal as well.

Resources: energy and environment until 1950

If J. S. Bach's 'Well-tempered Clavier' took a step away from the natural world's harmonies, then it is appropriate that the first item in his will comprises his shares in a mine.[20] Coal had been known as a source of heat for many centuries in the agricultural era but it was the knowledge of how to generate steam under pressure that brought into being a huge industry which was highly energy-positive, that is, every joule invested was repaid by access to many thousands more.[21] The concentration of the energy in coal (and much more so in oil) made portability possible and so the power of steam-driven (and then the internal combustion engine and electricity) could be taken to hitherto remote places. The tiny railways up to hill stations of the British Raj in India say 'steam and conquest' in each blast of the cylinders.

The ecology of the coal-mine and the oilfield are well known. The coal-mine affects land, air and water in its immediate vicinity, but the oilfield is more readily contained provided there no large accidents. Both oil and coal needed cheap forms of transport if they were to be used away from their geological bases: the railway in the one case and pipelines-plus-tankers in the other. Each took up land for installations and each produced emissions to air and water; oil refineries, in particular, have in the past given off cocktails of organic compounds to the air and water as well as using large quantities of water for cooling processes. Coal has been produced by open excavations as well as from deep shaft mines, and large quantities of unusable spoil are created as well as disturbed drainage and, if the surface is not restored, great scars across the landscape. Almost every shaft mine before the middle of the twentieth century had a smoking waste tip; if possible, railways or a conveyor system were used to take this material to be tipped away from the settlement, perhaps to an open and uncultivated hillside or even offshore. Management in the period to 1950 was, except in a few progressive places, driven by profitability, and the effect on the human population and the environment was generally regarded as part of the package that was crammed with cheap energy. But this management was successful in delivering energy: a solar-powered agriculturalist averages a throughput of 10 to 20,000 kilocalories per capitum per day (kcal/cap/day), whereas members of early industrial societies manage 70,000 kcal/cap/day and then, in full industrialisation, 120,000 kcal/cap/day. The access to resources made possible by the harnessing of fossil fuels was immense and although not on the scale of the post-1950 years, nevertheless worldwide in its reach and global in its effects. Few, if any, ecosystems failed to receive some transformation even if it was only in the form of aerosol fallout of, for example, lead (from its use as an additive in vehicle fuel) on to polar ice caps and radioactive particles (from testing of nuclear weapons) on to the high tundras of the north circumpolar zone. Many of the failed attempts to climb Mount Everest in the Himalayas left rubbish as well as the occasional undecayed body of a climber.

The export of industrial technologies to areas outside the core zones of Europe, North America and East Asia followed two main channels of impetus: that of trade and that of imperial conquest, one of which might follow the other. By 1750, many agricultural and remnant hunter-gatherers' societies had links with European or Chinese consumers and, in every continent, hierarchies had developed in which the richer nations imported from the poorer (slaves as well as environmentally derived materials) and, in turn, provided the poorer zones with products such as iron tools, guns and alcohol. Colonialism in the form of protection was an obvious follow-up in such places, and was often accompanied by the benefits of western technology, such as the railways, land drainage and irrigation on a large scale, and nineteenth-century medicine that at last knew about the role of bacteria in cholera and of mosquitoes in malaria. Cheap labour could produce cotton, tobacco, timber and sugar for industrial markets and, under colonial control, 'unused' lands could be converted to cattle ranching for beef export or to plantations for tea, coffee and cocoa. That the 'empty' lands had been the habitats of nomadic pastoralists or hunter-gatherers was usually ignored. An emerging theme is a continual increase in movement: of materials as, for example, rubber, bauxite and copper entered world trade in the second half of the nineteenth century; of people as they migrated from agricultural poverty to new sources of employment; and in support of which major canals such as that at Suez (1869) and Panama (finally finished in its recent form in 1914) were constructed.

There was, therefore, an unprecedented and major technological change, leading to revolutions in economies and in political structures, with immense social ramifications. Did environmental considerations play any part in these huge developments? This is apt to lead into a discourse on the 'why' of industrialisation which is too complex a task for this book once we abandon simple explanations such as charcoal shortages in the English Midlands in the seventeenth to eighteenth centuries.[22] Moral geographers, like Ellsworth Huntington (1889–1975), were sure that temperate climates with seasonal contrasts led to vigorous, hard-working individuals who were good at industrial production and leaders of trade and empire overseas. Images of the backward and lazy natives of the tropics were part of the fantasy world of environmental determinists of that time, now discredited. Any consideration of the complexities of both social and natural factors involved in early industrialisation is likely to conclude that there was a considerable degree of chance involved. What is not at issue are the changes in ideas and in material conditions that the new energy sources brought about. In the field of ideas, steam allowed people to entertain large-scale notions about the control of nature. It gave, as it were, a material form to the great collections of knowledge being gathered into extensive cathedral-like museums and into large volumes of print and pictures. Not least among these was the formulation of grand theories of the world such as Charles Darwin's (1809–82) crystallisation of ideas about organic evolution. Since the practically minded drivers of environmental change in the homelands and in the colonies were mostly western men of Christian traditions, Darwin's desacralisation of nature was a fertile bed that could be tilled by steam technology.[23] Crudely, it was morally acceptable to change the world of rocks, water and life and there were the means with which to do it. William Huskisson[24] wrote in 1824: 'If the steam engine be the most powerful instrument in the hand of man to alter the face of the physical world, it operates at the same time as a powerful lever in forwarding the great cause of civilization.'

Resources: feeding the industrious

Of all the organic products of human endeavour, food for our species is certainly the most important, although plant and animal materials contribute to many other flows of matter, as in animal fodder, clothing, tobacco and timber, for instance. As in the agricultural era, the main systems were rain-fed agriculture, irrigated lands, and pastoralism, with the latter undergoing an intensification to the form usually labelled ranching. Industrial output grew at about 3.5 per cent a year. from 1750 and per capitum by 2.3 per cent so that in the two hundred years currently under consideration, the total increased one hundred-fold. This has not been regionally uniform: if we take the British output at 1900 = 100, then the core industrial areas went from 2 in 1750 to 950 in the 1920s but the less developed countries having started at 127 finished in the 1920s at 220.[25]

All these systems underwent change because of the new technologies subsidised by fossil fuels, in particular, their mobile applications made possible by oil-based fuels. The nearer to industrial heartlands the systems were, the greater the changes but it is probably true to say that no food-and-fibre producers in the whole world have remained unaffected in some way, even if only by knowledge, the storage and transmission of which owed its existence to fossil fuels.[26] The theme of movement is exemplified by the mass carriage of materials during the nineteenth century by rail and sea in response to the demands of urban centres: meat from the High Plains to railheads in the American west, or by refrigerated ship from Argentina to Europe; cotton from India to England; wheat from Australia and Canada to the colonial metropolis: the list could be endless.[27] The production of food and fibre was, therefore, no longer purely a way of tapping the outcome of photosynthesis. The growth of scientific knowledge meant that plant and animal breeders could eliminate many of the chance elements in genetic recombination and tailor organisms to particular conditions: to produce cereals with a short straw as a defence against lodging, for example, or cattle better suited to tropical conditions than the indigenous varieties. Gradually, heavy machinery began to replace human and animal labour. The tractors which replaced horses (surprisingly late in the twentieth century even in western countries) freed land from producing food only for horses. Other machines allowed better tillage because they could plough deeper or break up soils into a finer tilth. Behind these direct applications was a massive secondary deployment of energy in, for example, chemical fertiliser manufacture. There was also the development of biocides, in which there was a great surge after World War II. All these metamorphoses were tied together with oil- and coal-using transport and all underlain by the accumulation of scientific knowledge which was, in turn, feasible only where energy was abundant.[28] The situation in which crops were confined to a particular continent or region, already breaking down in the agricultural era, was replaced by the dominance of New World species in the entire temperate zone.

Among those crops, the potato stands out as the food which sustained the rapidly growing industrial populations, but maize in southern Europe was important, as it also was in parts of Africa, albeit in less industrialised economies.[29] Environmental management at farm level thus became directly focused on each year's production, with rotations where necessary to avoid the build-up of diseases such as potato root eelworm, *Globodera rostochiensis*. No management, however, was successful enough to avoid outbreaks of a devastating fungus like potato blight (*Phytophthera infestans*, in a strain derived from South America, like its host) which caused the great famines in Ireland in 1845–49.[30] By 1950, however, potatoes produced in Ireland were, like most places, partly constituted of diesel oil.[31] A farm unsubsidised by fossil fuels could produce perhaps 1.0 kilocalorie per square metre per day (kcal/m²/day) of organic matter whereas by the 1950s, 5 to 10 kcal/m²/day were possible in the heavy energy-user countries such as the United States. If, as in the nineteenth century, even a fraction of an improved yield goes to urban workers as plant food rather than meat, then the feeding of population growth becomes possible. If meat and cereal supplies are secure then alternative crops become attractive. In the period 1879–1939 in Britain, these included pastures for fresh milk supplies to the cities (replacing unhygienic town dairies), industrial crops such as tobacco (uneconomical), hops, and sugar beet. The latter was popular because it took only 3.5 person-days per acre per year compared with wheat at 6.5 person-days/acre/year and produced both income and animal green fodder. Game crops appeal at times of affluence so partridges, pheasants and rabbits might be encouraged in odd patches of scrub and covert; if poor soils were present then a combination of horse manure and night soil could create an artificial soil 3 or 4 feet deep yielding profuse crops of mushrooms. The railways allowed the rapid marketing of fruit and flowers.[32]

The management and impact of irrigation farming are difficult to separate. The practices involve intervention in the hydrological cycle: water is stored and then released into a channel system. Some of it evaporates or is taken up into the crop ('consumptive use') but another fraction continues into the run-off, the 'return flow'. Added together for 1650, the water diversion for the world was 95 cubic kilometres per year, a total which rose to 226 in 1800, 550 in 1900 and 850 in 1950, with recent totals at about 3,000. Irrigation has for long been sufficiently important that, until 1950, it comprised about 90 per cent of the world's water withdrawals even in the face of industry and cities.[33] Though small-scale schemes have always been widespread, the larger projects have attracted much attention since colonial governments and their immediate successors were attracted by their ability to bring about massive transformations. In British India in the 1880s, 1 million acres (over 400,000 hectares) of 'scrub' were converted to wheat production in the lower Chenab and by 1939 British India held 116,000 square kilometres of canals watering 11.6 million hectares of land. Productivity increases tie the practice into industrial energy systems, so that water-lifting by bullocks is replaced by electric and diesel motors; intensive farming depletes nutrients and requires pesticides, so that there may be a need for 900–6,500 megajoules per hectare of energy input. There are changes in the physics and chemistry of the water and the soils, some of which are in the direction of improved fertility of the soils and less silty water. If evaporation exceeds precipitation, however, then a secondary salination occurs as minerals move up the soil profile. These compact the soil and cause waterlogging as well as increasing the pH to the 9.0–11.0 level. The impoundments are likely to get silted up quite quickly unless their watersheds are carefully managed and are also likely to acquire an anaerobic decomposition layer yielding hydrogen sulphide(H_2S). Downstream the clearer water may encourage aquatic photosynthesis that includes algal blooms. Obviously, management has to be skilful and, in larger schemes, has to encompass the whole project. Individual farmers lose control over the very basis of their livelihood and may also have suffered dislocation as the impoundment's water level rose. So many failures of large-scale projects have been due to their social context as much as to the influence of external environmental conditions. These factors have not prevented a huge growth of irrigation projects in the post-1950 period.

An analogous question about management runs through the history of pastoralism in the nineteenth and twentieth centuries. Those areas of the world (overwhelmingly semi-arid areas such as savannas and grasslands but also some mountain and high plateau areas) utilised by animal flocks that were moved on a daily or seasonal basis were usually regarded by traders and colonists as inefficient. This evaluation was compounded when scientific surveys detected soil erosion and overgrazing and, eventually, 'desertification'. Thus, from the nineteenth century onwards, European and North American influences were brought to bear upon animal herders in order to 'improve' their stock and change the life-ways of the pastoralist people or to replace them altogether with ranching. Both systems focus on a dominant species but, whereas pastoralism kills only a few animals and prefers sustained-yield products like milk, ranching focuses on slaughter products for a money market. Both take in large areas of land over which to spread the grazing pressure but ranching needs more labour. The ecological key for both is the conversion of primary production into animal tissue, in which not only the forage plants but the provision of water may be critical. In the case of cattle, 5–23 litres of water a day are required and lactating beasts need about 40 per cent more. Though the environments of pastoralists had seasonal regularities, usually including a pronounced dry season, they also have pulsed events such as prolonged drought or flash rainfall, much of which may escape short-term scientific survey just as measurements of soil erosion may be misleading when scaled upwards to whole landscapes. Concepts of stability and equilibrium are often misapplied in such environments.[34]

The development of European-style pastoralism was much enhanced by the transport of cattle to the Americas, starting with Hispaniola in 1493 and thence to the mainland at Vera Cruz (Mexico) some twenty-eight years later. Thus started the trend for colonists to run large herds of a single species and often a single breed, with cattle and sheep far outnumbering other species. Sheep, for example, were introduced into Australia in the early 1880s but there were 80 million of them by 1888. This history, however, disguises the fact that, in many places, cattle were introduced and then allowed to go feral, so that the beginnings of ranching in Texas, northern Brazil and Argentina, for instance, lay in the hunting of 'wild' cattle. These animals and many other more controlled herds were run on open ranges in the early to mid-nineteenth century. The takeover of apparently empty land was helped by the fact that some of it was occupied by hunter-gatherers with no concept of 'ownership' and other parts by indigenous pastoralists undergoing a downturn due to pulses of drought, disease or war as was the case on the East Africa savannas about 1900.[35] The transformation of pastoralism into ranching was given enormous impetus by the development in the 1880s of refrigerated steamships in which large quantities of chilled carcasses could be conveyed around the world[36].

The combination of demand for meat from urban-industrial populations, colonial power and scientific insights combined to push forward the idea of improvements in grazing in semi-arid environments. This was applied to both white-owned ranches and to the 'native areas'. The impacts of the former could be severe, as in Australia, where soil erosion, xerification of range, explosive populations of introduced species, such as the rabbit and the prickly pear, and the elimination of wildlife were part of the price of a 'golden age' of wool production in 1880–1900. Elsewhere, indigenous stock-raisers were often forced into smaller and often more marginal areas by colonists' land appropriation. Then their chances of avoiding higher rates of soil erosion, pasture degeneration and animal disease (and their human consequences) were poor. As a result, ideas of improvement of animal rearing were dominant in many colonial regimes. In Rhodesia (now Zambia and Zimbabwe) the plans (from about 1903 onwards) included the movement of selected bulls into native areas, fertilising pastures, reseeding, and rotational grazing but success on the experimental farm was rarely repeated in the reserves not least because Africans had no history of, and hence no cultural orientation towards, beef production.

One unusual example of pastoralism under change is that of reindeer in northern Scandinavia, in the area inhabited by the Saami people. Their pastoralism evolved from hunting in the seventeenth century to a form in which the animals ran wild for most of the year although they were owned and rounded up two or three times a year by settled forest agriculturalists and forest workers. After the depredations of World War II, the demand from the south for reindeer products coupled with the advent of the snowmobile allowed a few herders to accumulate the majority of the animals and displace many others into cities; keeping the reindeer from overgrazing the lichens of the forest and *tunturi* has been difficult due to interruption of their grazing patterns by developments such as reservoirs and tourist complexes.[37] There is a parallel with the Bedu of, for example, Syria and Saudi Arabia where the camel has been replaced by the Toyota truck, the 4×4 and even the light aeroplane, all providing for the better-off.

The cultural context of animal herding has changed as well as its technology. Perhaps the greatest change has been in the political impact of the state, whether colonial or post-colonial. Colonial regimes were notorious for making artificial boundaries that cut across pre-existing migration routes of seasonally migratory pastoralists and forcing them into overgrazing pastures or overusing water sources within the new political jurisdictions. (This may have meant that a cultural group who had been powerful were now rendered nugatory, a change not lost on colonial administrators.) Even post-colonial land reform imposed the sort of control that made it possible to alienate pastoralists' land in favour of agriculturalists, sometimes under the guise of sealing off areas where animal disease had broken out. Examples include Iran under the Shah, Israel, and British East Africa. As the semi-arid areas were drawn into worldwide systems of trade, further changes came about, combining economics and technology. In the United States, for instance, the penetration of railways into the High Plains imported both eastern capital and a demand for meat that brought the open range to full capacity. The range was then closed to facilitate more rational management, aided by the development of effective barbed wire, patented in Illinois in 1874. Even within the new units overgrazing was rife so that a move to high-quality stock was brought about by selective breeding, in which the new land-grant universities played a key role. Above all, there was territorial compartmentalisation which led to a mosaic of ground-cover types according to local stocking rates and management competence. The future of burrowing rodents, for instance, might depend on the attitude of a stock manager towards cattle numbers and coyotes. Ranching overall, however, seems to be culturally marked by cruelty towards animals.[38] At the heart of many environmental transformations lay demands for meat, aided by the spread of motor transport.

Translocation and transformation

A major difference between this era and the preceding millennia is the transport of food over large distances and its processing. Trade has always shipped food between people, and food has always been processed to preserve it. Industrialisation increased immeasurably the scale of both processes, with steamships and railways allowing the perishable products of New Zealand (like butter) to end up in Europe or Argentinian beef in canned and compressed form in North America. Refrigeration was first used for meat but proved to be useful for fruit as well. Canning drew upon metal ores and a great deal of energy in terms of heating, sealing and labelling. Cheaper energy meant that refining of comestibles could be more effective and so sugars with a higher sucrose level were cheaply made. Coal fires may have allowed more boiled water for tea and coffee and hence improved the general levels of health. On the other hand, food processing may remove minerals and vitamins and so energy has to be expended in producing, refining, packing, labelling and distributing food supplements. Even before 1950, cut flowers were sent by air from California to New York in winter. One lesson of the era is the way in which ecological transformations could be so thorough. In 1850s New England Henry David Thoreau describes an extensive swamp '. . .which first had been cut, then ditched broadly, then burnt over; then the surface paved over, stumps and all, in great slices; then these piled up every six feet . . . then fire put to them; and so the soil was tamed.' This was achieved by William Brown the farmer who was clearly following in the steps of the English rector of the eighteenth century who had, in a flush of innovatory enthusiasm in agriculture, planted the churchyard to turnips. His archdeacon chided him, saying, 'This must not happen again'. 'No sir, next year will be barley.'[39]

Whole environments: forests, recreation and warfare

The period from 1750 to 1950 was one of sustained impact upon the world's forest cover, though not as strong as that after 1950. Temperate zones and the tropics were affected: between 1700 and 1920, some 315 million hectares of forest and woodland in the temperate zone and 222 million hectares in the tropics disappeared. In the tropics, 56 million hectares of that total were transformed into grassland simply by felling and neglect. They became either cropland or grassland; at the same time 146 million hectares of temperate grassland became cropland. The trend was of acceleration: in the tropics once again, the change from 1700 to 1850 was a loss of 70 million hectares and from 1850 to 1920, 152 million hectares.[40]

Within these startling totals there are regional differences. In Germany, for example, forest loss led to strong governmental interest in the management of the resource. This was present in many of the smaller princedoms of the eighteenth century after the end of the Seven Years War in 1763. Following the publication of the influential *Principles of Forest-Economy* by Wilhelm Gottfried von Moser in 1757,[41] timber production became a principal aim of forests. Standard species of trees, orderly rows of monocultures and continual yields became the watchwords, and German foresters spread the techniques through many nations, including the tropics. By coincidence, 1757 was the start of the official British Raj in India, where a combination of fiscal policy and technology removed great areas of forest and other woodlands. In the first case, the removal of forest increased the amount of cropped land available for revenue. Thus, land was readily sold to (mostly white) landlords for conversion to plantations for tea, coffee and rubber. It was also cleared to feed expanding indigenous populations and to export timber, especially tropical specialities such as teak (*Tectonia grandis*). After about 1853, technology in the form of railways started to be a major influence. Fuel was an initial concern until coal became available from Raniganj after 1880, but construction was even more consumptive. In British India, there were 1,350 kilometres of railway in 1860 and 51,658 kilometres in 1910. In the Madras Presidency alone, that meant a yearly demand of 250,000 sleepers, meaning 35,000 trees.[42] The fading of the forests led eventually to a Forest Act of 1878 which fudged the question of common rights (hoping they would disappear), banned shifting cultivation and grazing, and moved the direction of management towards commercial production: the first Chief Forester was, in fact, German.[43] The same policies survived Independence in 1947. One other consequence of both forest loss and commercial orientation was lower populations of animals such as the tiger and the elephant, though imperial-style hunting added to both concerns.[44] In Africa, the savanna lands had islands of woodland, often wrongly interpreted as relics of former high forest in another example of regarding many tropical forests as pristine ecosystems.[45] More sophisticated twentieth-century science has teased out some of the detail of, for example, soil erosion, showing that short-term measurements are not necessarily a guide to long-term processes and that soil loss from one place may be soil gain lower down the watershed.[46] Nevertheless the rates of soil loss after deforestation in the nineteenth century increased by at least four times, often by ten and even by as much as seventy, if combined with (European style) settlement as in Michigan in the 1860s[47]. The new owners of large country parks often had an ingrained eye for profit and so faster-growing conifers were, in Europe at least, part of the planting schemes. The European larch and also the evergreen Japanese larch, and then the imports from North America: Douglas fir, Sitka spruce and lodgepole pine were set to be grown privately and then to become the staples of state forestry. Where a real exotic was wanted then the Sierra Redwood (*Sequoia gigantea*) was the obvious choice and given the common name of Wellingtonia in Britain demonstrated patriotism as well.

The cultural context of forests in this era is in some ways scale dependent. Locally, there were still few forests that did not share in the ambiguities of folk knowledge about them. Malevolent inhabitants were part of those habitats and so the disappearance of the trees had its virtuous side. At a national scale, forests and their history become part of mythologies: in England the oak stood for a whole people's virtues for it was venerable, patriarchal, stately, a guardian and therefore quintessentially English. Beech and elm were also well regarded. Even before then, conservatives like William Wordsworth thought that the larch (*Larix europaea*) was (in the Lake District, in 1810) a sign of increasing industrialisation.[48] The landscape designer Humphrey Repton (1752–1818) thought that any patron of his who wanted conifers was a parvenu. The German coniferous forests entered the German consciousness as part of a Nordic (and militaristic) history; the French saw them in the light of a passion for order, and in the United States they were part of a transcendental compact with the Creator.[49] In their untidy way, the British took against 'regimented conifers' when reafforestation became very important strategically after World War I.[50] Governmental attitudes were rarely irrelevant in the colonies. The pressure in Asia was often to clear land for sub-sistence farming and this might be aided by colonial policies which favoured agriculture over tree growth as short-term aids to a settled population. More rice and more rubber, as in Malaya and Indonesia, were preferable to keeping the forest trees even if, by the twentieth century, there was enough scientific evidence to suggest that in the longer term there might be environmental problems.[51] Wartime created pressures on forests as in Japan where post-1941 forestry emphasised cypress and *Cryptomeria* trees which created a 'dark forest' (*kuraku kanjiru*) instead of the preceding mixed forests. Traditional place names ('beech plateau') were lost and wild boar was no longer so tasty.[52] More widely still, it has been argued that forests developed as a 'dark other' in contrast to 'civilisation'.[53] The origins no doubt pre-date the industrial era but chime nicely with the perceived needs for wood products and for the clearance of forested land. Within each of these possibilities, local views and local construction can emerge which subvert any broad generalisations. But in all eras, the cultural constructions of nature are nowhere more evident (and commented upon by scholars) than in the case of trees, woods and forests.

The emergence of more widespread public attitudes to forests heralds the growth of outdoor recreations in which the environment is very important. The pleasures enjoyed by upper social echelons are by no means diminished but they are joined by the middle and working classes once those groups have more free time and some discretionary income. The delights of rural areas outside the city walls and of spring woodlands are celebrated in many lan-guages.[54] In the later nineteenth century the joys of days off spent in rural areas and by the sea began to spread through the often grimy world of industrialised regions.[55] In this, the role of cheap travel, especially by rail, was central and prefigures today's low-cost airlines. The role of gardens was now subsumed into a wider recreational context. Some cities made it possible for poorer people to have a small plot of land near their houses or flats in which to grow flowers or vegetables. Spare pieces of land near railways were often leased for such use, and their descendants can be seen today from the train between Schipol airport and Amsterdam's central station. In many cities, public parks were saved from the rapid developments caused by industrialisation. These usually depended upon philanthropy, when a landowner gave an area to a public corporation or an aristocrat opened his park or landscape garden to a wider set of users. Former hunting parks as far apart as London and China were transformed in that way. The ecology of a public city park tended to reflect the culture of the time: in the late nineteenth century of Europe and North America (and their imitators) this meant a fair degree of symmetry of flower beds and shrubbery, a few captive animals in a menagerie, some large trees and a nod in the direction of the wild in the shape of a non-circular pond.[56] In a different mood, changes in politics and warfare meant that cities no longer needed their encircling walls for defence. In Vienna, a ring road and imperial-looking buildings were the replacement, but in Lucca, for example, a belt of parkland was the successor use.

The pleasure that alpha males get from killing did not diminish. In England and southern Scotland, the control of the predatory fox (*Vulpes vulpes*) was ritualised during the nineteenth century so that 'the hunt' with horn calls, special dress, dedicated packs of dogs and negotiated access across farmland (preferably hedged and ditched but without wire fencing) and a parallel social life was integral to certain groups. It too was made possible by rail travel, as Anthony Trollope makes clear in his political novels, where the English Midlands are a favourite playground of Members of Parliament. The fox hunt harks back to a pre-industrial world of hunting whereas grouse shooting on the upland moors of England and Scotland was adapted to industrial-scale slaughter. Burning of the moors created heather monocultures and high grouse numbers; the birds were driven across a line of guns (breech loading, introduced from the mid-nineteenth century was the key technology) and large numbers could be killed by even short-sighted men. A day's bag into the thousands from one moor was the aim. Upper-class management and impact were variable: for fox hunting it was essential to maintain small coppice and scrub areas in which foxes can breed and sleep, and there is no strong evidence that fox populations have ever been seriously controlled by sport hunting with dogs. In the mid-nineteenth century, indeed, there was a shortage of foxes in England and so they were imported from France and Belgium and sold at Leadenhall Market in London. On the uplands, though, strict predator control was practised on grouse moors: the falcons and hawks were ruthlessly culled, no matter whether they fed on grouse chicks or not.[57] The mountains of Scotland needed less control, which did not prevent the near-extirpation of many species in case they took the occasional lamb or fawn, though the outstanding example, the golden eagle (*Aquila chrysaëtos*), is a preferential carrion-eater.[58] Grouse contrasts with its historical companion which was the development of red deer (*Cervus elephas*) shooting in the Scottish Highlands on estates where sheep ranching was less profitable and where the Royal example, often seen in the paintings of Edwin Landseer (1802–73) was persuasive. Here the purpose was to take only a few animals: especially the best stags. The servants might then cull the hinds, a system that broke down when there were many fewer servants – after World War I, for example. (Hence the current overpopulation of upland Scotland with red deer.) It is no wonder that hunting became a major recreation for those people running empires, especially the military officer when not actually campaigning against humans. So, in much of Africa and Asia, most of the larger species became 'game', and targets for sport hunting. The more likely that the animal might fight back then the higher its status in trophy terms: the Indian tiger perhaps above all, so that it was shot from the safety of a small castle on the back of an elephant. Even herbivores could be worthwhile prizes if they were known aggressors, such as the rhinoceros and African buffalo.[59] One other zone of conquest was, however, very much socially and financially confined: exploration. Though imperial, scientific and spiritual reasons might be put forward for climbing in the Himalayas, finding the sources of the great rivers or reaching the South Pole, a common underlying theme is the imposition of western notions of dominance over nature. In hunting and exploration, as in business, a type of Social Darwinism favoured the dominant male.

Parallel to these high-profile recreations, there was from the 1880s onwards a series of working-class recourses to the outdoors. The killing desires were focused on fish which, in Europe, were often the coarse fish species and were put back for another day. In North America, the salmonids were not reserved for the upper classes and all were liable to be killed for trophies or food. A kind of working-class imitation fox hunting took the form of hare coursing, provided there was the right kind of grassland to support the hares (*Lepus timidus*). But the mass recreations involved the sea and the countryside. Sea bathing and sea air as antidotes to the illnesses and oppressions of industrial cities were in Britain an obvious follow-on from her maritime preoccupations but became popular elsewhere quite quickly, with resorts developing in, for example, New Jersey, central Japan, Brittany and on the Baltic. New towns developed with good rail connections, suitable for a day out or perhaps for longer stays. The management of the sea became important: access to it was important so that promenades and walkways had to be built, no matter that storm damage might increase; a sandy foreshore was desirable and so groynes trapped the sand, no matter that areas down-drift were then starved and more vulnerable to storms. If the intertidal zone was shallow, then a pier (normally made of wrought iron) might be necessary and could provide municipal profits for other schemes. Sand dunes were often manipulated: they might be in the way of development or they might be kept in the usually forlorn hope that trampling would not lead to blow-outs; the result was normally the loss of a cheap and effective coastal protection habitat, foreshadowing the later loss of mangroves in the tropics.

Inland, rural areas attracted thousands on Sundays, Bank Holidays and feast days.[60] Provided there was adequate public transport, then strolling, picnics, serious hiking, cycling, and 'just being there' were popular. In general, perhaps, wilder ecosystems were the most attractive, so that the Japanese Alps, the Appalachians, the English Lakes and moorlands were among the most visited. Alpine rock climbing and winter sports were for a wealthy group, not the plebs. In much of continental Europe, forests were favourite places even if they had been converted from beech to conifers by followers of von Moser; the British not only never took to such environments, they actively campaigned against them, at least until widespread car ownership came about after the 1950s. For moorlands, 'regimented blankets of trees' were derided but the term 'over-grazed wet deserts' was never heard outside the ecologists' laboratories.[61] In crowded nations, some conflicts were inevitable. Farmers complained of gates being left open and so stock were lost or allowed to mate promiscuously; animals were attacked by loose dogs; and some landowners simply hated not being able to keep the working classes out.[62] Walkers complained that they were excluded from vast areas of land that were kept for grouse or which were part of the catchment areas of water-supply impoundments. In both places (in England and Wales at least), armed keepers enforced exclusion perfectly legally with some resolution coming only in the year 2000.[63] The environmental impact of mass recreation before the 1950s is hard to judge because so much of the evidence in terms of erosion, for example, has been overtaken by later pressures. Photographs certainly show that, in popular areas of western countries, boots were having an impact, and there is evidence that in regions of coniferous forest the incidence of forest fires went up.

In the western world and its colonies recreation is full of social constructions that reflect a particular set of cultures at a defined time in history. There is no shortage of interpretive work but all, perhaps, have in common the fact of choice. The industrial worker may toil in a mill or a foundry or on an assembly line and there is only the grind of repetitive tasks. For recreation, though, other imperatives can be expressed, whether through the creation of flower beds, the excitement of the Ferris wheel or the calm of lying in the grass listening to the wind and the bees. The imagery of a rural idyll was especially strong in the interwar period, thus leading to much conversion of land to suburbia; it also gave the illusion of participation in a now-past Golden Age of harmonies disturbed by the machine and allowed the myth of wild areas as sites of re-creation. So, a spiritual dimension could be added to the simpler feelings of just being somewhere other than the workplace: the great prophet was Henry David Thoreau (1817–62) in his New England retreat.[64] At the grandest cultural and spatial scales, the value of landscape (primarily as a visual experience not an ecological one) led to the creation of National Parks, starting with Yellowstone (United States) in 1872, with England and Wales just creeping in in 1949 after many other countries. This movement, with a variety of conservationist labels to apply to land areas and with a strong nature-protection element in many countries, became worldwide in the post-1950 era but its origin and cultural reference points are those of the later nineteenth century and first thirty years of the twentieth century.

Last in this treatment of total environments is warfare. Deliberate and accidental effects of conflict were present in the agricultural era and these were characteristic of the eighteenth century and most of the nineteenth century as well. A time-span from 1750 to 1950, however, contains some major changes: from armies in red coats and shakos marching along the roads and tracks of Europe, North America and South Asia to the aerial devastation and radioactive legacy wrought on Hiroshima and Nagasaki. Navies are involved, too: shipbuilding had a strong impact on timber harvesting and management until iron hulls took over after the 1850s. Shores were then converted to large shipyards because vessels became dramatically bigger. Air power took a hand with the flattening and grassing of large areas and the use of coastal sands and salt-marshes for bombing practice.

An eighteenth-century army on the march may devastate much in its path and consume anything edible but it is relatively small and transient: few rural places now carry scars from the Napoleonic Wars. Burning of forests was more likely to last and was carried over from Classical times to North America; it needs a Mediterranean vegetation or high densities of conifers, preferably accompanied by a dry season, to succeed. The burning of forests (especially in anti-guerilla campaigns) was helped by air power, as when the French bombed forests in Morocco in 1921–6. The construction of large dams and dyke systems involved vulnerability to destruction as when large parts of the western Netherlands were flooded during 1944 or when 4,000 villages and millions of hectares of cropland were inundated along the Yellow River in China in 1938 in an attempt to hold up the Japanese advance. Mechanised armies retreating are apt to destroy everything to deny it to the enemy, as with the scorched-earth policies of both the Red Army and *Wehrmacht* east of the Oder in World War II.[65] They fail if their fuel supplies are cut off: no oil, no war.

One of the ecologies of industrialisation is the concentration of energy and materials and this is shown vividly on the Western Front in World War I. The slogging across largely static battle lines meant that narrow bands of terrain were reduced to simple ecosystems (albeit high-energy zones) of men, iron and steel, mud, horses and lice. (The paintings of Paul Nash convey this like no other depiction.) Similar transformations were brought about on smaller Pacific Islands in World War II when Japanese resistance turned whole islands like Iwo Jima into a pockmarked soil and scrub ecology. As elsewhere, the portable successor to Greek fire, napalm, consumed plants as well as people. Although steel remains in soils as shells and mines, most of these areas have become farm and forest once again except where deliberate efforts to preserve the past have been made. The threat of biological warfare allowed the authorities to use a small Scottish island (Gruinard) as a testing ground for anthrax and its decontamination has only recently been accomplished.[66] Nearly all warfare had secondary effects. Tanks need steel and so need energy as coal for smelting iron ore. More coal production consumes more timber underground. Soldiers and munitions workers need calories so grassland is ploughed (perhaps 6 million hectares in North America in World War I); horses, too, have to be fed and they were important right through to 1945. By contrast, fish stocks have a chance to recover if the fleets are kept out of mined zones and factory destruction reduces emissions to air and water. The sea was, however, a dumping ground for old shells and bombs: those thrown into the channel between Scotland and Ulster have a distasteful habit of being washed ashore. Old and unexpended munitions are a problem in many land areas, too, especially shells and mines which can inhibit productive land uses.

In general, combat zones and aerial warfare before 1950 produced devastation that can be overcome. The great exception was radiation which, as we shall see for the post-1950 era, makes the greatest threats because of the longevity of radioactive particles.

Dead matter into living symbols

In his famous book, *The City in History*, Lewis Mumford (1895–1990) wrote that, 'The chief function of the city is to convert power into form, energy into culture, dead matter into the living symbols of art, biological reproduction into social creativity'.[67] While the city is not the only place for which to discuss the use of energy, minerals and water, it is a good place to start for, without the ability cheaply to transform inorganic substances, such as rock and clay, and to manage water, there would have been no cities of the nineteenth and twentieth centuries and these are, after all, among the great phenomena of those centuries. The growth of cities and urban agglomerations under the influence of industrialisation in developed countries and the subsequent explosion of their populations in the less developed lands are testimony to a massive change in the environment of many humans.[68] By 1800, Britain led the world by having 30 per cent of its population urbanised and, indeed, between 1750 and 1800 some 70 per cent of European urban expansion was in England. But 1800 to 1900 was the time of the new industrial city in Europe, North America and Japan, with a few pockets elsewhere. So, by the 1950s, the world became 34 per-cent urbanised, with Africa at 18 per cent and Europe at 61 per cent. A major change of the twentieth century, accelerating as time progressed, was the shedding of rural populations in the poorer countries into their cities.[69] Still, in Africa about 70 per cent of the population was engaged in agriculture, compared with 14 per cent in North and central America combined.

A common metaphor for the city's metabolism, opposed to Mumford's upbeat summary, is to regard it as parasitic. Not even the medieval city could supply its own energy needs though it might have produced some of its food, and both were usually supplied from relatively near at hand. But the quantities of hydrocarbons needed for the people and factories of industrial cities came from the mines that also supplied the railways and steamships. The pull of the city became strong enough to draw in coal from long distances, so that the coal yards of Chicago, for example, were a dominant feature of the urban topography. Most big cities had gas lighting in the streets by about 1820. There were emblems of a second kind of demand: of organic materials for food and clothing and structural purposes such as timber for roofing. In Chicago, the stockyards fed meat to the city and then processed it eastwards as well. The various flows may have become long-term stock if they became structures such as buildings or sewers, but may equally have had a short-term utility as food and fuel before becoming wastes.[70]

The city's growth may be managed by urban planners, so as to minimise some environmental impacts but many of the latter are inevitable. (Table 4.1). At the local level some soil may remain in parks and gardens but the dominant

TABLE 4.1 Environmental impact of the city

Input to city	'Upstream'	Transformation within city (maintenance rather than construction)	'Downstream'	Output from city
Atmosphere	Unmodified unless by other cities upwind	Air velocity changes around buildings Air turbulence increased over roofline of city Loadings increased of: particulates – increase thunderstorm frequency heat radiated from buildings gases, esp. CO, CO_2, SO_x, NO_x PAN + NO_x = photochemical smog as in California, Japan lead, fluorides Precipitation – water is shed quickly from paved and built areas snow melts relatively quickly		Air flow resumes normal characteristics Particulate loadings may increase precipitation immediately downwind Particles fall out downwind Photochemical smog may damage organisms NO_x, SO_x fall-out as acid rain
Ground and surface water	River systems modified for flood control and water storage – reservoirs channelisation Watershed cover modified for water yield and quality Groundwater tapped for urban use	Evaporation of water from many sources Incorporation of water in organic and inorganic mass stored in city Calefaction of waste water, esp. from power generation Water as waste carrier. Affects local water quality – in solution in suspension Waste water to run-off. May add to flood peaks		Flood peakiness of river enhanced Contaminated rivers affect fauna, flora. Decreases with dilution Contaminants may be toxic to life, affect aesthetic quality of river Flood hazard increases engineering of river, e.g. channel modified for faster flow, cut-off channels Water may be led off for aquifer recharge

TABLE 4.1 continued

Input to city	'Upstream'	Transformation within city (maintenance rather than construction)	'Downstream'	Output from city
				Water may be purified in sewage treatment works and reused or put into river for use lower down
Land and surface forms	Little effect except as consequence of other processes, e.g. water management, road construction. Sand and gravel, brick clays extracted to build city and adjacent infrastructure. (Also may affect seabed.) Other rocks used for urban–industrial construction	Construction of embankments etc. across flood plains may cause ponding back at times of peak flow. Slope stability – infiltration changes may hasten slides; buildings constructed in unsuitable zones, e.g. Hong Kong, some Japanese cities high hazard when earthquake risk also present: zoning for land use may mitigate damage. Japan, California. Subsidence – extraction of groundwater or removal of e.g. coal or other solids, e.g. Mexico City, Venice, Long Beach. Surface changes – removal of vegetation increases silt yield and causes local flooding. Stream-course engineering – to manage and avoid floods, improve navigation. Coastal cities – dune-blowing, river channel dredging, storm protection construction, piers, seawalls		Silt transported away more rapidly to sea. Removal or alteration of sediment transport may affect landforms at some distance, esp. increasing coastal erosion hazard. Solid wastes must be disposed of: circum-urban fill sites sought, e.g. quarries, gravel pits. Contaminants from tips may affect groundwater. Necessity for sewage treatment unless river or sea volume sufficiently large to make it appear unnecessary. Eutrophication of water unless N_2 and P removed by multi-stage treatment

Food	Urban demand may exert strong influence on agricultural patterns Transport network reflects necessity to move food to city	Waste dump accumulation Coastal landfill for industrial sites, housing, airports, e.g. San Francisco Bay Some food stored, most consumed and transformed to organic wastes	
Plants and animals	City acts as roosting/nesting zone for rural feeders, e.g. starlings, pigeons Fire-prone vegetation may need to be managed to reduce hazard to city	Vegetation more or less all changed: managed vegetation often grass with shrubs and trees ('urban savanna') unmanaged vegetation dominated by weedy species, incl. quick-growing trees and shrubs some species cannot cope with contaminated atmosphere (esp. lichens, mosses) some species adapt to city life, e.g. songbirds, fox, badger, escaped pets	Escaped pets may establish themselves in rural areas (e.g. Australian parakeets in S. England)
Energy	Construction of conduits, e.g. pipe and transmission lines, transport networks to import energy; generating	Energy resources into power at generating sites (steam, electricity). All forms into wastes (heat; particles, gases)	Gradual dilution of effects of energy transformation and use

materials become asphalt, brick, glass, concrete and metals. These change the values of heat reflectance, radiation and absorption, and the roughness of the surface. The city provides a habitat for adapted plants and animals, notably the rat. Regionally, the city generates large quantities of heat and emits gaseous and solid wastes. Some of these are scavenged from the air quite close to the city whereas others travel long distances laterally and may also migrate into the upper atmosphere. Thus, there are truly global connections becuase the combustion of hydrocarbons generates carbon dioxide, a major component of the enhanced greenhouse effect (see pp. 145–6). Dust and aerosols may also become globally distributed via the upper atmosphere: when lead was a common additive to motor fuels then it rained out on to polar ice in aerosol form; the cities were the main sources.

The city as a cultural phenomenon and milieu has evoked many millions of pages of writing. The city has been the site of freedom to think and behave differently and to foment revolution. It has also been the place where inequalities have become manifest and so the idea of there being a permanent 'other' lower down the hierarchy of esteem may have contributed to a world view in which the non-human is somewhere towards the base. And yet it was in towns, argued Keith Thomas,[71] that new attitudes towards cruelty to animals arose, especially as industrialisation took hold. Cities are heterogeneous, too, so that new ideas about the city and its environment are likely to take root in one place and not another: they may be 'pushed' by bad conditions as well as 'pulled' by the example of good open space, the presence of trees or the reining in of motor vehicles. Cities developed a particular character which may have attracted the kind of people who thought environmentally as well as commercially though, in the United States, the dominant thinker of the pre-1950 period, Aldo Leopold, wrote firmly out of a rural background.[72] Detroit qualifies as the genesis site of the mass-production motor car after 1900: consider the environmental impacts of that centre of innovation. Glasgow in Scotland was a key innovation centre in developing open-sea steamboats after 1818, with enormous consequences for colonial land use, especially when allied to major projects such as iron ships (1839 onwards) and the Suez Canal (1869). Environment in Peter Hall's historical treatment of the city is confined to necessities such as water supply and amenities like parks in nineteenth-century Paris; not even Stockholm in its social-democratic heyday seems to have contributed to considering the city as an integral component of human–nature interaction.[73] At one level of cultural abstraction, the skyscraper connects the worlds of the subsoil and the sky [74] though this might signify conquest rather than a sense of belonging, for some city dwellers disliked 'nature': it was less comprehensible to them than their urban homes. Part of the attraction is the city as spectacle: after 1900 the night-time 'bright lights' formed part of its energy consumption. The advent of war changes a city's ecology: Germany and Japan after 1945 are obvious examples but even lesser destruction and deprivation will bring about the cropping of gardens and the keeping of pigs, goats and rabbits. The city of Turku in Finland had four pig-geries in 1940 and eighty-seven in 1945; wood cutting near the city created open areas and, because bombing created waste land, then this became the waste dumps rather than designated landfill sites.[75]

The city dweller has always been a high consumer of water. Everybody needs about 2 litres per day to drink. Further, even in the urban areas of less devel-oped countries, use can run at about 200 litres per person per day if there is a piped supply; if not then about half that becomes available. The rich use much more: 300 to 400 litres per person perday is common. The other great urban-industrial use is in industry itself: to produce 1 kilolitre of motor fuel takes 7 to 34×10^3 litres of water. One tonne of steel requires 8 to 60×10^3 litres, a tonne of worsted cloth 266,000 litres, and a tonne of paperboard 60 to 376×10^3 litres. These need not be consumptive figures, in the sense that the water can usually be treated and reused but, in the period before 1950, such conservation-minded attitudes were rarer than now and, although historical data on water use are hard to find, it is safe to presume that between 1800 and 1950, the upper figures at least represent the amount of new water demanded by each batch in the process.[76] The industrial era was host to two major developments in water use: the start of hydropower generation (in Wisconsin in 1882) and the explosive expansion of irrigated agriculture. The former leads eventually to large dams. In 1900 the world held 700 dams over 15 metres high whereas by 2000 the figure was over 50,000; at the Revolution in 1949, China had a mere eight. Now, the reservoir volume for the world is five times that of rivers and would cover the area of California. Impoundments also serve the devel-opment of irrigation schemes for crops. Agriculture is responsible for about 70 per cent of the human interventions into the hydrological cycle: a proxy measure of its growth is the amount of 'consumptive use' of water, that is, that proportion that does not go back into the run-off downstream of a project. In 1800, the world total was 226 cubic kilometres, in 1900 550 cubic kilometres and in 1950 it had doubled to 1,080 cubic kilometres, of which 850 cubic kilo-metres were in Asia. In the eighteenth and nineteenth centuries, the area under irrigation doubled every thirty years.[77]

No resource development on such a scale and with such penetrations into all aspects of human life was without impact and equally in need of manage-ment. Historical investigations inevitably emphasise the large scale but every-where there has been a phase when small-scale use of wells, springs, shallow aquifers and small dams has predominated. What is surprising is the occa-sional loss of technology: the skills of Roman aqueduct builders seem to have been lost until the eighteenth century and the substitution of iron pipes for lead or bored timber affected cities only in the nineteenth century, so that water supply in, for example, London was intermittent until the 1880s.[78] Cheap energy in the form of pumps and drills enlarged the resource but the large dam symbolises the industrialisation of water management. Its uses could be for urban-industrial supply, irrigation, power generation or flood control, and sometimes all at once. The impacts are well known and have not in general diminished since 1950: the drowning of land under impoundments is primary, with much social dislocation, especially in poorer countries. Downstream from the dam, the chemical and physical qualities of the water are usually changed so that the river regime becomes artificial and often controlled by the agency in charge of the dam. Poor management of water levels or miscalculations of water loadings have led to catastrophic dam failure in most places at one time. Water supplied to users may follow two basic paths. It may be returned to the hydrological cycle in liquid form quite quickly, though possibly now carrying the traces of its use: sewage, fertilisers, silt and other evidence of summer nights. By contrast it may be taken up into storage in the form of materials such as food (think of the cucumber or melon) or evaporated into the atmospheric part of the cycle. Evaporation water is generally free from contaminants but return flows contain both suspended and dissolved materials, which change the ecology of the water into which they are disseminated and may threaten human health as well. Much is known about the effects of silt on the photosynthetic rates of plants, for example, as it is of high levels of nitrogen and phosphates upon algae. Warm water and nitrates make for 'blooms' of fast-reproducing algae which are then decomposed by bacteria and take up oxygen, leading to the death of fish. Offshore, dinoflagellates bloom in 'red tides' in which their toxic exudates poison most other organisms. Only in the twentieth century did most industrial regions start seriously to clear up their nineteenth-century legacy of gross water contamination with the wastes from chemical industries, which shed high volumes of alkalis and acids alike into rivers and shallow estuaries. Similarly, sewage treatment for many urban agglomerations was still absent or at best confined to taking out some of the more solid objects: the River Tyne in north-east England was said to be too thick to drink but too thin to plough. In 1867, a Parliamentary Committee reported on the way in which small boys set the Bradford canal on fire for amusement.[79]

Since water is so central to life itself, let alone to industry and agriculture, how was it that such poor management came about, both in terms of supply and of waste-water treatment? A partial answer has to be sought in terms of shifts in responsibility as the concentration of populations increased. In a dispersed or low-density population, every family can be responsible for its own flows, provided they do not impinge negatively upon others and that they obey any common (but local) laws that govern water supply and disposal. This does not work at high densities but there is always a lag between the perceived need for public control (through a municipality or a company to which the responsibility is delegated) and its acceptability. In the case of industry, capitalism requires that costs be externalised as much as possible, and responsibility for external flows disowned wherever possible. Water may cross political boundaries and therefore much co-operation between jurisdictions is needed for management. There is a further cultural level beyond the political economy, however, which has seen water as a free good. There has been the feeling that despite the technology needed to supply water and treat sewage, that to pay for these services is not somehow necessary, in a kind of hangover from pre-industrial days. Only since 1950 in 'western' nations has there been any convincing shift in that cultural attitude. One result is the entry of water into the arena of struggle for control within societies: the hydrological cycle and the flows of money become inextricably mixed.[80]

Fishes and whales

Moving to salt water, we come to an inorganic medium populated with organisms of all kinds. Human impact during the industrial era was mainly on fishes and whales, though other ecologies were disturbed, especially along coasts. The imagery from the Indian Ocean tsunami of December 2004 is a reminder that industrialisation has not totally wiped out small-scale artisanal fisheries, though virtually all of these depend on diesel engines and on modern fibres for netting. The industrial revolution in fishing came with applications of steam. On land, refrigeration of railway trucks allowed fresh fish to reach the growing urban markets of inland cities. At sea, cargo ships could perform a similar service. The leap from salted fish to fresh fish was made by the conversion of the fishing fleets themselves from sail to steam. Initially, steam tugs took sailing vessels to sea against the wind: the earliest record seems to be from 1860 [81] but widespread steam-powered vessels followed in short order. In the 1870s menhaden (*Brevoortia tyrannus*) was caught off the north-eastern United States by steam trawlers and then factory processed for its oil. In Britain, 1,573 trawlers were built between 1881 and 1902 and so, about fifty years after the rest of the economy, the deep-water fisheries were industrialised. By 1910, Japan had caught up and by 1938 was the world's leading fishing nation in terms of landings, having made a massive investment in deep-sea vessels. Steam also powers on-board winches that haul up nets from ever greater depths, and industrial processes provide the nylon nets which, after 1945, proved to be stronger and more durable than organic fibres.[82] The horse-powered dredger could be replaced by steam and so channels into harbours were deeper and more stable than ever before.

The seas are great and wide but fish tend to swim in schools or lie on the bottom and have cyclic variations in population numbers. Many spawn in estuaries subject to reclamation for industrial or urban uses. Between these two pincers, fishes were caught in greater numbers (aided by the sonar developed in World War II) without regard to their reproductive capacity and growth rates.[83] Thus, soon into the industrial era in fisheries, some stocks began to decline. The North Sea, unsurprisingly, was one of the first localities, with plaice (*Pleuronectes platessa*) stocks showing signs of stress in 1890. Cod, herring, hake, haddock and ocean perch all showed declines through the first half of the twentieth century around the Atlantic and various parts of the Pacific, especially between Seattle and the Aleutians; the response was occasionally to try conservation measures but more often to look for another stock further afield and to accompany trawlers with factory ships that processed the fish soon after the catch and then refrigerated them in bulk. Conservation was inhibited everywhere by the fact that outside immediate national waters, the seas were a common resource, open to all. But for many years, optimism prevailed: the President of the Royal Society of London in 1883, Professor T. H. Huxley, asserted at an international fisheries exhibition that away from the confines of the North Sea, '. . . the multitudes of these fishes [cod, herring, pilchard mackerel] is so inconceiveably great that the number we catch is relatively insignificant; and, secondly, that the destruction effected by the fishermen cannot sensibly increase the death rate . . .'[84] He might have been right about the numbers but not about the death rate; yet, as late as the 1950s, the world's oceans were being touted as the obvious solution to its nutritional shortages, especially those of animal protein.[85]

In 1772–5, Captain Cook toured the southern polar regions and put forward the notion that 'the world will derive no benefit' from that region. Yet between 1778 and 1830, the southern fur seal of South Georgia[86] was brought to the verge of extinction, as have been many other fur seals; they are easy prey once on land. This vignette pans into a wider frame with the history of whaling. Like fishermen, whalers under sail ventured some distances from their home ports since pelagic whaling came in around 1650, and, by the time of the advent of steam, there were familiar with most of the seas between 80 °N and 55 °S. Steam came into whaling after 1857 and factory ships from 1907. Like nylon nets in the 1950s, there was a technological turning point. In this case it was the patenting of the explosive harpoon by Sven Føyn in 1873. In spite of moving to smaller species, by 1920 at least six great whale populations had collapsed; in 1933, nearly 58,000 whales were killed. Blue whales probably numbered about 200,000 before whaling began, but are perhaps 3,000 today.[87] Humpbacks fell from 125,000 to 20,000 in the same period, which includes a period of attention to conservation promulgated by the International Whaling Commission set up in 1946 that is discussed in the next chapter (p. 194). But in 1950 both whaling and fishing were essentially regulated mainly by markets and the attempts at the imposition of regulations which were usually regarded as unwarranted government interference.

The cultural framework of fishing is complex and, in recent years, much investigated since several ways of reducing the effort have been undertaken. Basically, sea-fishing is still largely a hunter-gatherer exercise: the fish are not genetically selected by humans and they are chased. Because the seas and oceans are so vast, those involved were always reluctant to accept any sense of limits and so there is a parallel with frontier resources like timber, grasslands and some minerals. The plenitude seemed to be lasting since there were always new places to exploit and nothing was known about the historic changes that human groups had brought about.[88] The tenacity shown by men engaged in a dangerous occupation extended to whaling. The commercial arguments are so thin that there must be a deep-seated 'need' to engage in this form of slaughter (as in the annual kill of pilot whales in Tórshavn harbour in the Faroe Islands) as a way of exhibiting the kind of testosterone levels normally taken care of by football and beer.

An age of minerals

Industrialisation was, indeed, a second Iron Age, and the refinement of iron ores into various kinds of steel is the frame of that entire construction. Iron is not the only metal of interest and many others have found uses in industrial processes and manufactures: copper and aluminium are just two from a long list. Other Earth resources were sought in large quantities as well: stone for the impressive parts of eighteenth-century and nineteenth-century cities, for instance; clay for bricks; sand and gravel for aggregate once concrete became a major building material. Limestones formed essential fluxes in iron smelting and also became the foundation of many road beds and landfill projects as well as fertiliser. From the 1930s uranium played its part, first in the making of atomic weapons and then in the spin-offs of civilian power in the later 1950s. Some of these materials were virtually ubiquitous (iron ore, for example, though very variable in quality) whereas others are found only in a few places, giving them strategic value. Most are found on the sea-bed, though only a very few in concentrations that make their recovery feasible. In the years to 1950, only sand and gravel were dredged in any large amount.

One encapsulation of the nineteenth and twentieth centuries is therefore to talk of them as the time of 'holes and heaps' when mines and quarries produced most of the minerals and the reject material from overburdens and from refining was piled up in waste tips. The main aim of management was the reduction of cost. Technical innovation played its part but cheap land on which to set up processing plants and their associated transport networks meant that estuarine salt-marshes and sand-flats, riverside meadows and depastured hillsides were often converted to the new uses. Once in the rivers, many toxics were entrained in the silt fraction and so were spread on riverside flood plains at times of high water and then debouched into estuarine silts and muds; many are still there, working their way downstream as, for example, slugs of arsenic or lead. The air around deep mines was full of soot and acids from the coal used for winding and sorting gear, and near metal smelters (nickel and copper provide particular examples) the landscape was devoid of plants and animals downwind for several kilometres. Regional effects were dominated by acid precipitation, as the sulphur compounds in oil and coal were scavenged out as a weak solution of sulphuric acid.[89] Some 1,000 years of copper mining around Falun in Sweden (until 1993) meant that the soils had so much sulphur in them in that lakes showed no recovery from acidification even as the open-air roasting of ore decreased.[90] Bulk low-cost material, such as coal or ores, also convert environments through their demands on transport; large areas of flat land are needed for ore terminals and marshalling yards. If ores, for example, came from colonies or in large ore carriers at low cost (for example, from Sweden to Britain) then supplies seemed illimitable and cheap so that notions of re-use and recycling were often confined to times of exceptionally high use or interruptions of supply, of which wartime is the main instance. Images of housewives giving up their aluminium saucepans to make aircraft during World War II capture this process.

The subset of energy-containing materials is central to the industrial era. The move from stored but recent photosynthesis (usually called biomass) to geologically ancient materials with much higher energy concentrations pro-duced many sets of environmental consequences. In the case of coal, for example, it is deep mined or quarried,[91] transported, burned and turned into ash, with the emission of gases and solids. Oil is extracted on land or at sea, transported by pipe or ship to a refinery and mostly burned with gaseous emis-sions; some of it is turned into other industrial and pharmaceutical chemicals. Natural gas is simpler, for it needs no refining and produces no solid wastes. Out of all the impacts of land conversion, transport, pipeline construction and maintenance, one perhaps stands out: the need for water in many of the bulk processes of, for example, oil refining and electricity generation. In many places, this was supplied by seawater or major rivers, so that their adjacent lands and intertidal zones were favourite places for 'reclamation' and their waters subject to heating (and hence to the acquisition of a different fauna) by the outflow streams. Lastly, and crucially, all this combustion (in the getting and transporting of energy as well as in its end uses) started to push up the con-centration of gases such as methane and carbon dioxide in the global atmos-phere. The whole question of wastes is central to an industrial era. Smaller cities, smaller-scale industries, lower population densities and 'organic' mate-rials create a volume and a type of waste that is relatively easily transformed by local ecosystems. Energy-containing compounds merely carry the energy, and the carbon is unwanted. But with higher energy intensities, the volume of dis-carded solids and gases is bound to rise because every energy conversion is inefficient. Wastes, therefore, constitute a significant ecological flow, the more so if a 'new' substance is concerned, that is, one which is unknown in nature, such as aniline dyes formulated in the nineteenth- century chemical industry and for which (like later pesticides) there was no 'natural' breakdown cycle. Though these all rose in that period, the main priorities after the eighteenth century were the product of human excrement and domestic heating, neither of which was amenable to town-based recycling systems in the nineteenth and (most of the) twentieth centuries, even in thrifty resource-conscious societies like that of the German Democratic Republic ('East Germany') between the late 1940s and the 1970s.[92]

The cultural context of mineral extraction and use is very diverse, and gen-eralisations difficult to sustain. In the case of mining, what had hitherto been the occupation of slaves and other forms of forced labour, now became the basis of whole communities who fiercely defended their conditions against change and – especially – closure. On a wider canvas, whereas mineral industries were always prepared to consider the virtues of thrift (mostly achieved by technological innovation), energy suppliers were normally inclined to look for more and to be optimistic about the supply holding up for ever. Before 1950, using less was not much discussed, for reserves of coal and oil seemed to be virtually infinite, natural gas was little touched, and much of the world was still dependent upon renewable biomass. The populations of the developed world enjoyed having slaves in the form of machines and acquiring many more personal possessions, and there seemed no reason to suggest that there should be limits to that kind of progress. 'Scarcities could be relieved or almost indefinitely postponed . . . by a rise in prices' was the 1956 verdict of a mineral company president whose cultural context clearly included neo-classical economics as a way of thought.[93]

PROMETHEUS' NEXT BOUND

The consequences of those two centuries are with us all the time, and one effect comprises the difficulty of understanding it all in both its empirical diversity and depth of change in ideas about 'the habitable world'. In terms of a chronicle, three trends are outstanding: the industrialisation of crop production, the increase in the volume of movement of goods and people, and the use of fossil fuels.

The main worldwide environmental impact of the industrialisation of crop production can be summarised (Table 4.2) in the conversion of forest and grasslands into croplands, at regional and at continental scales. (The data are often aggregated so that the 1950–2000 period of chapter 5 cannot be separated out.) The loss of woodlands and forests was by no means all to cropland:

TABLE 4.2 World land transformation 1700–1950

	10^6 ha			% change		
	1700	1850	1950	1700–1850	1850–1920	1920–50
Forests & woodlands	6,215	5,965	5,389	−4	−4.8	−5.1
Grassland & pasture	6,860	6,837	6,780	−0.3	−1.3	0.5
Croplands	260	537	1,170	+102.6	+70	+28.1

Note: The 1700 condition of many of these lands was not 'natural' in the sense that human manipulation of them had been thorough, sometimes for a long period.

Source: extracted from J. F. Richards, 'Land transformation', in B. L. Turner et al. (eds) *The Earth as Transformed by Human Action. Global and Regional Changes in the Biosphere over the Past 300 years*, Cambridge: Cambridge University Press, 1990, Table 10–1, p. 164. More detail for cropland 1700–1990 can be found in Table 3b of N. Ramankutty and J. A. Foley, 'Estimating historical changes in global land cover: croplands from 1700 to 1992', *Global Biogeochemical Cycles* 13, 1999, 997–1027.

some of it was eventually reforested, and other lands remained in a more or less derelict condition. The grassland conversion was mainly to cereals, however, with a smaller proportion being managed so as to support denser herds of sheep or cattle. In the end, the period 1700–1850 saw an immense growth in cropland, borne along especially by North America and Russia, whereas in 1850–1920 it was south-east Asia, Oceania and Latin America that expanded the most, with the first experiencing the conversion of much dry-land farming into irrigated crop production. Worldwide, about 136 million hectares (Mha) were made into cropland in 1700–99, 412 Mha in 1800–99 and 658 Mha in 1900–99. A broad generalisation has the nineteenth century converting the temperate zones' natural grasslands (and some forests) to crop-land, and the second half of the twentieth century transforming the tropical forests to food and fibre production. Overall, a cropland area of 265 Mha in 1700 became 1,471 in 1990 and a pasture total of 524 Mha expanded to 3,451 Mha; in 1700 less than 6 per cent of the world's land surface was in agriculture but by 2000, the equivalent figure was 32 per cent. All through the eighteenth and nineteenth centuries, the clearance of forest for shifting cultivation was about ten times that used for wood harvests; by 2000 it was only twice that area. The wood harvest itself was 1.07 petagrams ($Pg = 10^{15}$ grams) per capitum in 1920 but only 0.61 Pg/cap in 1961, showing the effects of popula-tion growth.[94] In Europe the conversion of land-use types was based on the demands of cities which created zones of intensity of transformation. In the less developed peripheries, the growing cities also created such zones but they were set within an economy in which exports to the industrial core were drivers of the economy. Such a country might well have frontier industries where a moving belt of environmental transformation tore out forests, grazed grasslands heavily or mined any mineral easily accessible to low levels of tech-nology, such as alluvial gold. If permanent settlement followed then the demand for food went up as well. Crops, though, might well now include lux-uries such as tobacco and sugar as well as items so embedded in people's lives that they seemed essential, such as cotton clothing.

The development of cotton, tobacco and sugar for the delight of industri-alising and resource-consuming populations started well before the nine-teenth century but they and many other products were subsequently being moved on a scale unimaginable before steam power: production and con-sumption move apart in an unprecedented way.[95] This characteristic of mass movement of materials has had many environmental consequences and only two will be mentioned here. The first is the environmental alterations needed to facilitate large-scale movement in bulk: in particular, the construction of canals and railways and then, in the post-World War I era, roads. The canals are instructive in the sense that they were built on a variety of scales. The early examples were usually local, to connect, for example, a coal-mining area with a seaport or with an area of iron ore.[96] Then, national systems were put together to connect stretches of navigable river so that vessels might penetrate far into land masses: the Rhine and the St Lawrence alike were partially canalised to allow either big barges or ocean-going ships to provide cheap transport into continents. Then come larger canals with a strategic purpose, such as the Kiel Canal, to allow German vessels (especially warships) to avoid the narrow passage of the Kattegat. World-altering canals at Suez and Panama made it cheaper to transport bulk goods such as ores, cereals and rubber from areas of production to the manufacturing zones of the world. Thus, the second consequence: the processing of materials. In the case of metal ores, the changes are obvious and the role of energy supplies equally so. But consider also food which is shipped in raw form (though very likely refrigerated in the case of meat and fruit) and then processed elsewhere. This uses energy in the actual physical changes in the tissues (for example, cooking, pulping, juicing) and also in any transport between processing stages (think of wheat becoming flour then going to a bakery, being sliced and wrapped and then driven to a retail store) and then may well cause further energy consumption in the manufacture and packaging of dietary supplements because the food processing had eliminated useful minerals and vitamins. Both the shipping and the processing have spin-off consequences for the environment: some are large scale and profound, as with the penetration of biota through the major canals with the potential for their populations to explode in new environments. Fruit and meat may bring unwanted micro-organisms or even hairy spiders. There were many other 'accessory' changes in the years to 1950: the aeroplane, for example, was beginning to absorb land and fuel, though on nothing like its subsequent scale. Motor vehicles, likewise, are lower in significance except in the United States and in the major road-building schemes of the 1930s in Germany. But the leap into movement as a key accompaniment to an industrialising world is very important and cannot be confined to resources, because it involved people and ideas as well.

Underlying all this extra movement was the application of hydrocarbons. Fossil fuel use totals about 0.11 calories per square metre per day globally, and the weather accounts for about 100 calories per square metre per day so it is quantitatively not a huge component; this does not mean, however, that it does not have a considerable impact. This impact can be described in two parts: the direct impact of energy use and its emissions, and the secondary impact of those processes it makes possible. For direct impact, there are fallouts such as dust and carbon particles from coal combustion and nitrogen compounds from oil distillates such as diesel fuel. Carbon monoxide is more of a threat to human life than to the environment generally, but all hydrocarbon fuels produce waste carbon dioxide which finds its way into the upper atmosphere, from where it is scavenged only slowly. The current concern over the 'greenhouse effect' (perhaps better called the Enhanced Greenhouse Effect) is therefore rooted in the accumulation of carbon dioxide (CO_2) in the planetary atmosphere which began to rise from industrial emissions, and so a level of 280 parts per million (ppm) in 1750 became about 300 ppm in 1900 and 320 ppm in 1950, with early twenty-first-century levels at 350 ppm.[97] Carbon dioxide is not the only trace gas whose emissions rose in the same period: nitrous oxides and methane are the outstanding additions to the list and, in the case of methane (CH_4), the molecules are much more effective at retaining radiation than CO_2 and stay longer in the atmosphere. Overall, the atmospheric CO_2 level increased by 30 per cent between 1750 and 2000, and the methane concentration by 150 per cent. More discussion of these phenomena comes in the next chapter but here it needs to be recorded that there were 200 years of rise during which its impact, actual and potential, was ignored outside a few restricted scientific circles.

MORAL, INTELLECTUAL AND MATERIAL

In 1849, Thomas Macaulay universalised his English subject matter to assert that, 'Every improvement of the means of locomotion benefits mankind morally and intellectually as well as materially.' This can be interpreted as a slogan for the coalescing tendencies of the two centuries being examined here. In 1800, the world was rapidly becoming Europeanised: Paris fashions may have had a limited market but they functioned as a vanguard for many other introductions in the period to 1950. Without question, the North American and European technological introductions and spread dominate. To single out one set of innovations is risky but perhaps the most important was the 'improvement' in communications in the sense that they became faster and available on a much wider basis, more comfortable and affording greater privacy to some at least.[98] The steamship with a regular schedule, the railway with its demands for standardised time, the electric telegraph, radio, powered flight, film, television (though more widely after 1950) and especially printing allowed the transmission of goods, people and ideas on scales never before envisaged. Print, especially, became a validating medium like men in pulpits in an earlier era. All these proliferations, backed up by firearms and quinine, led to more imposition of the dominance of western ways of thought and action. Primary among those was trade, which carried raw materials and finished goods as well as large amounts of food and other comestibles, such as tea and coffee, and was in general responsible for the accumulation of capital: in those two sentences lie the germs of thousands of environmental alterations. Add to trade the often parallel penetration of colonial acquisition and rule in which many production techniques from the metropolitan countries were imposed on the dominated lands, and the recipe for large-scale ecological change is determined. In the 1930s there was an African Research Survey the basic aim of which was to apply scientific methods in the cause of co-ordinating and, if possible, standardising the colonial policies of Britain, France, Belgium and Portugal. It was abandoned in 1940 but had developed a somewhat contrary stance in adopting ecological attitudes which called into question the usual colonial rejection of traditional ecological knowledge.[99]

This shows that, as so often, a hegemonic nexus calls forth opposition. There were independence movements in the colonies which disrupted the orderly flow of products and, in western circles, new thinking on the treatment of animals, for example, began to take hold. This might have been mostly a projection of human social concerns on to animals but it pointed up changing attitudes. The extension of moral concerns beyond any immediate circle was also one consequence of new flows of information and of technology: the gradual (and incomplete) abolition of slavery was one outcome, and the various conventions on the conduct of warfare another, both with environmental implications. Also normal in most parts of the world was slavery, fundamentally made redundant by the advent of fossil fuels.[100]

One of the great themes of industrialisation is concentration in the material sense: in and of the city, of chemicals and of wastes. This has an analogy in the convergence of display in numerous 'World Fairs' in the nineteenth and twentieth centuries. These claimed to show the best of every country's arts and sciences (both including technology) and so were dominated by western nations. Trade was a central focus but, interestingly, several of them held world congresses of science as well. In 1904, at the St Louis World Fair, there were addresses by Ludwig Boltzmann (1844–1906, a founding father of particle theory) and Ernest Rutherford (1871–1937, likewise of radioactivity science); the newest section of this International Congress of Arts and Sciences was ecology. Science looked to be a unifying force in the world, now that it had made such immense collections of knowledge as in the museums of nature and of humankind in Paris, Berlin, London and Washington DC. One example of its unifications was the Linnaean system of plant and animal nomenclature, universally adopted, though subject (like European Union legislation today) to nationalistic derogations from time to time. Meteorology moved to the city and the laboratory from the 1840s onwards, away from its individualistic and rural amateur base to worldwide networks of comparable data.[101] The first man (Svante Arrhenius, 1859–1927) to quantify the influence of the concentrations of carbon dioxide in the atmosphere upon temperatures at the Earth's surface did so in the late nineteenth century, though he may well have crystallised ideas of eighteenth-century pioneers.[102] In such a world it is not surprising that universalist claims were made for ideologies, of which the outstanding example is the work of Karl Marx (1818–83) for whom the accumulation and deployment of capital were the central facts of the world. The closest parallel in understanding the human individual was Carl Jung (1865–1961) for whom psychology had to reach out beyond European systems of thought.

The environmental consequences are pervasive and follow two main trends: the demands of western economies and their aspirants, and the growth of population. One worldwide effect is the spread of species. Crop plants and their pests are obvious examples: the import of sugar cane into Australia brought with it cane beetles which provoked the introduction in 1935 from

Hawaii of a fast-breeding toad (*Bufo marinus*) which, in turn, has become a large-scale pest. It is toxic to most other species and also ineffective because it cannot jump as high as the fronds where the beetles live. In a piece of nostalgic symbolism, trout were exported to many parts of the British Empire (for example, the Madras hills in 1863, Tasmania in 1864, and thence to mainland Australia and New Zealand, 1898 South Africa, and 1947 the Falkland Islands).[103] The spread of commercial wine-making from its European heartlands and the interchange of stocks meant that diseases of the grape vine could be transferred from one continent to another. Hence California contributed to European production declines in the form of a fungus (*Oideum* spp.) in 1851 and then to virtual wipe-out in 1860 by the aphids of the genus *Phylloxera*. California also contributed a mid-century mildew *Peronspera viticola*. The inadvertent transfer of organisms is exemplified by examples such as the movement of species through great inter-ocean canals, as at Suez and Panama, and of micro-organisms using humans as carriers. The greater numbers of humans and their higher densities improved the chances of epidemics: Muslim pilgrims and the British army alike spread cholera in the nineteenth century; the pandemic of influenza two years after World War I, and which affected a fifth of the world's population, is cited as something likely to happen again only on a larger scale. Land-use changes also spread diseases: mosquitoes are encouraged by forest clearance, irrigation, standing water and house-building so that malaria, yellow fever and dengue are spread. Irrigation encourages schistosomiasis. Only tick-borne diseases have failed to colonise the humanised spaces, with the exception of Lyme disease in the twentieth century which followed wild deer into secondary vegetation.[104]

Deliberate transfer of species has not led to any universal distributions that could be called worldwide in the sense that influenza decimated populations as far apart as the Inuit and the Indonesians. The European house sparrow (*Passer domesticus*) was taken to North America as a reminder of home, and several exotics escaped from captivity in the parks of the rich to colonise their new homelands: small species of deer, such as the Japanese sika (*Cervus nippon*) and the muntjac (*Muntiacus reevesi*), have been favourite introductions. Timber trees from favoured environments, such as the west coast of North America, have become widely grown but even Sitka spruce (*Picea sitchensis*) and lodgepole pine (*Pinus contorta*) have limits to their tolerances. Even more widespread, however, was (and is) the planting of various species of *Eucalyptus*, an Australian genus. Look by roadsides in Egypt, India, Portugal and California alike and one will soon come into view. The transfer of species meant that hybrid species, such as a cross between the European and Japanese larches (*Larix eurolepis*), became possible and attractive to plantation owners.[105] In Australia itself, the European immigrants tended to believe that the native fauna would be replaced by introduced species anyway and so the Tasmanian marsupial wolf had a bounty on its head in 1912, a closed season only in 1929, and protection in 1936 by which time it was extinct. In

New Zealand, white clover would not grow until suitable bees had been imported in 1839 and then sheep were able to become a major crop on ecosystems converted to meadowlands.[106]

One way in which the findings of the physical sciences (by far the best at prediction) suffused the world was in the preparation of chemical compounds with precise effects upon defined targets. The general cure-all for plant diseases of dusting with sulphur or one of its compounds was part folk knowledge but also came out of German science. Likewise, a more precise formulation of industrial fertilisers, in terms of their need to contain nitrogen, emanated from German laboratories. The winner element in chemical death until the 1960s is chlorine which was built up into a variety of organic compounds of which the most famous was DDT, a member of a group known as chlorinated hydrocarbons (CHCs); the public was unaware of their environment effects until the publication of Rachel Carson's famous book *Silent Spring* in 1962. Together with several similar products, the enzyme chemistry of nerve transmission in insects (DDT was largely developed as an anti-malarial in World War II) was disrupted by this substance. What was not predicted was that the group would have very low breakdown rates outside the target organisms and so became subject to biological amplification whereby very low concentrations in, for example, water became lethally high further up a predator–prey food chain. One very noticeable effect was the accumulation of CHCs in fish- and flesh-eating birds whose eggshells underwent thinning so that they broke during incubation. Many species became almost extinct until western governments began to prohibit the use of CHCs. This story might be thought of as a example of fragmentation rather than coalescence except that residues of CHCs are distributed throughout the world's oceans (where they are still available for uptake) and thence, via aerosol formation at the water–air interface, can get rained out on land surfaces and on, for example, the Antarctic ice. There is a case, therefore, for regarding this industrial residue as having a global distribution. An even stronger claim can be made for waste gases that, having been led off into the atmosphere, are only slowly scavenged out and remain long enough to affect the processes of the atmosphere. Even though present only at 'trace' levels, the likelihood that gases which contribute to the effectiveness of retaining radiation within the Earth system will affect climate, seems high. In 1750, the concentration of carbon dioxide was about 280 parts per million (ppm) and in 1950 it was 310 ppm, with a straight line between the two estimations. Other industrial-era emissions such as nitrogen oxides (270–285 ppm) and methane (650–1,100 ppm) followed the same trend. Once in the atmosphere, their distribution is virtually homogeneous spatially: no matter who emits, everybody will receive consequences though these will be spatially variable. This globalisation is certainly material, is the subject of much intellectual activity, and is without doubt moral in its implications. Not all of it stems from the 'means of locomotion' but the centrality of motors of many kinds cannot be evaded.

THE COLLAPSE OF CONTINUITY

Since the use of the natural world must always reflect what humans think about it (as well as allowing for any unconscious effects), developments such as the Enlightenment and the amplifications of modern science have enormous relevance for the non-human parts of the planet. In our context, the Enlightenment of eighteenth-century Europe and North America is ambiguous. Its philosophers and practitioners swore by the idea of a universal human nature but at the same time preached tolerance, diversity, reason, and the encouragement of science and technology. The notion of humans as individuals who did not belong to anybody else becomes possible as the private replaces the public and so there is less sharing of goods and thereafter the production of more possessions from more 'natural' resources.[107] If everybody has their own stockpot, then there must be more iron mining; if city workers live in houses, then more bricks mean more clay-pits and deeper coal-mines. One intellectual stance became that of the literary critic, F. R. Leavis, in the 1930s who envisaged a formerly organic society broken up by industrialisation. If he had looked beyond the horizon of south-eastern England then he would have had to admit that the ending of slavery in the Atlantic in the nineteenth century was also the fragmentation of a type of organic society, with implications for land use and land cover where large plantations were broken up.

One evolution can be seen in the intellectual construct of modernism. Applied firstly to the arts in the period between (roughly) 1890 and 1914, it has come to express implications for the rational use of resources (as in town and country planning) and the styles of that rationality. The grid plans of some cities and the straight-line architecture of the classical skyscraper are examples of modernist architecture, as are Picasso's cubism or the geometric lines of Mondrian's paintings. Behind the rigid, however, there is a pervasive current of '. . . fragmentation, introversion, and crisis . . .'[108] which has resonance across a broad spectrum of the arts, politics and science. The core of these ideas is in the many spheres in which any element of continuity is broken up by the realisation that the world and its representations are discontinuous. The atom is not a new idea in the nineteenth century but its central role in physics and then, indeed, its fragmentation into smaller particles with behaviour based only on probability comes from the twentieth century, as does the delineation of movement on film by capturing sixteen individual frames per second in 1903. The early pictures of Piet Mondrian show the representation of, for example, trees breaking up into fragments, and the pointillism of Georges Seurat makes every brushful of paint a series of dots. With words, James Joyce's *Ulysses* (conceived in 1907) was a series of pieces resolved only by the magnificent and moving soliloquy of Molly Bloom which shifts tenses as often as a restless sleeper. In the concert hall, Schoenberg and Webern's atonality provides little clue about the next note. Politically, the invention of the concentration camp sequesters whole non-combatant populations 'for their own good', a nostrum often credited to the British during the second Boer War (1899–1902) but more correctly the inspiration of Valeriano Weyler y Nicolau, a Spanish army officer in a Cuban war of independence, who constructed three *campos de reconcentraciòn* in the province of Pinar del Rios in 1896.[109] This separation of function is also behind the analytic approach to planning which results in suburbs, industrial zones and shopping malls, each with a single purpose.

Such separations are very obvious in the case of the sciences, where its language became incomprehensible to the untrained (rather as medieval Latin had been), and so a whole group now acquired a primary identification as 'scientists', a word coined only in the later nineteenth century.[110] In no time at all, the 'republic of science' was cantonal, each area with its own language incomprehensible across the border. Science contributed to the climate of fragmentation because it was endlessly analytical, seeking to break things into ever smaller parts in attempts to model processes and predict their future behaviour. The success of Linnaean taxonomy can be seen in that light.[111] Its success was incalculable in terms of many of the phenomena of physics and chemistry but less so with complex biological entities, such as 'communities', which were the forerunners of 'ecosystems', pointing to an element of intellectual construction in both.[112] In environmental terms, however, the success of pharmacology in disease control, helping to contain death rates, had considerable consequences. Equally important, perhaps, was the ability of technology in about 1900 to impose the fractured time of the stopwatch on human movements and thus make an assembly line for all kinds of goods. Not least among these was the motor car, the ascent of which into private ownership is one of the most environmentally pervasive technologies of the twentieth century. Henry Ford's assembly line also broke up employment: in 1914, heavy industry in the United States had an annual labour turnover of 115 per cent.[113] That statistic demonstrates a more general result of industrialisation: the demise of communities based upon a shared locality and long-standing connections to the land, and their replacement by associations based on particular interests, whether these were of the workplace or of leisure time. The German terms *Gemeinschaft* and *Gesellschaft* are often used of these two states and it has been argued that the former is more likely to be 'tender' towards nature than the latter, which is so often urban and detached from sky, sea and soil.

The fragmentation of ecosystems follows the technology which allows social drives to express themselves. Many natural and solar-powered, but humanised, systems were replaced apparently totally but around the edges and in the interstices, so to speak, there are usually small pieces of the precursor orders. Small patches of woodland, a pond, some predators, quite a lot of scavenger species and a few weeds remind us of a previous ecology. The enclosure of the common fields in Europe as feudalism waned was the leading edge of much breakdown of common-property resource use. Though much anathematised in later years,

this way of managing the nature of yield was often successful and highly democratic in terms of local control, locally agreed upon and locally enforced.[114]
Communities which knew that, if they depleted a resource base, then nobody
would ride to their rescue had a strong incentive to live within their limits and
to erect buffers against hard times. One of the coalescent forces of industrialisation was the emplacement of such groups within a wider nation-state context
which might bale them out. At the same time, the boundaries of the nation
state and its component units might well not be the best for resource management, a situation still true of, for example, water in the Middle East and many
of the oceans. States such as Tokugawa Japan, which had deliberately cut itself
off from the world for 200 years (1603–1868), had faced the consequences of
famine, tsunami and earthquake without any outside assistance, yet faced
more difficult decisions when confronted with an industrial world in the form
of steamships with large guns.

The diminution of the public in favour of the private had many ecological
consequences. One of these was the need for the public bodies to acquire land
for wider use. Instead of common land, there was now public-access land
under the name of 'park' in one form or another. On the private areas the
industrial scale of some uses made other users unwelcome: the mass shooting
of grouse on moorlands in England and Scotland (p. 128) meant that walkers
were the targets in the nesting and shooting seasons; likewise in colonial plantations and forests, any compromise with the previous land-management
systems was grudging at best and often denied. On the worldwide scale, the
gap opened up between the developed nations and the 'less developed' countries in which the former had high per capitum energy use and resource consumption and low rates of population growth. The 'developing' countries
used less energy from commercial sources, and had higher rates of population
growth among which there were many poor people.[115] It was thought that the
rich preserved more nature (unless it could be the basis of tourist income)
because they could call in resources from their economic periphery of poorer
dependants. The poor often had no choice and they either converted long-
standing systems to cash crops for exports, no matter what the environmental cost, or transformed fragile ecologies into subsistence systems at great risk
from environmental hazards such as floods, landslides, and even tsunamis.
The same industrial revolution which brought the world closer together, as
Walt Whitman's paean extols, also estranged the human winners and losers,
to say nothing of the non-human members of ecosystems. It begot multiple
worlds, not all of which either understood, or had sympathy for, one
another.[116]

In all these changes, two features need final emphasis. The first is the impact
of colonialism in so many regions of the world beyond the temperate zone:
both wet and dry environments were subject to production systems and environmental management whose origins were in temperate areas and which
might be imposed (though not always) irrespective of the different conditions.

Thus, shifting agriculture was often stigmatised as 'primitive' and 'wasteful' even though it was usually a rational response to certain soil–vegetation combinations. The second is the response to fragmentation of ecologies in places with very rapid change in the nineteenth century and early twentieth centuries. The United States is a prime example, with many public policies developed and brought in by foresters, such as Gifford Pinchot (1865–1946) and President Theodore Roosevelt (1858–1919), in the wake of rapid ecological transformation of forests and grasslands. But note the instrumental theme in all these instances: nature for its own sake may get a national park here and there yet in the United States, the National Parks Service in the 1950s promulgated a 'parks are for people' policy.[117] Any ontological continuity between humanity and its co-existing entities had been lost.

Representing industriousness

Invented in China, transformed in Europe by Johannes Gutenberg (c.1400–68), printing is the key representation of the age, including the dissemination of paintings by engravers. After the eighteenth century, the numbers of books began to explode, accompanied by many printed ephemera, such as newspapers. One function of the book was to be a compendium in which all knowledge could be recorded, though this ambition rather faded as the nineteenth century progressed. The details of the world were recorded as science as in the great multi-volume tracts of Alexander von Humboldt's simply titled *Kosmos* (1845–62) or in works of travel such as the radical (though backward-looking)William Cobbett's *Rural Rides* in nineteenth-century England (1830), or the slightly aloof French traveller, Alexis de Tocqueville (1805–59), in both Britain and North America. The stimulus to imagination of the new way of life inspired novels and poetry: the Romantic poets were definitely against it (Wordsworth's view of railways in the Lake District is a good example), Elizabeth Gaskell set up a lasting divide in *North and South* (1855), and D. H. Lawrence (1885–1930) was in little doubt about the morality of coal-mining as distinct from pheasant-rearing.

Threading through this emphasis on print is a change in consciousness through the use of words. 'Industry' becomes a thing in itself rather than a human attribute, with Adam Smith one of the first writers to use it for a collective of institutions, in his *Wealth of Nations* of 1776. 'Industrial Revolution' was coined in the 1820s by French writers, and art's transition from the work of the artisan to 'imaginative truth' was of the 1840s. Between 1780 and 1880 perhaps fifty words in English acquired new or much altered meanings which indicated a shift in ways of seeing and representing the world. The extension of literacy in the 'western' world and the emphasis given to it in social and economic betterment inevitably make something printed the symbol of the industrial era: let us light upon the US dollar bill. At an abstract level, the neoclassical economic theory developed after Adam

Smith is distinguished (among other ways) by its absence of meaning taken from other spheres of life. It becomes a closed and self-referential web of concepts that dispelled morality from human livelihood and deprived all local systems of any meaning.[118] It thus placed no obstacles in the way of accumulation of wealth which brings in further materials and money which spin off into further environmental manipulation. The influence of Britain in the nineteenth century extended to the exhaustion of soils in the American South, the pushing of the American wheat belt towards erosion, the conversion of native vegetation to pasture in Argentina and Australia, the continued extraction of sugar from the West Indies, and the way in which Swedish forests were managed. Where specialisation of rural production prevailed, the ecologies of areas of cotton, wheat, tea and coffee were much the same even on different continents.[119]

A WASTE LAND?

The relationships of science, technology, culture and environment are complicated. But nobody can deny the transformations in all of them that came with the industrialisation of the nineteenth and twentieth centuries. In some ways, the new energy sources presented a flood of abundance which echoed Charles Darwin's contemplation in 1859 of the progress of life-forms from 'famine and death' to the production of the higher animals: 'There is grandeur in this view of life.'[120] In keeping with this expansive attitude there came about an acceleration of speed and accessibility that rippled out to very far corners of the world. Between 1880 and 1914, there were sweeping changes in technology and culture which created new modes of thought about such fundamentals as the relationships of time and space, not least the theory of relativity. At a hands-on level, there came the telephone and telegraph, X-rays, the cinema and the bicycle, the moter car and the aeroplane.[121] Yet words do not capture well the linkages and interdependence of specialisation, atomisation, instrumental rationality, the growth of knowledge and its provisional nature. The connections to environment are at one level obvious: 'whatever we want, we want it now' might have been chanted in the streets had that been the customary cry of demonstrators. It was more subtle as well. In 1834 the Statistical Society of London averred that the 'first indispensable body of information' was 'physical geography' and, of course, 'the means of its modification'.[122] This outlook was applied to many of the colonies which supplied not only vast new territories for economic exploitation but new mental vistas as well. In the fields of deforestation, soil erosion and species extinction, colonial administrators and those whom we would now call scientists evolved new ideas which would now be called environmentalist. The French led the way with the British and Dutch some distance behind and extra-colonials, such as George Perkins Marsh, having rather less influence outside the US and perhaps Italy, where he was Ambassador.[123]

The nineteenth century was when humans started to know a great deal more about invisible things but this was greatly outbalanced by the rise of the visual towards a status equalling that of the printed word in shaping world views. Images can destroy certitudes, reveal ignorance and challenge the limits of comprehension; on the other hand, they might often be retouched so that interloping people or thoughts are shaded out.[124] They also impose a frame on what is included and what is left out. We think most often of photography but painting, too, may have captured shifts in world view, moving from scenes that were part of a pre-established order to those that were the creation of a single artist's perceptions: the difference between the landscapes of Gainsborough and those of Constable has been used as an example.[125] The idyllic qualities of rural landscapes were matched by heroic industrial paintings of a boosterist nature although beneath there might be another current. It is possible to see in some of J. M. W. Turner's sea pictures the sheer inability to do anything but yield (even in a steamboat) to nature's forces (*Snow Storm – Steamboat off a Harbour's Mouth* [1842] is a good example), echoed in A. E. Houseman's lines of 1900, that he was

> I, a stranger and afraid
> In a world I never made.

The mood of an industrialised world was caught brilliantly by T. S. Eliot's *The Waste Land* of 1922. The poem is a series of fragments and quotations, allusive as well as direct, and multilingual. It reminds one interpreter[126]of '. . . landscapes, where billboards loom out of farmlands, and a jungle of modern tenements surrounds the splendour of the Acropolis', and by many others the poem is held to sum up modernism in its disruptive aspects, the intrusion of fossil-fuel-powered economies into a world formerly glued by the sun. The world had been 'hopscotched' in the sense that not all of it was landed upon (this is a historian of ideas writing and not an environmental historian) but in a way that emphasised separations. These judgements from other disciplines reinforce any conviction that, in 1950, there was more fragmentation than coalescence in the world. After 1950, the symmetry changes as 'globalisation', as the is term currently used, is more apparent.

If we look for fundamental factors in the metamorphoses of 1750–1950, a number of generalisations [127]can be brought out:

- An increasing capacity for collective action. This is most evident in the power of the state but backed up by corporate entities that get ever larger. These processes were dominated by states and companies in Europe and North America.
- The extension of state and corporate power in the shape of colonial regimes. Even frontiersmen [sic] logging forests, running cattle ranches

and mining alluvial gold have the mechanisms and power structures of the state backing them up, usually from a warm office.

- Since about 1500 there has been a world market not only for precious items, such as silver and gold, but also for cheaper commodities, such as cotton, sugar, timber, tobacco and iron ore. After intensification of land use in the industrialised heartlands, the search for these products brought about similar processes in the tributary regions which were then tapped.
- Coal and oil meant that a nation was no longer dependent on its surface area for energy sources: it had those underground to add to any it might acquire by trade or conquest. Hypothetically, that might lead to lower environmental impact. In fact, it proved to be a positive feedback loop.
- Population growth was a major contributor in most regions of the world both as occupiers of the land and creators of demand for its products, as well as those of the sea. The basic data for world population are a rise from 610 million in 1750 to 2,500 million in 1950, a continuation of the take-off beginning about 1650, with accelerating rates of growth in many areas towards the end of the period.
- The accessibility of technology to so many more people. The state has battleships, the individual can have vulcanised rubber boots. In the early decades of this period, the ingenuity of individuals was the major driver of innovation but, with time, basic science preceded the engineering achievements. Overall, technology removes barriers but ideologies, ideas and demands drive the resulting modifications in land use and land cover that are the visible reminders of environmental change.[128]

The underlying ecology then manifests certain redirections, One is of concentration and intensification: the sewage of a family farm is easily processed by the local river but that of a city of half a million is a different prospect. As the population rises in the monsoon zone, people must either convert more hillside to rice terrace or produce more crops per year. Either way, more energy per unit area has to flow through the agro-ecosystem, which is the basis of intensification. A second shift is the introduction of new substances into environmental flows. All natural and many low-intensity human-directed ecosystems have mechanisms for breaking down complex molecules so that they can be recycled. But some of the new outputs from the chemical and pharmaceutical industries have no such pathways and, indeed, may have been designed not to break down. Clear examples exist in the form of the CHC biocides discussed above and in the residual radioactivity of isotopes such as plutonium-239, the safe disposal of which requires it to be kept away from all forms of life for at least 250,000 years. There is a chiasma of both transitions in the intensification of content: tobacco is grown for higher nicotine levels, sugar refined for more sweetness, and the increasing availability of meat. The growth of vegetarianism in the west in the 1880s to 1890s seems not to have affected the demand for the product that transformed whole regions and attracted people to them: in the 1870s, Anthony Trollope was in Australia and noted that 'the labouring man . . . eats meat three times a day in the colonies, and very generally goes without it altogether at home'.[129]

This world is still here even though many people in the world have yet to experience the abundance of goods and services and hence have access to the levels of energy throughput that industrialisation has brought to the richer regions. It is here in the habits of thought engendered by two hundred years of expansion of production and consumption. There is an expectation that there can be more of everything, that any limits are only temporary and also that solutions to problems are most likely to be found in science and technology.

NOTES

1. If nuclear fusion is to be successful outside the laboratory, a method of containing the very hot plasmas has to be found. Magnetic fields and lasers are the top candidates.
2. Joseph Vissarionovich Stalin (1878–1953) was Georgian and his family name was Djugashvili. He also invented his birthday, nationality and education as well. S. Sebag Montefiore, *Stalin. The Court of the Red Tsar*, London: Weidenfeld & Nicolson, 2003. 'Steel' is a generic term for a variety of alloys of iron, carbon and other additives, such as tungsten and wolfram.
3. The best history of energy use is V. Smil, *Energy in World History*, Boulder CO: Westview Press, 1994. For the United States, see also D. E. Nye, *Consuming Power. A Social History of American Energies*, Cambridge MA: MIT Press, 1998.
4. D. S. Landes, *The Unbound Prometheus: Technological Change and Industrial Development in Western Europe from 1750 to the Present*, Cambridge: Cambridge University Press, 1988. Pig iron was the initial product of the smelting of the ore and so was the base material for other forms of iron and for steel.
5. Not so evocative, I suppose, for anybody born too late to travel during the era of steam-hauled trains.
6. In approximate numbers, 1 kilogram of lignite (brown coal) contains 1,800–4,500 kilocalories; hard coal 550–7500; oil 9,500–10,500. Natural gas comes by the cubic metre and each contains 8,500 kilocalories.
7. R. G. Wilkinson, 'The English industrial revolution', in D. Worster (ed.) *The Ends of the Earth*, Cambridge: Cambridge University Press, 1988, 80–99. Biomass energies are still important in nations now characterised as 'poor'.
8. For example, see M. Osborn, ' "The weirdest of all undertakings": the land and the early industrial revolution in Oldham, England', *Environmental History* **8**, 2003, 246–69.
9. Quoted in the chapter on Glasgow in P. Hall, *Cities in Civilization. Culture, Innovation and Urban Order*, London: Weidenfeld & Nicolson, 1998.
10. The term seems to have originated in France c.1827 but its significance and emplacement are due largely to Friedrich Engels in the 1840s. The term and its mythological resonances are explored in detail by D. C. Coleman as chapter 1 of his collected essays *Myth, History and the Industrial Revolution*, London and Rio Grande OH: Hambledon Press, 1992, 1–42.
11. There is a huge literature and a constant debate about which engine was seminal, the relative price and availability of coal and wood, and the social acceptability

of factory working. Here I shall accept a simple narrative: that of Landes op.cit. 1988.

12. This is much simplified from A. Grübler, 'Technology', in W. B. Meyer and B. L. Turner (eds) *Changes in Land Use and Land Cover: A Global Perspective*, Cambridge: Cambridge University Press, 1994, 287–328, Table 1.

13. I have tried to put this together in my *Changing the Face of the Earth* (Oxford: Blackwell, 1989, 2nd edn) and see also J. McNeill, *Something New under the Sun. An Environmental History of the Twentieth Century World*. London and New York: Allen Lane, 2000. The treatment in J. D. Hughes, *An Environmental History of the World*, London and New York: Routledge, 2001, is unusual but remarkable in the way it puts the industrial material in a long-time framework. The huge book is B. L. Turner et al. (eds) *The Earth as Transformed by Human Action. Global and Regional Changes in the Biosphere over the Past 300 years*, Cambridge: Cambridge University Press, 1990.

14. C. McEvedy and R. Jones, *Atlas of World Population History*, Harmondsworth: Penguin Books, 1978; M. Livi-Bacci, *A Concise History of World Population*, Oxford: Blackwell, 2001 (first published in Italian 1989); C. Haub, 'How many people have ever lived on earth?' *Population Today* **23**, 1995, 4–5.

15. P. Bairoch, 'International industrialization levels from 1750 to 1980', *Journal of European Economic History* **11**, 1982, 269–333.

16. These and almost any other data on population can be obtained from the website of the Population Reference Bureau.

17. J. M. Grove, *The Little Ice Age*, London: Methuen, 1988; R. S. Bradley, K. R. Briffa, J. Cole, M. K. Hughes and T. J. Osborn, 'The climate of the last millennium', in K. D. Alverson, R. S. Bradley and T. F. Pedersen (eds) *Paleoclimate, Global Change and the Future*, Berlin: Springer, 2003, 105–41.

18. M. Davis, *Late Victorian Holocausts. El Niño Famines and the Making of the Third World*, London and New York: Verso, 2001.

19. The data are for 1997 and suggest that the year's combustion involved 44×10^{18} gigacalories, that is, about 400 times the twentieth-century global net primary productivity (NPP). Coal from plants is less than 10 per cent efficient in terms of solar energy conversion; oil and gas are less than 0.01 per cent efficient. See J. S. Dukes, 'Burning buried sunshine: human consumption of ancient solar energy', *Climatic Change* **61**, 2003, 31–44.

20. My paper, 'Bach's butterfly effect' *Environmental Values* **5**, 1997, 210–19, wrongly assumed that Bach's shares were in a coal-mine. In fact, the mine in Saxony was a silver mine, not one for coal. But silver probably formed the medium of exchange that allowed mines to be developed

21. Not a unit then in use: 'horsepower' was the benchmark but it would read oddly in that sentence. Most machines that used the coal were highly inefficient, with only a small proportion of the energy in the coal actually moving machinery or providing light. There are relevant data throughout V. Smil op. cit. 1994.

22. There is, of course, a huge literature on the reasons for the process. A summary set of accounts with often unusual perspectives is found in J. Goudsblom, E. Jones and S. Mennel, *The Course of Human History. Economic Growth, Social Process, and Civilization*, Armonk NY and London: M. E. Sharpe, 1996.

23. Marx and Freud also contributed world-scale theorisations.

24. A former British colonial secretary and president of the Board of Trade. Most remembered for his death in 1830 as the first ever victim of a railway accident, at the opening of the Manchester–Liverpool railway; a somewhat ironic fate in the light of his earlier pronouncement.

25. A. Grübler, 'Industrialization as a historical phenomenon', in R. Socolow (ed.) *Industrial Ecology and Global Change*, Cambridge: Cambridge University Press, 1994, 23–41.

26. I suppose there may be some remote groups unvisited by a development-agency worker who was trained in a building lit by electricity and who arrived by 4×4 vehicle? But very few. After the tsunami of December 2004, Indian government representatives would not for some time land on Sentinel Island in the Andaman and Nicobar Islands for fear of the traditionally hostile reaction of what the British press called 'Stone Age tribes'.

27. Overviews of this material are given by A. M. Mannion, *Agriculture and Environmental Change*, Chichester: Wiley, 1995, 3.4–3.5; D. Grigg, *The Transformation of Agriculture in the West*, Oxford: Blackwell, 1992; D. Grigg, *The Dynamics of Agricultural Change. The Historical Experience*, London: Hutchinson, 1982; a conceptually very important exploration was D. and M. Pimentel, *Food, Energy and Society*, London: Edward Arnold, 1979 but there is no better account than V. Smil, *Energy, Food, Environment. Realities, Myths, Options*, Oxford: Clarendon Press, 1987. Most of these books bring the story into the late twentieth century and hence are relevant to the next chapter as well.

28. Consider how much science could have been carried on without electricity. The symbolism of any movie of *Frankenstein* is highly revealing.

29. W. H. McNeill, 'American food crops in the Old World', in H. J. Viola and C. Margolis (eds) *Seeds of Change*, Washington DC: Smithsonian Institution Press, 1991, 43–59.

30. As always, there is a social context: the British government could have sent timely food aid to Ireland.

31. A famous pronouncement by the American ecologist Howard Odum: 'for industrial man no longer eats potatoes made from solar energy; now he eats potatoes partly made of oil'. In *Environment, Power and Society*, New York: Wiley Interscience, 1971, p. 116. In both phrases the role of mineral nutrients is overlooked.

32. J. Thirsk, *Alternative Agriculture. A History from the Black Death to the Present Day*. Oxford: Oxford University Press, 1997.

33. M. I. L'Vovich et al., 'Use and transformation of terrestrial water systems', in B. L. Turner et al. op. cit. 1990, 235–52.

34. See many of the papers in D. R. Harris (ed.) *Human Ecology in Savanna Environments*, London: Academic Press, 1980. There are regional overviews of the twentieth century in M. D. Young and O. T. Solbrig (eds) *The World's Savannas. Economic Driving Forces, Ecological Constraints and Policy Options for Sustainable Land Use*, Paris: UNESCO/Carnforth: Parthenon Publishing, Man and the Biosphere vol. **12**, 1993.

35. For an even earlier example of these interactions see I. Pikirayi. 'Environmental data and historical process. Historical climatic reconstruction and the Mutapa state 1450–1862', in W. Beinart and J. McGregor (eds) *Social History and African Enviroments*, Oxford: James Currey/Athens OH: Ohio University Press/Cape Town: David Phillip, 2003, 60–71.

36. P. C. Salzman and J. G. Galaty, 'Nomads in a changing world: issues and problems', in P. C. Salzman and J. G. Galaty (eds) *Nomads in A Changing World*, Naples: Instituto Universitario Orientale, series minor no **33**, 1990, 3–48.

37. T. Ingold, 'The day of the reindeerman: a model derived from cattle ranching and its application to the transition from pastoralism to ranching in northern Finland', in P. C. Salzman and J. G. Galaty (eds) op. cit. 1990, 441–70; J. L. Fox,

'Finnmarksvidda: reindeer carrying capacity and exploitation in a changing pastoral ecosystem – a range ecology perspective on the reindeer ecosytem in Finnmark', in S. Jentoft (ed.) *Commons in a Cold Climate: Coastal Fisheries and Reindeer Pastoralism in North Norway: the Co-Management Approach*, Paris: UNESCO/Carnforth: Parthenon Publishing, Man and the Biosphere vol. **22**, 1998, 17–39. [Is there something about reindeer that makes commentators adopt long titles for their essays?]

38. Ingold op. cit. 1990, including the observation about cruelty.

39. D. R. Foster, *Thoreau's Country. Journey through a Transformed Landscape*, Cambridge MA and London: Harvard University Press, 1999, p. 39 (a journal entry by Thoreau for January 28, 1853); the churchyard story is in a paper the substantive findings of which are shot through this chapter: E. L. Jones, 'Environment, agriculture and industrialization in Europe', *Agricultural History* **51**, 1977, 491–502.

40. There is no better source of both data and interpretation than M. Williams op. cit. 2003; these data are from Part II. An earlier and more condensed version of the history can be found in his contribution to B. L. Turner et al. (eds) op. cit. 1990: 'Forests', ch. 11, 179–201. Some of the data have been updated by newer studies: see G. C. Hurtt et al., 'The underpinnings of land-use history: three centuries of global gridded land-use transitions, wood-harvest activity, and resulting secondary lands', *Global Change Biology* **12**, 2006, 1208–29. The overall magnitudes are much the same.

41. Its title in German is *Grundsätze Forstoekonomie*.

42. In the United States at least, the introduction of creosote in the 1920s (derived from crude oil) saved many forests from being logged out for railroad ties (sleepers).

43. B. Weil, 'Conservation, exploitation, and cultural change in the Indian Forest Service, 1875–1927', *Environmental History* 11, 2006, 319–43.

44. M. Gadgil and R. Guha, *This Fissured Land. An Ecological History of India*, Delhi: Oxford University Press, 1993.

45. J. Fairhead and M. Leach, *Misreading the African Landscape. Society and Ecology in a Forest–Savanna Mosaic*, Cambridge: Cambridge University Press, 1996. The rainforests further south were correctly interpreted much earlier by E. W. Jones ('Studies on the rain forest of southern Nigeria', *Journal of Ecology* **44**, 1956, 83–117) when he said that the whole of the (then) continuous forest had been inhabited and cultivated at some time or other.

46. R. L. Heathcote, *Back of Bourke. A Study of Land Appraisal and Settlement in Semi-arid Australia*, London and New York: Cambridge University Press/Melbourne: Melbourne University Press, 1965; A. B. Smith, *Pastoralism in Africa. Origins and Development Ecology*, London: Hurst/Athens OH: Ohio University Press/Johannesburg: Witwatersrand University Press, 1992; M. Stocking, 'Breaking new ground', in M. Leach and R. Mearns (eds) *The Lie of the Land. Challenging Received Wisdom on the African Environment*, Oxford: James Currey/Portsmouth NH: Heinemann for the International African Institute, London, 1996, 140–54.

47. J. Dearing and R. T. Jones, 'Coupling temporal and spatial dimensions of global sediment flux through lake and marine sediment records', *Global and Planetary Change* 39, 2003, 147–68.

48. S. Daniels, 'The political iconography of woodland in later Georgian England', in D. Cosgrove and S. Daniels (eds) *The Iconography of Landscape. Essays on the Symbolic Representation, Design and Use of Past Environments*, Cambridge: Cambridge University Press, 43–82.

49. O. Jones and P. Cloke, *Tree Cultures. The Place of Trees and Trees in their Place*, Oxford and New York: Berg, 2002, ch. 2.

50. J. Tsouvalis, *A Critical Geography of Britain's State Forests*, Oxford: Oxford University Press, 2000.

51. Williams op. cit. 2003, ch. 12.

52. J. Knight, 'When timber grows wild. The desocialization of Japanese mountain forests', in P. Descola and G. Pálsson (eds) *Nature and Society. Anthropological Perspectives*, London and New York: Routledge, 1996, 221–39.

53. R. P. Harrison, *Forests: the Shadow of Civilization*, Chicago: Chicago University Press, 1992.

54. The inspiration for Ludwig van Beethoven's *Pastoral Symphony* (Opus 68, 1808) came from land and water within the narrow limits of some 15 to 20 kilometres either to the north of Vienna, or at Baden or Hetzendorf. At those places there were conceived and written (or at least sketched) ten great works recording Beethoven's impressions face to face with nature. The path known today as the '*Beethovengang*', which leads to the brook of the 'Pastoral' symphony, is in the valley of Wildgrube.

55. There is much writing about individual places in the nineteenth century but few general studies of any weight. The massive multi-volume federal study of outdoor recreation in the United States (informally known as the ORRRC Report) had only twenty-eight pages in its final volume given over to some history: 'Historical Development of Outdoor Recreation', in *Outdoor Recreation Literature: a Survey*, Washington DC: Government Printing Office, 1962, ORRRC Study Report vol. 27, 101–29. In Britain, the work of J. Allan Patmore has introductory historical material but no detail: see *Land and Leisure in England and Wales*, Newton Abbot: David and Charles, 1970; *Recreation and Resources. Leisure Patterns and Leisure Places*, Oxford: Blackwell, 1983. There is a whole chapter in C. Harrison, *Countryside Recreation in a Changing Society*, London: TMS Partnership, 1991.

56. In a bid for large-scale connections, a comparison was once made between the physiognomy of city parks and savannas, with the implication that humans were re-creating the environment in which they had evolved. Hm.

57. It is now illegal in Britain to kill birds such as the merlin and the hen harrier. The merlin (*Falco columbarius*) is probably tolerated because it will prey on rats (which eat grouse eggs) but the hen harrier (*Circus cyaneus*) is still shot and poisoned. Prosecutions in such areas are hard to bring and often not successful in front of a local magistracy. There is more (much more) in I. G. Simmons, *The Moorlands of England and Wales. An Environmental History 8000 BC to AD 2000*, Edinburgh: Edinburgh University Press, 2003.

58. I. G. Simmons, *An Environmental History of Great Britain*, Edinburgh: Edinburgh University Press, 2001.

59. See, for example, J. M. MacKenzie, *The Empire of Nature. Hunting. Conservation and British Imperialism*, Manchester and New York: Manchester University Press, 1988; E. J. Steinhart, 'The imperial hunt in colonial Kenya, c.1880–1909', in M. Henninger-Voss, *Animals in Human Histories. The Mirror of Nature and Culture*, Rochester NY: University of Rochester Press, 2002, 144–81.

60. In most developed countries until the 1950s, Saturday was a half-day, not a full holiday. I worked in Selfridges in London's Oxford Street in the summer of 1956 and we closed on Saturdays at lunchtime; some stores in Germany still do so. The exodus on Sundays was regularly condemned by the churches in Protestant regions, less so in Roman Catholic subcultures.

61. Jones and Cloke op. cit. 2002, p 32 quote from a 1992 newspaper article: 'We British . . . have looked at these coniferous plantations and decided we do not like them. We have brewed up a frantic symbolism of revulsion around them.'

62. M. Shoard, *The Theft of the Countryside*, London: Temple Smith, 1980.

63. The situation varied greatly from nation to nation. Sweden's generous laws of access to all but enclosed private land are well known (though currently under threat in the south) and other countries vary considerably. Few are as restricted as England, though here there is a dense network of *de jure* public footpaths, zealously guarded by citizen associations. The water catchments have been gradually opened up. For Britain, see M. Shoard, *A Right to Roam*, Oxford: Oxford University Press, 1999.

64. In the United States, Thoreau has been romanticised a great deal. There is a realistic account of his ecological and land-management context in D. R. Foster, *Thoreau's Country. Journey through a Transformed Landscape*, Cambridge MA and London: Harvard University Press, 1999.

65. J. R. McNeill op. cit. 2000 is one of the few overview books to have a decent treatment of warfare.

66. The publications of the Stockholm Institute for Peace Research (SIPRI) are invaluable. See especially A. H. Westing, *Environmental Hazards of War: Releasing Dangerous Forces in an Industrialized World*, Newbury Park CA: Sage, 1990 and his earlier *Warfare in a Fragile World: Military Impact on the Human Environment*, London: Taylor and Francis, 1980. Also, S. D. Lanier-Graham, *The Ecology of War: Environmental Effects of Weaponry and Warfare*, New York: Walker & Co., 1993.

67. L. Mumford, *The City in History: its Origins, its Transformations, and its Prospects*, New York: Harcourt Brace, 1961, ch. 18.

68. See the chapter by B. J. L. Berry, 'Urbanization', in B. L. Turner et al. (eds) op. cit. 1990, ch. 7.

69. Given the conditions in which many of them lived, 'shedding' has a curious ambivalence.

70. I. Douglas, *The Urban Environment*, London: Edward Arnold, 1983.

71. K. Thomas, *Man and the Natural World. Changing Attitudes in England 1500–1800*. London: Allen Lane, 1983.

72. The key work is A. Leopold, *Sand County Almanac*, New York: Oxford University Press 1949.

73. P. Hall, *Cities in Civilization*, London: Weidenfeld & Nicolson, 1998. Another mega-book on cities has even less, focusing mostly on art: P. Conrad, *Modern Times, Modern Places*, London: Thames & Hudson, 1998.

74. S. Pile, 'Cities', in S. Harrison, S. Pile and N. Thrift (eds) *Patterned Ground. Entanglements of Nature and Culture*, London: Reaktion Books, 2004, 246–8.

75. R. Lahtinen and T. Vuorisalo, '"It's war and everyone can do as they please!" An environmental history of a Finnish city in wartime', *Environmental History* **9**, 2004, 679–700.

76. These numbers are mostly from the 1970s and 1980s and will show some effects of more conservation-minded attitudes. For the period before World War II, let us guess, the upper figures are likely to be nearer the historical situation. Generally speaking, as plant is renewed and technology is improved, unit consumption drops.

77. Every book on world resource and environment processes and problems has a chapter on water. Most draw for contemporary material on the UNEP and WRI data. The chapter in B. L. Turner et al. (eds) op. cit. 1990 is by M. I. L'Vovich and G. White, 'Use and transformation of terrestrial water systems', 235–70; a

shorter version is in W. B. Meyer, *Human Impact on the Earth*, Cambridge: Cambridge University Press, 1996, ch. 5. The work of Gilbert White from the 1950s onwards on water has been an inspiration to many ever since. A very useful general work is P. H. Gleick (ed.) *Water in Crisis: a Guide to the World's Freshwater Resources*, New York and Oxford: Oxford University Press, 1993. Also, G. Petts (ed.) *Man's Influence on Freshwater Ecosystem and Water Use*, Wallingford: International Association of Hydrological Sciences, 1995.

78. H. Cook, *The Protection and Conservation of Water Resources. A British Perspective*. Chichester: Wiley, 1998, ch. 2.

79. Quoted in B. W. Clapp, *An Environmental History of Britain since the Industrial Revolution*, London and New York: Longman, 1994, p. 75.

80. E. Swyngedouw, 'Water. Circulating waters, circulating moneys, contested natures', in S. Harrison, S. Pile and N. Thrift op. cit. 2004, 119–21.

81. This was the *Heatherbelle*, a paddle-steamer tug from Sunderland in north-east England which towed two smacks offshore about 8 to 16 kilometres; by 1864 there were twenty-four such rigs working off that coast.

82. D. H. Cushing, *The Provident Sea*, Cambridge: Cambridge University Press, 1988.

83. See the classic R. J. H. Beverton and S. J. Holt, *On the Dynamics of Exploited Fish Populations*, London: HMSO, 1957 Fishery Investigations series 2, vol. **19**, Facsimile reprint, London: Chapman & Hall, 1993. Also J. R. Beddington, R. J. H. Beverton and D. M. Levine (eds) *Marine Mammals and Fisheries*, London: Allen & Unwin, 1985.

84. Quoted in many sources and in my *Environmental History. A Concise Introduction*, (Oxford: Blackwell, 1993), p. 112.

85. '. . . here at the beginning and the end is the great matrix that man can hardly sully and cannot appreciably despoil.' M. Graham, 'Harvests of the seas', in W. L. Thomas (ed.) *Man's Role in Changing the Face of the Earth*, Chicago: University of Chicago Press, 1956, 487–503 at p. 502.

86. At 54° south, this is the same latitude as York (England) and the Queen Charlotte Islands; scarcely polar.

87. K. Dorsey, 'Whale', in S. Krech et al. (eds) *Encyclopedia of World Environmental History*, London and New York: Routledge, 2004, vol. **3**, 1324–7. There is an interesting list of the equipment and supplies of a sixteenth-century whaler, pp. 1325–6; M. Klinowska, *Dolphins, Porpoises and Whales of the World: the IUCN Red Data Book*, Gland, Sw., 1991; R. Gambell, 'World whale stocks', *Mammal Review* **61**, 1976, 41–53.

88. W. J. Bolster, 'Opportunities in marine environmental history', *Environmental History* **11**, 2006, 567–97.

89. The literature on emissions from industry is vast. A good guide is the work of J. O. Nriagu, which nearly always has a historical dimension. See his 'Global inventory of natural and anthropogenic emissions of trace metals to the atmosphere', *Nature* **279**, 1979, 409–11; with J. M. Pacnya, 'Quantitative assessment of worldwide contamination of of the air, water and soil with trace metals', *Nature* **333**, 1988, 134–9; and a summary, 'Industrial activity and metal emissions', in R. Socolow (ed.) op. cit. 1994, 277–85; *idem*, 'A history of global metal pollution', *Science* **272**, 1996, 223–4 and for the total effects of a smelter, J. O. Nriagu , H. K. Wong, G. Lawson, and P. Daniel, 'Saturation of ecosystems with toxic metals in Sudbury basin, Ontario, Canada', *Science of the Total Environment* **223** 1998, 99–117.

90. A. S. Ek, S. Lögren, J. Bergholm and U. Qvarfort, 'Environmental effects of one thousand years of copper production at Falun, central Sweden', *Ambio* **30**, 2001, 96–103.

91. Sometimes known in English as 'strip-mining' or elsewhere 'open-cast'.

92. S. Baumgartner, 'Thermodynamics of waste generation', in K. Bisson and J. Proops (eds) *Waste in Ecological Economics*, Cheltenham: Edward Elgar, 2002, 13–37; V. Winiwarter, 'History of waste', ibid. 38–54. For the United States, see the work of M. V. Melosi, *Effluent America: Cities, Industry, Energy and the Environment*, Pittsburgh: University of Pittsburgh Press, 2001; *idem, Garbage in the Cities: Refuse, Reform and the Environment: 1880–1980*, Pittsburgh: University of Pittsburgh Press, 2005, revised edition.

93. D. H. McLaughlin, 'Man's selective attack on ores and minerals', in W. L. Thomas (ed.) op. cit. 1956, 851–61. The use of 'attack' is interesting and unusual. The equivalent essay on fuels is largely concerned with supplies (especially in the USA) and disavows the possibility of forecasting (E. Ayres, 'The age of fossil fuels', op. cit. 367–81).

94. Extracted from J. F. Richards, 'Land transformation', in B. L. Turner et al. (eds) op. cit. 1990, 163–78 and from later and more refined work by the LUCC project of the IGBP. See, for example, N. Ramankutty and J. A. Foley, 'Estimating historical changes in global land cover 1700–1992', *Global Biogeochemical Cycles* 13, 1999, 997–1027; K. K. Goldewijk, 'Estimating global land use change over the past 300 years: the HYDE database', *Global Biogeochemical Cycles*, 15(2), 417–34, 2001.

95. M. D. I. Chisholm, 'The increasing separation of production and consumption', in B. L. Turner et al. (eds) op. cit. 1990, 87–101.

96. E. A. Wrigley, 'The supply of raw materials in the industrial revolution', *Economic History Review* **15**, 1962, 1–16.

97. D. Raynaud et al. in Alverson et al. op. cit. 2003, ch. 2.

98. R. Sennett, *Flesh and Stone. The Body and the City in Western Civilization*, New York and London: W. W. Norton, 1994.

99. H. Tilley, 'African environments and environmental sciences', in W. Beinart and J. McGregor (eds) *Social History and African Environments*, Oxford: James Currey/Athens OH: Ohio University Press, 2003, 109–30.

100. O. Bernier, *The World in 1800*, New York: Wiley, 2000.

101. V. Janković, *Reading the Skies. A Cultural History of English Weather 1650–1820*, Manchester: Manchester University Press, 2000.

102. H. Rodhe, R. Charlson and E. Crawford, 'Svante Arrhenius and the greenhouse effect', *Ambio* **26**, 1997, 2–5. The first publication was in in Swedish in 1896, followed by refined calculations in German in 1903 and 1906. A likely pioneer was Arvid Högbom (*fl*. 1894): see E. Crawford, 'Arrhenius' 1896 model of the greenhouse effect in context', *Ambio* **26**, 1997, 6–11.

103. C. Rangely-Wilson, 'Trout of the empire', *The Field*, December 1999, 128–31. (The waiting-room fruits of a middle-class medical centre.)

104. M. N. Cohen, 'The epidemiology of civilization', in J. Jacobsen and J. Firor (eds) *Human Impact on the Environment: Ancient Roots, Current Challenges*, Boulder CO: Westview Press, 1992, 51–70.

105. In this case the hybridisation was accidental after the planting of specimens of European and Japanese larches close to each other on the Duke of Atholl's estate at Dunkeld in Scotland in 1885.

106. T. R. Dunlap, 'Australian nature, European culture', in C. Miller and H. Rothman (eds) *Out of the Woods. Essays in Environmental History*, Pittsburgh: University of Pittsburgh Press, 1997, 273–89; A. W. Crosby, 'Biotic change in nineteenth-century New Zealand', ibid. 263–72.

107. Privacy is one result: people die in private (especially if legally executed) and those declared insane are no longer an entertainment spectacle. Or at least were

not until the kind of 'reality television' which can only be a step away from contriving these sights.

108. *The Fontana Dictionary of Modern Thought*, London: Fontana Press, 2nd edn 1998, p. 540.

109. W. R. Everdell, *The First Moderns. Profiles in the Origin of Twentieth-Century Thought*, Chicago: University of Chicago Press, 1997. He notes that the camps were given their name 'at the same time as cubism and quantum physics, by the same civilized Westerners'. (p. 117). Everdell's breadth of understanding and felicitous writing make it a most persuasive read.

110. A. Desmond, *Huxley*, London: Penguin Books 1997, reports that it was used as a term of abuse on a visit by Huxley to the United States in 1876 but that by 1888 the Royal Society used the word. Huxley, apparently, avoided it.

111. Carl von Linné (1707–78), known as Linnaeus, systematised plant and animal scientific names in the system still largely in use today, though classification by DNA structure is likely to replace the earlier emphasis on anatomy.

112. It is claimed that between 1947 and 1959 there was a paradigm shift in American ecology which saw 'real and integrated' biological communities becoming artefacts of the human imagination. Connections with other aspects of culture (existentialism would be an obvious transference) are poorly made by the scientists themselves. M. G. Barbour, 'Ecological fragmentation in the fifties', in W. Cronon (ed.) *Uncommon Ground. Toward Reinventing Nature*, New York and London: W. W. Norton, 1995, 233–55.

113. However, Henry Ford tried to create a bird reserve on his farm and built hundreds of bird houses. He then imported 600 pairs of songbirds from Europe at one go: when released they all flew away.

114. A very influential paper decrying common usage was G. Hardin, 'The tragedy of the commons', *Science* **162**, 1968, 1243–8 and much reprinted since. Counterarguments which conclude that rational use can come from communally run systems, include a vigorous denunciation in P. Dasgupta, *The Control of Resources*, Oxford: Blackwell, 1982, ch. 2. There is a good discussion with examples in J. Rees, *Natural Resources. Allocation, Economics and Policy*, London and New York: Routledge, 2nd edn 1990, ch. 6. Later work includes F. Berkes (ed.) *Common Property resources: ecology and community-based sustainable development*, London: Belhaven, 1989; R. A. Devlin and R. Q. Grafton, *Economic Rights and Environmental Wrongs: property rights for the common good*, Cheltenham: Edward Elgar, 1998.

115. The nomenclature shifts with fashion and notions of correctness. The use of 'The South' to denote the poorer nations was not in use until well after 1950. Most common at that time would have been *DC* and *LDC*, with the implication that development was a kind of linear trajectory; then 'developing' replaced *LDC* and eventually 'rich' and 'poor' were often applied, to the annoyance of those who objected to its purely material parameters. Then we have 'Low Income Economy' and its variants. 'Underdeveloped' came into general use after President Truman's use of it in 1949, where it replaced 'backward areas' then current; by 1958 the FAO of the UN was using 'Less Developed Regions'. This is set out by L. Dudley Stamp, *Our Developing World*, London: Faber & Faber, 1960, ch. 12, which is an updated version of some lectures he gave in 1950 in the USA and published as *Our Undeveloped World* in 1953. He talks of the USA (ch. 11) as a 'poor little rich girl' and noted that 'The architects of the Dollar Empire . . . are compelled to adopt and adapt most of the devices of their predecessors [as imperial powers].' (p. 169).

116. D. S. Landes, *The Wealth and Poverty of Nations*, London: Little, Brown/New York: W. W. Norton, 1998.

117. R. W. Sellars, *Preserving Nature in the National Parks. A History*, New Haven and London: Yale University Press, 1999.

118. A. Hornborg, 'Ecology as semiotics. Outlines of a contextualist paradigm for human ecology', in P. Descola and G. Pálsson (eds) *Nature and Society. Anthropological Perspectives*, London and New York: Routledge, 1992, 45–62.

119. E. R. Wolf, *The People without History*, Berkeley, Los Angeles and London: University of California Press, 1982 and with an updated Preface 1987.

120. All from *On the Origin of Species*, 1859.

121. S. Kern, *The Culture of Time and Space 1880–1918*, Cambridge MA: Harvard University Press, 1983.

122. P. Gay, *The Cultivation of Hatred*. Vol. III of *The Bourgeois Experience. Victoria to Freud*, New York and London: W. W. Norton, 1993.

123. R. H. Grove, *Green Imperialism. Colonial Expansion, Tropical Island Edens and the Origins of Environmentalism, 1600–1860*, Cambridge: Cambridge University Press, 1995.

124. B. M. Stafford, 'Presuming images and consuming words: the visualization of knowledge from the Enlightenment to post-modernism', in J. Brewer and R. Porter (eds) *Consumption and the World of Goods*, London and New York: Routledge, 1993, 462–77.

125. C. Klonk, *Science and the Perception of Nature. British Landscape Art in the Late Eighteenth and Early Nineteenth Centuries*, New Haven and London: Yale University Press, 1996.

126. R. L. Schwarz, *Broken Images. A Study of 'The Waste Land'*, London and Toronto: Associated University Presses, 1988, p. 13.

127. Adapted and amplified from J. F. Richards, 'Land transformation', in B. L. Turner et al. (eds) op. cit. 1990, 163–78.

128. Interesting contrasts can be found between writers who think that technology drives change from the boiler up, so to speak, and those who say that social factors, such as demand and the desire for control, bring forth the technology. For examples see J. Mokyr, *The Lever of Riches. Technological Creativity and Economic Progress*, New York and Oxford: Oxford University Press, 1990 and G. D. Snooks, *The Dynamic Society. Exploring the Sources of Global Change*, London and New York: Routledge, 1996. For further discussion, see some of the essays in M. R. Smith and L. Marx (eds) *Does Technology Drive History? The Dilemma of Technological Determinism*, Cambridge MA: MIT Press, 1990.

129. A. W. Crosby, *Ecological Imperialism. The Biological Expansion of Europe 900–1900*, Cambridge: Cambridge University Press 1986, reprinted 1993, 2nd edn 2004, p. 300 in 1993 edn.

CHAPTER FIVE

A post-industrial era?

FIGURE 5.1 *Joseph Beuys's* The End of the Twentieth Century, 1982–83 *(detail). Basalt, clay and felt. Bayerische Staatsgemäldegalarien. Pinakothek der Moderne. Photograph by Caroline Tisdall.* © DACS 2008

Joseph Beuys was a radical artist who was very concerned with environmental matters. He often tried to use 'natural' materials such as felt and animal fats and was aware of the role of energy in natural and human-led systems. His largest 'static' work was *7,000 Oaks*, begun in 1982 in Kassel, Germany. His plan called for the planting of seven thousand trees, each paired with a columnar basalt stone approximately 4 feet (1.2 metres) high above the ground, throughout the greater city of Kassel in Germany; the last tree was planted in 1987. Beuys intended the Kassel planting and sculpture to be the first stage of a worldwide project in effecting environmental and social change.

The sculpture shown is portable enough to be exhibited around the world, as is appropriate for an artwork with such wide ambitions. The illustration does not quite convey its size: the area covered is perhaps five or six times the area shown in the photograph. A common interpretation is that the blocks of basalt represent the Earth and the 'plugs' are representative of the wounding of the planet and its repair – the excised parts are replaced.

Beuys is one of the generation of artists who has moved into a kind of hybrid relationship with 'natural' materials. This *oeuvre* is partly studio based (as above) but moving out into a wider frame with *7,000 Oaks*, and his work seems to be a precursor of the 'landscape art' of the kind that uses a 'soft' approach (including temporary structures) rather than importing technology-based artefacts: Andy Goldsworthy and Martin Hill would be good examples. In this, they all covey a tenderness to the natural world not characteristic of twentieth-century modernism and so fit better with the pluralities of the post-modern movements in the arts. To what extent they bring about changed attitudes in a wider context is perhaps unknowable: on the one hand, 'high culture' can seem radically divorced from the real life of the very poor or the really rich but, on the other hand, the artist may often possess antennae that are more sensitive to coming atmospheres and thinking.

See M. Gandy, 'Contradictory modernities: conceptions of nature in the art of Joseph Beuys and Gerhard Richter', *Annals of the Association of American Geographers* **87**, 1997, 636–59.

I saw it on TV

Television is a product of the 1920s but becomes an integrated part of most humans' lives after 1950. Its use emphasises that in the period from 1950 to 2000 it was possible for a very wide viewing public to learn a great deal more about humans and their environments, especially after satellite transmission reached remoter places. The visual content was remarkable but the price was a loss of immediacy: the world became mediated through the screen's everyday presence. The two-dimensional digital world of the visual is apparently tailored to an individual but, in fact, is pre-programmed by the providers of the images.[1] Television's view of the past is also distorting in the sense that it turns past fiction into present fact (notably via the docu-drama) and, although its representation of the past may be more self-conscious than that of our precursors, it is no less selective.[2]

The lineaments of the period to 2000 is still here so that the tenses of the verbs in this chapter often convey the continuation of that fifty-year period forward from the horizon of the year 2000 into the present. Through all this time, the material world was often subject to exponential rises in rates of interaction driven by human actions. Some of this was simply due to population increases but other demands increased because of the high demands from the wealthier people of the world. One outstanding linkage is the end-use of energy by humans in ways which are luxuries in the sense that we lived for millennia without them, and which are far enough from a source to be an apparently inefficient use: the appetite for meat is one of them. To eat as a top carnivore means that much photosynthetically fixed energy is dissipated through the herbivore stage and also transformed into heat as the carnivores run around; at the table, meat is about 2 per cent efficient. Another extravagance is the internal combustion engine in which only refined fuel can be used and in which only a small proportion of the energy in crude oil actually works the motor; the efficiency is about 10–12 per cent.[3] In both cases, a lot of heat is given off without performing any useful function and there are gaseous emissions of many kinds of which two at least (methane and carbon dioxide) are implicated in global changes in the composition of the upper atmosphere. There are associated changes in land cover as animal farming is favoured, often with heavy inputs of fossil fuel, and as roads allow the penetration of powered vehicles into most places. The use of vehicles (mostly privately owned and the bigger the better) then dictates land-use pattens: about 1.3 per cent of Britain is road surface.[4] So soils, plants, animals, water and atmosphere are all subject to alteration and all on scales greater than hitherto.

THE CULTURAL ECOLOGY OF THE WORLD AFTER 1950

The term 'post-industrial' (and the appended question mark in the chapter title) need some exploration. It is normally used of societies in which the actual manufacture of goods carries less monetary value than that of services. (The acronym PIE [post-industrial economy] will be used here.) The goods are still available but they are made in low- and middle-income places. Thus, the question mark, because many parts of the world have nowhere near the same mix of making and consuming as the 'western' nations such as those of western and northern Europe, North America and Japan. Do not ask somebody who was in an Ethiopian relief camp in the 1980s if they felt themselves to be a member of a post-industrial world.[5] But because the western ways are innovative and so much copied and because the west controls the advocacy of these ways of living, most countries aspire to them.[6] Overall, the ingredients of post-industrial life ways seem to be (a) the transition from coal to oil as the main fossil fuel in use, with natural gas increasing rapidly in popularity; (b) an economy in which services (financial, educational, medical) generate more income and provide more employment than manufacturing; (c) movement is ever more possible, including cheap air travel and also access to private means of transport; and (d) more leisure time, coupled with higher disposable incomes and better health, much of it made possible by the use of machines as 'slaves'. All of them lead to environmental interactions, not least for energy demands where consumption by individuals (in the home, in vehicles and as food) dominates public discussion more than its supply to industry. 'Inefficient' labour is replaced by cheap fossil fuels with both social and environmental effects.

Evolution and dispersal

The PIE has a technological foundation which undergirds its access to resources, its disposal of them and its outlook upon the non-human components of the globe. To some extent, this is a development from the previous era (the greater use of refined oil products to fuel motors, for instance) but there are novel elements as well. New energy sources are discussed below but there has been an emphasis on a shift wherever possible to electricity as a source of power.[7] The ease of getting electricity to most places has facilitated a digital revolution with an emphasis on the computer, and its storage in battery form has helped along the miniaturisation of many machines, especially in the field of communications. These last are now instant, cheap and often visual in nature and are meshed together by a global network of satellite relays. Two features become apparent. The first is that many changes are driven by technology: we have the means, so let us find an application for them. Previous eras may have seen social wishes allowed by technological developments; now the technology is powerful enough to create its own world. The second is a reflection on the role of war in accelerating trends: the digital computer, targeted biocides, antibiotics and radar might all have come to their present degrees of penetration and impact, anyway, but it seems at least arguable that World War II provided the impetus for an accelerated evolution after 1950.

There is a predictable air to maps of the world that show income, birth rates, longevity, energy consumption and all the other conventional measures of a society. When combined, the results show a division in the world between those who are prosperous and those who are not and, between, regions that have been climbing out of widespread poverty (Thailand would be an example) and those who are regressing into it, such as Zimbabwe. In neither of those two cases can the environment be said to be causal though, in both, its state is interwoven with the economic changes that have occurred. Worldwide, there are core areas and a constantly shifting periphery, as in the nineteenth century. In the case of the post-1950 world, without doubt the initial impetus lay with the United States. From there, the combination of corporate capitalism, high resource use, with rapid throughput of materials, and a raised standard of material life for many (though never for all) has spread. Trade has been a major artery as always but the influence of film and especially television is a difference from the preceding era. There has been resistance to the American package in the shape of state socialism (in the USSR and its satellites, in a diluted form in the post-war Labour governments of Britain, in central America, North Korea and in Cuba, and above all in China) but most of these are now firmly in the camp of 'free trade' and neo-liberal thinking. This means that resources and their environmental linkages are deemed marketable and can have a price. Much of the resistance to such ideas, however, has also been initiated in the United States. Hence, the story of the environment in both material and cultural terms since 1950 can often be epitomised by examples from that nation, from whence came the idea of this phase as a Great Disruption.[8]

Environmental relationships

If there has been a problem using all the information gathered in the period 1750–1950, then it is greatly increased in the years thereafter. The miniaturisation of all kinds of instruments and the digital chip means that huge quantities of data can be stored, albeit expensively since they seem to have to be remounted every few years, which was less of a problem with the book. Everybody of note is now a specialist and, in academia, generalists are called 'Renaissance men' [sic] with a certain air of patronage and a distinct lack of promotion.

As throughout the entire Holocene, the number of humans has been a basic element in environmental relations. In 1950, the annual growth rate was 1.5 per cent and between 1962 and 1972 rose to just above 2.0 per cent before dropping back to about 1.2 per cent by 2000. This worldwide figure conceals the kind of variation that shows some African countries to have rates of 3 per cent, meaning a doubling time of twenty years. Two-fifths of the world's population, however, now live in countries with a replacement level of population growth. Nevertheless, the 2,500 million (2.5 billion) people of 1950 hit 6 billion in 1999 and just under 80 million are being added to the total each year. One element in this growth has been increased life expectancy: worldwide in 1950 this was fifty-six years but this had become sixty-six years by 2000, and over sixty even in Low Income Economies. Again, this conceals substantial variations, and expectancy is falling in countries such as Botswana and Zimbabwe: whereas it was over sixty years in Botswana in 1980 to 1990, it fell to forty-three years in 2000. The reason here is the incidence of HIV/AIDS: worldwide, two million people died of it in 1998, with 90 per cent of the deaths in sub-Saharan Africa; in Botswana and Zimbabwe, about one-quarter of the adults were infected.[9] Because population growth is so clearly an element in humanity's environmental relationships, outspoken attitudes towards it have often been adopted by environmental writers and campaigners. The high growth rates of the 1950s and 1960s, for example, led to a great deal of neo-Malthusian advocacy, with numerous campaigns and think tanks aimed eventually at reducing growth rates in poorer countries but also in the rich nations where the consumption patterns were held to be destructive.[10] The main public phase of this argument was finished by the United Nations Cairo Conference of 1994, and it is significant that in the United Nations Environment Programme Millennium Survey of 'environmental experts in over 50 countries', 'population growth and movement' were ranked seventh in the problem list, named by 22 per cent of respondents (first came climatic change at 51 per cent).[11] Note, however, that mass migrations would also be included and so, in Africa especially, there are notable environmental consequences of famines and armed conflicts. It has looked as if population growth is a social trend that loses its public high profile after a while. One attitude would be that this was a feedback process and that, indeed, the high visibility afforded to it has resulted in the means being provided to most people to control their fertility. Pessimists have been heard to mutter that Malthus might be disapproved but is not disproved.

Whatever the raised consciousness of population growth until the 1970s, it was overtaken by concerns about climate, which have persisted to the present day. Discussion of climate in scientific and popular media is usually bound up with scenarios of future change; here in this chapter, only the history will be given. Against the background of the Little Ice Age (LIA) and its end in about 1900, the twentieth century has been a very warm period: as the Intergovernmental Panel on Climate Change (IPCC) puts it, 'An increasing body of observations gives a collective picture of a warming world and other changes in the climate system.'[12] A brief summary might focus on an increase in global average surface temperature by about $0.6° \pm 0.2$ °C over the twentieth century with a particular upward kick in 1995–2000. So 1990 becomes the warmest year and the 1990s the warmest decade since 1861, that is, during the period of reliable instrumental records. Therefore, the increase in temperature in the northern hemisphere in the twentieth century is likely to have been the largest of any century in the last 1,000 years. One remarkable fact is that between 1950 and 1993, night-time daily temperatures over land showed an increase of about 0.2 °C per decade, which lengthened the frost-free season in many mid- and high-latitude regions. Snow cover and ice extent have decreased, with snow cover having fallen by about 10 per cent since the late 1960s, and a reduction of about two weeks has taken place in the period of ice on rivers and lakes in the higher latitudes of the northern hemisphere. Mountain glaciers have retreated and the sea ice of the Arctic has shrunk in extent by about 10–15 per cent since the 1950s and by 40 per cent in thickness. Warming and melting have led to an average worldwide sea-level rise of 0.1–0.2 metres in the twentieth century, and the heat content of the oceans has also risen. There is only a finite amount of water on the planet and so, if precipitation has increased in one set of regions (such as the northern-hemisphere land regions and the tropical land areas), then it decreases over others such as the subtropics. Extreme high precipitation and temperature events have gone up in frequency, and droughts have been worse and come more often in parts of Asia and Africa. El Niño phases have become more frequent, persistent and intense since the mid-1970s. The IPCC is certain that these changes could not have occurred by chance. There have also been poleward and altitudinal shifts in the ranges of some plants and animals, declines in the populations of some biota and the earlier flowering of plants, emergence of insects and egg-laying in birds. All these have been documented for aquatic, terrestrial and marine habitats, and are considered to be independent of direct human activity.[13] The role of the land surface, and especially of agriculture and forest cover, came somewhat later in investigations of influences but their importance and their links to environmental history started to grow in the late 1990s;[14] after a period of apparent invisibility, the continuing role of fire was recognised. About 200 to 400 million hectares of savanna are burned annually, as are 5 to 15 million hectares of boreal forest, yielding perhaps 2.0–4.0 petagrams of carbon to the atmosphere. In Canada, the annual release of carbon from fires is comparable to the combustion of fossil fuels.[15] Some important aspects of climate appear not to have changed: some southern-hemisphere regions seem not to have warmed, and phenomena such as tornadoes, thunder and hail are still much the same in places where they have been measured.

All these changes have been linked to the use of fossil fuels. The energy use of this era is still dominated by coal, oil and natural gas but there have been significant shifts within the repertory and additions to it. The crossover in the substitution of oil for coal between 1960 and 1965 is important, denoting a world ever more dependent upon oil supplies, even though oil may be transported over distances as great as 12,000 kilometres. In absolute terms, nevertheless, coal is still being produced in ever-larger quantities especially with the surge in industrialisation in China. Both oil and coal can be easily, if not efficiently, converted to electricity. In the post-industrial era, the demands for electricity were sometimes met by hydro-power generation (HEP): between 1945 and 1985 nearly 120 plants with a capacity in excess of 1 gigawatt were opened in over thirty countries. Although the dominant producers are the United States, Canada and Russia, HEP is important for many smaller countries and also in China. In Africa and Latin America, HEP may often be the major source of electricity and so the large HEP scheme has often been a staple of international development agency funding in Low-Income Economies. Worldwide, HEP generates about 20 per cent of world supplies of electricity.[16] The limited scope for ever more dams, along with periodic concerns about the size of coal and oil reserves, have meant that yet other sources of energy have been developed. The most important of these in the post-1950 era has been nuclear power. This uses the fission reaction developed first for military purposes, and the separation of the two has often been incomplete. From the first commercial reactor in 1956 to 400 plants (in operation or advanced planning) in 1973 represents a period of remarkable growth. About 17 per cent of world electricity was generated from this source in the 1990s though, after 1973, there was a withdrawal from new plants except in dependent countries like Russia and its former republics, France (which has the highest dependence at 78 per cent), Belgium and Japan. In 1998, there was one less installation in the world than five years earlier so that, with an average life of only eighteen years before closure, the capacity of the industry levelled off in the 1980s. For reasons which are detailed under Management and Impact (below), nuclear power has lost popularity and acceptance since the 1970s, and so alternative methods of energy generation had a high profile: a renewed interest in biomass, wind turbines, tidal power and solar cells forms a category of 'equilibrium' or 'renewable' energies, though many depend upon large fossil fuel inputs to start them up, as indeed does nuclear power.[17]

The technologies of these economies are in part continuations of the previous industrial decades and in part novel. The centrality of steel has been supplemented by the use of copper in electrical wiring, for instance. One ubiquitous set of materials since 1950 has been plastics, themselves offshoots of the refining of oil and, even in the worst situations (in refugee camps, and after earthquakes), plastic sheeting is one of the first materials to appear. Trees are still the providers of paper, a demand propelled by increasing literacy. In most of these mass-material uses a different resource flow has arisen, namely that of re-use and recycling. This is not especially new in poor countries, where the picking-over of the city rubbish tip is normal but has become an increasing practice among the rich countries as a reaction to, for example, the over-packaging of almost all consumer items. The great technological novelty is miniaturisation. In one literal sense this means that instrumentation in the environment (as elsewhere on Earth) can be placed in satellites or at the bottom of the oceans and thus accumulate measurements of a precision and a quantity hitherto impossible and only manipulable because of the digital computer. At the same time, biotechnological procedures have allowed manipulation of living material at the molecular level so that engineering of the genetic characteristics of plants, animals, fungi, bacteria and viruses has become commonplace, albeit contested in some circumstances. The next step along this road is nanotechnology, using individual atoms as building-blocks. Environmental consequences of all these changes are likely to be great but are impossible to forecast accurately. The United Nations Environment Programme's millennial overview recognised this in retrospect when it remarked that 'new issues rarely appear without warning' but went on to list a series of 'environmental surprises since 1950'.[18] These were classified into unforeseen issues, unexpected events, new developments and changes in trends. (Table 5.1).

TABLE 5.1 Environmental surprises since 1950

Type of surprise	Examples(s)	Date
UNFORESEEN ISSUES	CFC-induced ozone depletion	Hypothesised 1974. Confirmed 1985
UNEXPECTED EVENTS	Major oil spills	*Torrey Canyon* 1967; *Amoco Cadiz* 1978; *Exxon Valdez* 1989
	Accidental toxic events	Minamata disease 1959; PCB poisoning 1960s; Sveso dioxin leak 1976; Bhopal leak 1984; Basel chemical fire 1986
	Air pollution	London smog 1952; Indonesia forest fires 1997
	Nuclear accidents	Windscale 1957; Urals 1958; Three Mile Island 1979; Chernobyl 1986
	Biological invasions	Zebra mussel 1980s; mesquite in Sudan 1950s
NEW DEVELOPMENTS	Pollution	Pesticide effects in birds 1962; Love Canal toxics; contaminants in Arctic ice 1970s
	Acid precipitation Climate change	Growing concern from 1972
	Tropical deforestation	Satellite images from 1980s
	Widespread consequences	Effects of Aswan High Dam 1970s; shrinkage of Aral Sea, 1980s
CHANGES IN TRENDS	Climate change: increase in occurrence of severe weather and El Niño events	1997/8 El Niño very severe; 1988 summer very hot
	Oil crisis from producer worries about depletion	1970s
	Fisheries collapse, e.g. Atlantic cod	1990s

Source: UNEP, *Global Environmental Outlook 2000*, London: Earthscan, 1999, p. 336.

Some of these surprises result from new technology, such as the increased use of personal and domestic aerosols; others from the intensification of already extant processes such as fossil fuel use, others from synthesised compounds with no quick breakdown pathways. The 1970s oil crisis is the only one which did not happen, so to speak. There is a school of thought that considers any global warming is due entirely to 'natural' processes and that evidence linking it to fossil fuel emissions is unreliable.

Because of the quantity and immediacy of information available, perceptions are intimately connected with material events and processes. The 'oil crisis' of 1972–3 was partly driven by the producers' perception that their stocks were running out, which turned out not to be true. Less materially, the 1969 'Earthrise' photograph from *Apollo II* became iconic as an image of a fragile but unitary ecosystem.[19] Ecological science urged that many social investigations and concerns (population, pollution and resource depletion, for example) could be seen as related to one another. Wider economic and political contexts gave them shorthand labels, such as sustainable development, climate change and biodiversity. The scale of focus also shifted because better knowledge and more concerned people showed that many substances had, not only local, but national, international, worldwide and sometimes global pathways and storage pools. This kind of knowledge enhanced the realisation that environmental events always have cultural contexts: in the Cairo earthquake of 1992 many illegally built high-rise concrete blocks of flats collapsed, and government help was tardy and poor, whereas local Islamic groups were very quick and helpful; the Yangtse floods of 1998 converted high but 'normal' rainfall into disaster because of all the reclamation and engineering that had been enforced by a growth-minded central government, especially by lowering storage in the river basin through the reclamation of lakes and fluvial islands and the filling of impoundments with silt.[20]

In very general cultural terms, there has been a continuing belief in the power of science and technology to solve problems: rather than use less energy, for example, high-technology ways of sequestering carbon under the oceans have been considered, and money is invested in the possibilities of 'clean' energy from nuclear fusion: still forty years off, as it has been for the last forty years. Thus 'growth' is still considered a prime economic aim, though it is increasingly interrogated with questions like, 'what kind of growth? who benefits? and what are the external costs?' The gaps between rich and poor are still very wide, and all kinds of environmental consequences are tied into those discrepancies, not the least of which is the way some rich nations tried to export their toxic wastes for disposal in poorer places without effective legislation. Movement of thousands of kilometres is involved and this is just one sample of a wider PIE phenomenon. Transport became cheaper after 1950 and so baby sweetcorn can be flown from Thailand to Europe, and Europeans can fly to Thailand for the weekend. Neither is without environmental consequences. Through all this, the pervasive power of the visual media has increased, pushing forward rising expectations of material anabasis.

Management and impact

As in previous chapters, this section considers the main resource processes in their environmental context where that means not only the material flows but some of the matrix of ideas which surrounds them. In this era, the quantity and diversity of environmental attitudes are so great that there is even more selection than usual: many advocates of particular positions will be dissatisfied at the treatment given to their convictions.

Resources: energy and environment after 1950

None of the energy sources discussed in the previous chapter fell out of use after 1950 and, in fact, they all increased in use in absolute terms. Biofuels also continued to be significant in many low-income economies, and their continuing supply a matter of controversy in semi-arid regions where the relative importance of climatic shifts, grazing pressures and population growth has been fiercely debated. One hallmark of the post-industrial economy was nevertheless the development of new energy sources and, in particular, that of energy from nuclear fission. It is also the environmental linkages of nuclear fission which brought about evaluation and disagreements as never before. The principal problem is that of the emission of ionising radiation which may enter the environment in two main ways, assuming that the generating plant is functioning properly. There may be deliberate release to the atmosphere, as happened in the years before the test-ban treaty in 1962, or controlled release to the atmosphere or water (mostly marine) systems, all in small quantities. Then there is the nightmare event of uncontrolled releases when a plant malfunctions, as at Windscale (now Sellafield) in 1957, Three Mile Island (United States) in 1979 and Chernobyl (Ukraine) in 1986, along with a number of others, of which incidents in Japan have attracted most attention. The isotopes released have quite long half-lives (typically of twenty-five years or more) during which they are carcinogenic to humans even if not immediately fatal through radiation poisoning. Small wonder, therefore, that other new sources, such as electricity from solar collectors or photoelectric devices have been extensively tried as sources of energy. In general terms, however, older sources reworked with new technological knowledge, such as wind power and small-scale hydro-power, gained higher profiles. The use of hydrogen cells for motive power kept bobbing up only to be deflated by the explosive potential of hydrogen. In high-income economies, land can be given over to crop production for 'biofuels', using rapeseed and other oils; in Brazil, surplus sugar is also converted to a vehicle fuel.

The attitudes taken towards new energy generally avoided the obvious course of using less. Western industrial processes have generally become much less energy intensive in the post-1950 period, but that applies to few other sectors. Instead, hi-tech solutions of radioisotope disposal were sought, involving underground storage or even contemplating the proposal to fire waste-bearing rockets into the sun. Large-scale development of wind turbines can generate arguments on the basis of landscape change or impact on wildlife. These are not apparently ineluctable, however: the general acceptance in Denmark and Germany contrasts with the wrangles in Britain. Overall, energy sources and their environmental connections have been well aired in high-income economies, though less so in countries where a higher per caput availability would bring clear benefits.

Resources: agriculture

All the techniques of the industrial era were still present but intensified in bringing energy to the land: machinery, fertilisers, biocides, plant and animal breeding (including recent developments in direct genetic modification), plastic ground covering, polytunnels and glasshouses, do not exhaust any list. At one remove there is a great deal of scientific knowledge which is only possible with large inputs of electricity: think of all those lecture rooms without any windows. Though most applied in the best soils of the high-income economies (the fenlands of East Anglia are a case in point), many emerging economies followed the path of moving where possible to more crop per unit area per unit time, that is, to intensification. This usually means a cash-crop economy, producing a uniform product for an outside market. The eventual end-user is thousands of kilometres away, and with edible materials, air transport was increasingly used. So we get crop monocultures coupled to high energy consumption and multiple emissions to the atmosphere. The import of the technology and the economics into less-regulated countries means that, for example, pesticides are used without concern either for non-target species or for field operators and, in a general sense, the whole process follows the nineteenth-century course of uncontrolled emissions and the dispossessed rural population, whose bodies' energy input is no longer needed, migrate to the towns and cities with entirely predictable environmental and social results.[21] The post-1950 components of a process which started in the nineteenth century have been dominated by the growing of soy beans for cheap animal feed: its meal became three-quarters of all high-protein livestock feed in the high-income economies, ending up as milk, eggs and meat. Animal production in feedlots produces large quantities of slurry: it is, in fact, an industrial undertaking, resulting in about 60 per cent of river-pollution incidents in western countries as well as adding to the nitrogen enhancement of soils with its eventual effects on freshwater systems. (Before the 1930s, animal wastes were a resource rather than a problem.) Soy consumption in Europe and North America produced environmental consequences in Brazil as well.[22] Fast-food empires developed after the 1950s and rely on scientific animal nutrition, genetics and disease control. The control of this type of industry and, hence, of its environmental impact have been in a small number of companies: in the United States, the yearly production of 3.7 billion chickens comes from fifteen primary breeders. Concentration is similarly great in the seed industry where advanced biotechnology ensures that about ten to twenty seed- and plant-technology companies dominate the world's seed trade: nature is seen as a vast organic Lego kit for genetic characters.[23]

Cultural resistance is growing. The Food and Agricultural Organization of the UN (FAO) plan, that every place should grow its 'best' crop and export it falters when the subsidies made to agriculture in the richest nations are taken into account. By contrast, non-governmental organisations (NGOs) and a few national governments argue in favour of more local self-sufficiency and the revival of the primacy of local knowledge and local varieties of crops. In the richer world, a move towards growing plants and animals without chemical fertilisers or long-lived biocides and marketing it close to the producing site (lowering the number of 'food miles') has a hold in middle-class eating patterns, spurred on by medical attitudes to animal fat, refined sugar and salt; the Italian 'slow food' movement is also part of this more 'organic' approach, as is the buying of 'Fair Trade' products from tropical countries. The PIE pull can be deduced from the example of a tea bag in Britain, with tea from Sri Lanka, India and Kenya, wood products from North America and Scandinavia, polypropylene from Belgium, and assembly at four different sites in England. In Britain around the turn of the twenty-first century the cost of 'food miles' was c.£9 billion per year compared with £6.4 billion for agriculture. The number of lorries carrying food doubled between 1974 and 2002.[24]

As farmers in well-watered areas discovered the usefulness of occasional added water, so the distinction between irrigated agriculture and rain-fed equivalents became less apparent. Even in the British Isles, not noted for their aridity, sprinklers are a common sight in summer. The immense increase in the irrigated area of Asia, has been largely responsible for the rise in the absolute area of cultivated land on a world scale, by 1990 amounting to 17 per cent of cultivated land. The amounts of water used, as a proportion of global water stock, are minuscule but there may well be regional constraints. In the Middle East, for example, a few rivers supply a high proportion of the irrigation water and they are subject to much political quarrelling. A nation such as Israel must therefore re-use all its water as much as possible and, for irrigation, to drip-feed water to individual plants rather than saturate soils with no plant roots in them. The centre-pivot distributor, which creates such remarkable landscape patterns from the air, is normally 80 per cent efficient but, even so, geological aquifers, such as the Ogalalla strata of the High Plains of the United States, must eventually be drawn down to an uneconomic level and another way of using that land be found. Indeed, some related shifts took place.

In her novel *That Old Ace in the Hole* (2004), Annie Proulx takes on the hog-raising units which are replacing open-air agriculture on the High Plains in Oklahoma and Texas. Immense numbers of pigs are reared indoors, producing a great deal of slurry and a very characteristic smell. The rearing of domesticated animals on a 'zero-acreage' basis has a long history (probably from the Neolithic onwards in Europe) but ubiquitous energy supplies make it possible on a large scale. Thus, although Danish bacon is a well-known export, a journey through Denmark is visually pigless since the animals are all housed. In the United States, beef cattle spend their time on concrete bases ('feedlots') and not in fields: to eat grass-finished steak is an eccentricity of the rich. Dry-lot dairies and battery chickens are analogous. So, while the animals in these systems save energy by moving little, the wider economy invests it in feed concentrates, water supplies, animal transport and vetinerary pharmaceuticals. (The meat tastes of very little.) Meat is so much in demand in the PIE world that many of the species hitherto undomesticated have been considered either for actual domestication (that is, altering their genetics) or for a form of ranching using native grazing resources. Thus, the North American bison is now used for meat production, unthinkable fifty years ago when its numbers were minuscule; red deer are farmed in Scotland as well as hunted; and in Africa there have been many attempts at 'game farming' (usually of antelope species) which provide employment and also food in a context which diminishes the pressure to extend the cereal farming that extirpates the wildlife which is usually an excellent source of income.

No account of the ecology of present-day nutrition can ignore the importance of food processing. In PIEs the emblem is the ready meal on the supermarket shelf. Its energy relationships, the packaging and its disposal, the trip by car and indeed the impossibility of paying for it if there is no electricity to power the tills are all part of a resource ecology dependent on abundant energy. Some retail chains talk of their energy-conserving buildings and low-consumption lighting but these are insignificant beside the throughput of energy-intensive materials. In the course of processing, many chemicals may be added and other constituents (for example, fibre) taken away, so that a very largely artificial substance is created, often high in salt and sugar. Governments may rail against these types of product, especially for children, in the context of rising obesity in high-income economies but rarely connect them outwards in their wider ecology.

Whole environments: forests, recreation and warfare

There are two major trends in the history of forests in the last fifty years: deforestation, especially in the lowland tropics, and afforestation in countries where land was released from an intensified agriculture (to the tune of 16 million hectares in Europe and North America since 1950) or where semi-arid land is converted to fast-growing conifers and eucalypts. The rapid deforestation is both legal, as in Brazil where it has been an element of government resettlement policy, and illegal, as in Indonesia where trees are 'mined' for a quick cash return to those with power.

The deforestation rates were initially estimates extrapolated from local surveys but, after about 1980, satellite imagery has made it possible to refine the data. The precursor context is of clearing about 11 million hectares per year in the 1920s and 1930s, of which 70 per cent was in the tropics. After 1950 the rate of clearance accelerates and is mostly in the tropics. Net of afforestation, about 318 million hectares of tropical forest disappeared in 1950–80 out of a total world loss of 336 million hectares.[25] After 1980 the rate rises again though more accurate data now suggest slightly more modest rises. In 1990–7 5.8 ± 1.4 million hectares were lost every year, with 2.3 ± 0.7 million hectares visibly degraded.[26] The regional pathways to deforestation vary considerably, depending upon the environmental history of the region, the combination of causes bringing about the land-use change, and the feedback structure which may bring about rapid degradation of ecosystems and often the impoverishment of human societies as well. In Latin America, for example, the proximate causes of deforestation concentrate upon the construction of roads followed by migrant settlers with an agricultural economy together with pasture creation for cattle ranching: motors and meat are both implicated to a high degree.[27] In South-east Asia, the equivalents are timber logging (often illegal), and conversion to plantations. In west and central Africa, timber logging by private companies and the intensification of shifting cultivation are dominant.[28] In India in the 1960s clear-felling was followed by the planting of exotics so that industry was the main beneficiary, at the expense of subsistence cultivation.[29]

It is scarcely possible to sum up a cultural context for such diverse places. Poverty, both local and national, greed by local and national loggers, the demand for wood products by countries unwilling to consume their own resources, the market for cheap beef, the building of extravagant capital cities in the forests, and the lack of recognition of the short-term irreversibilities of tropical ecologies have all contributed. The major changes in worldwide opinion have related to such factors as the role of the forests in regional and perhaps global climatic patterns of today, as well as their ability to sequester carbon from the atmosphere.[30] The value of the forests as long-term producers of income and of novel products, such as pharmaceuticals, has also become more widely recognised. The carbon sequestration argument is one of the reasons why the afforestation rates of tropical countries arouse interest. Tropical afforestation could be a useful contributor to the mitigation of global emissions of carbon, by about the same amount as slowing deforestation.[31] It is, though, more regional concern that leads afforestation in developing economies, such as those of China (30 million hectares since 1950) and even Brazil with 4.9 million hectares in 1990. But afforestation in low-income economies is still only proceeding at a rate of about one-fifth that of deforestation. It can nevertheless be part of a development strategy though some species are claimed to alter the local ecology detrimentally: the desiccation of local agriculture by eucalypts in India is one such contention. Cessation of summer cattle grazing in upland Norway has fragmented the habitats of many plant species, especially as woodlands have come back but also as soil fertility has diminished.[32] The interaction of the natural and the human directed can be seen in the phenomenon of die-back in central European forests (*Waldsterben*) in the 1960s which was quickly attributed to air pollution. But recoveries after 1980 seem not to be associated with better air quality and so other and deeper historical factors may have been implicated: the simple explanation of emissions-caused die-backs was too modest.[32] Beyond these instrumental considerations, there are still forests as paradises, as spiritual places, as mythological landscapes and, indeed, as gendered landscapes but most of these valuations can fall to the chainsaw: when Governor of California (1966–74), Ronald Regan said that when you had seen one redwood, you had seen them all.[34]

Large-scale writing on environmental change tends to the serious: today's pleasure is often overlooked even if medieval hunting parks are not.[35] After

1950, however, there was an explosion in recreation travel and activity which cannot be ignored. For convenience, two kinds of activity will be distinguished: intra-national recreation and, in particular, that part of it which is outdoors; and foreign travel.[36] When it comes to their impact upon ecosystems, the results may be similar though, in catering for a foreign tourist, many places seem willing to undertake more intensive environmental manipulation than for their own citizens. Both types may also focus on cities and produce the same kinds of effects.[37]

Outdoor recreation on a day or weekend basis (with added mid-week use by greying populations in the high-income economies) often focuses on wild areas, frequently misconceived as 'natural'. Longer trips are especially desired to 'wilderness' areas where a perception prevails that human activity has been absent or minimal. In such areas, some of the technology of the PIE is abandoned: 'Take nothing but photographs, leave nothing but your footprints' is sometimes the aspiration. (Large designated areas may be too big for the mobile phone but not out of sight of the global positioning system [GPS].) Unless regulations are tightly conceived and enforced, much technology is often brought into wild places, representing an influx of fossil fuel energy. This may be embedded in simple machines, such as the mountain bike, or in more complex arrangements such as quad bikes, trail bikes and 4x4 vehicles. For all those users, the challenge of 'overcoming' adverse conditions is paramount. The acme of such imports is in winter sports, where trees are felled for runs, pistes are graded, lifts of all kinds are essential and large amounts of electricity may be used to create artificial snow. The effects on plants, for example, suggest that plant cover and species richness declined by about 10 per cent and that machine-graded pistes had five times the amount of bare ground even after re-sowing. Artificial snow lies longer than natural snow and so snowbed plants (such as the alpine snowbell, *Soldanella alpina*) are favoured over early-flowering species like gentians. These effects became more pronounced at higher altitudes but were already obvious in areas such as Italy and Austria where about 40 per cent of ski runs use artificial snow. This requires 4,000 cubic metres of water per hectare (in France enough to supply a city like Grenoble) and 25,000 kilowatt-hours per hectare of electricity. Climatic warming prognoses suggest that in thirty to fifty years' time reliable snow will be found only about 300–600 metres higher than it is at present.[38]

Recreation in artificial but outdoor conditions was and is dominated above all by golf. No other sport occupies and manages such large rural areas. Worldwide, there are over 25,000 golf courses, with almost 10 per cent of these being located in Britain. Golf courses average between 50 and 60 hectares in size; every golf course consists of highly managed areas (the greens and tees), less intensively managed areas (the fairways) and non-playing areas (less intensively managed habitat or 'rough'). The non-playing areas generally represent between 25 and 40 per cent of the total area of the course.[39] Environmental impacts have been cited that included habitat loss, water-resource depletion, chemical contamination of soil and groundwater from pesticides and fertilisers, and increasing urbanisation. Studies in the United States showed little concern on those but, in low-income economies, there is no doubt of the land and water use and high rates of biocide application. Land appropriation led to civil conflict and death in the Philippines in the 1990s. In particular, competition for water seems a likely area of conflict for golf developments. This is especially so in the Mediterranean where a general shortage of water is the latest in a series of environmental impacts which contributed to the ecological image of the region as being a set of degraded landscapes.[40]

Foreign travel may focus on existing settlements, especially historic cities like Venice, and there create problems of water availability for seasonal or diurnal influxes of visitors and likewise those of air quality management. The annual pilgrimage to Mecca is an example of adaptation by the authorities to an immense seasonal influx; city managers are in general well placed to minimise environmental impacts of these events. The response to demands for beach and rural recreation has often been less well arrayed, with hotels and similar built structures clustered in favoured locations without much regard for their water supply, sewage disposal and road connections. A strip north of Sousse in Tunisia in the 1970s, for example, was built on sand dunes which became destabilised, which made the hotels prone to a coating of blown sand. Water was pumped away from nearby smallholdings and so a source of fresh food was undermined; sewage piped offshore made sea bathing unwise. Similar examples can be found worldwide, and their vulnerability to marine perturbations is seen in hurricane seasons (for example, in the Caribbean) and after tsunamis, as on 26 December 2004. In savanna Africa, tourism not only clashes with indigenous cultures, but the sheer density of safari vehicles is said to affect the behaviour of some animal species, notably the large predators which everybody wants to photograph.

The boom in recreation is entirely underlain by fossil-fuel use. Without cheap fuel to run motor vehicles and jet aircraft, it would exist as it did in the pre-1950 era, only for the rich. (In 1993, 1.2 billion passengers departed on international flights, that is, one for every five people on Earth.) The democratisation of recreational travel brings pleasure to many but its environmental costs are rarely added up. Thus, the cultural context is largely dominated by the precepts of economics: people have the time, the disposable income and the access to technology, and so a market arises. Like so many, it is one in which the environmental costs were either externalised or, as it were, discounted into an apparently infinite future. To which can be added the political strength of those whose enterprise has created local and regional wealth. Deeper questions can be asked: is recreation simply a positional good in which going to more remote or expensive places than other people is the major satisfaction?[41] Or, has the need to escape a modern life-style (either with greater hedonism or with the 'desert island' syndrome of total simplicity, for a while) provoked a desire for movement and escape?[42]

Some forms of recreation have always, it seems, revolved around violence. Killing wild animals was for long the preserve of an aristocracy but, to some extent, was democratised in the nineteenth century. Angling for coarse fish has more adherents than any other outdoor recreation in many countries and the pastime can induce over-fishing even in the sea.[43] Hunting for small species of deer has been a widespread seasonal activity in France, Italy and the United States, for example, resulting in high injury and kill rates, of the hunters. The pleasures in killing have only partly translated themselves into warfare though some candid memoirs confess to the pleasures of killing and destruction of an official enemy. Post-1950 wars in, for example, Indo-China included the usefulness of small diesel engines to the military and civilian economies of North Vietnam.[44] Post-1950, this pattern is amplified by the addition of chemicals, especially the employment of defoliants in Indochina by the United States during 1961–75: some 10 per cent of South Vietnam was sprayed from the air (72 million litres on to 1.7 million hectares) with the result that 1.25 million cubic metres out of 8.5 million cubic metres of timber was destroyed. The defoliant was a mixture of growth-hormone herbicides which included 2,4,5-T which was contaminated by dioxin (TCDD). This is a teratogenic agent and many deformed children were born with effects that are still present. Much of the high forest recovered, not least because shelling and bombing implanted metal in the trees and made them commercially unattractive. Indeed, the official view was that low and secondary forest suffered the worst.[45] Hence, mangroves were vulnerable to spraying, with one pass killing an average of 36 per cent of the trees. Recovery has been very slow and exposed coastlines to damage by storms or made them easy to convert to shrimp farms which have turned out to be unsustainable.[46] Residues of metals in trees are reminders of the 10 million active mines in the soils of sixty-four countries of the world, providing some sterilised wilder terrain in some and a source of danger in all, led by Egypt (23 million), Iran (16 million) and Angola (15 million).

Where oilfields were the scene of conflict, they added immensely to environmental damage. The First Gulf War, fought in Kuwait and Iraq in 1991, is the best example. The oil wells were blown so that the oil sprayed the local landscape and could be ignited, producing a great deal of heavy black smoke. The oil may also find its way into rivers and the sea, causing anxiety about fisheries, for example. The long-term worries have been about hydrocarbons and heavy metals staying in sediments but being steadily released to form a constant flow of contaminants. As far as can be seen, these worries have not been fulfilled and the effects overall were less than was feared in 1991 and shortly thereafter.[47] Internal conflict in Iraq resulted in the drainage of the marshes of the Tigris and Euphrates. Only about 7 per cent of the original area (about, 20,720 square kilometres) was left after Saddam Hussein's actions. Perhaps 30 per cent could be revived but there are many obstacles, such as the supply of water down the Euphrates which is dominated by

Turkey, and local habits recently acquired of electro-fishing entire water bodies.[48]

One worry about the Gulf struggle of 1991 was that the cloud of smoke from the oil fires would coagulate over India and prevent the onset of the monsoon. This unrealised fear was a subset of what was the most influential war-environment scenario of all time (it was developed in the 1980s), a scenario in the sense that it has not happened but could. The hypothesis was that a 5,000-megaton nuclear exchange between the Soviet Union and the United States would produce so much fine particulate matter in the upper atmosphere that photosynthesis would be inhibited by a coalescent black cloud and that this 'nuclear winter' would last for a period measured in years. The consequences of the nuclear bursts would themselves be immense and lead to total social and economic breakdown but this would be compounded by the lack of sunlight for an indeterminate period but one certainly long enough to cut out food supplies in the northern hemisphere and perhaps in much of the south as well. It was an apocalyptic hypothesis and not unexpectedly somewhat controversial (though the scientific ideas at its root were mostly quite robust); it is credited in some circles, however, with exerting great influence on the leaders of the United States and the Soviet Union in coming to greater understandings and mutual reduction of forces. The scenario is now not much discussed but the weaponry that might achieve it still exists.[49]

The nuclear-winter idea is, in some senses, an intensification of many other pathways in which war and the preparation for it affect the environment. Residual radioactivity persisted in Japan long after 1945 and it has also built up in the bodies of circumpolar people following bioaccumulation in Arctic-tundra vegetation and its transmission via animals such as caribou and reindeer. In the Pacific, weapons testing disrupted the lives of many in the Marshall Islands and, at every site in the western world (and no doubt in Russia and China as well) where weapons-grade material was processed, rogue escapes and accumulations of radioisotopes have eventually come to light. Even small conflicts now damage water supply and sewerage systems, create dust and destroy crops; and, in times of war, nobody much worries about resource consumption and the disposal of wastes: victory is all, as always. What is new after 1950 is the sheer scale of delivery of high explosive using large bomber aircraft and Cruise missiles, and the potential for nuclear confrontations in, for example, South Asia and the Middle East even if the United States and Russia are at present standing back from each other. The 'military-industrial complex' syndrome of which Dwight Eisenhower spoke in 1961 is still going strong.[50]

Special places

It has been a continuing theme in this book that wild plants and animals have occupied a special cultural role in human societies, one apart from (or as well as) their usefulness. In many cases, this was an aristocratic phenomenon but, since the 1950s, there has been an enormous growth of interest in wild species. Cynics are apt to say that the level of interest increases as the quantity decreases but there is perhaps a more complex relationship. It seems as if there are two main strands to the recent state of affairs. The first is scientific in origin: within the wild species of the world there is an enormous reservoir of genetic variation and biochemical molecules. If biotechnology is to fulfil one of its potentials, then the reservoir of genetic material will allow the insertion of many desirable qualities into all kinds of species: lions may be bred to lie down with lambs, for example. The second is basically that of pleasure: to see the non-human in its self-hood brings pleasure and satisfaction to many people. There is a third strand, namely that human life and livelihood depend upon the integrity of the non-human world, that is, that it fluctuates within certain foreseeable boundaries. The role of non-human species in this whole complex is now called 'biodiversity conservation'. No matter that the whole endeavour receives a great deal of attention: in 1996, some 3,314 species of vertebrates and 5,328 species of plants were in danger of extinction. By 2000, there were 3,507 and 5,611 respectively. About 15,000 species from all taxonomic groups were considered endangered at the turn of the twenty-first century.[51]

The way in which environmental management and impact work is highly diverse in itself. The most noticeable in many instances is the attempt to put a fence (literally or regulatory) around a 'natural' area in order to sequester it from other economic activities. These undertakings are most successful when there is a lot of money to be made from the sight of the wild species, as in the game parks of Africa, or when there is a lot of international interest, and hence pressure, to devote resources of time and money to the perpetuation of a species or habitat. The giant panda's survival in China is of this type. The devotion of a special-interest group is also likely to succeed, no matter that sometimes its primary interest is in killing the species concerned: the western European migratory geese populations are much enhanced by the influence of wildfowlers in protecting otherwise vulnerable habitats. Sheer remoteness may also make it easier to declare a 'reserve' but experience suggests that such places (parts of Alaska come to mind) can be undeclared when an important resource (especially oil) can be exploited.

The period 1950–2000 saw the production of many impressive-looking maps of protected areas in the world, with comparative percentages of designated areas but few measures of success through time in keeping, so to speak, the wolf at the door. The degree of internal ecological change, the external pressures for change and the intensity of the management effort all need representation as well as the line of the boundary fence.[52] Other parts of the cultural context display no shortage of interest. The membership levels of conservation organisations became very high, and selling outdoor clothing and binoculars, for example, is a very big industry. Many people are content with a bird-table within sight of a window, but others endure stormy seas in smallish boats to see penguins or whales. There is, of course, a great deal of lip-service since, if faced with the choice between the loss of central heating and the presence of blue tits, most westerners would take little time to make the decision. The interest veers towards the instrumental in the sense that a strand of 'you never know when they might come in useful' is woven into most western attitudes, and some biologists have evoked the metaphor of the canary in the mine: wildlife as an indicator of the 'health' of the planet. Basically, it seems as if the more carbon that passes through human economies, the fewer wild plants, animals and habitats there are. This is brought home again by attitudes to waste materials. There was pressure either to export them or to keep them in a cycle so as not to pollute. But both have relied on energy use so that the emission of gaseous carbon has increased; all production used more energy when 'planned obsolesence' became common in western economies.[53]

One more historical layer was started in the post-1950 period. 'Restoration ecology' aims at recreating habitats from a former time, mostly from an agricultural and pre-fossil-fuels era. Woodlands are popular sites because their composition in previous periods can be affirmed by pollen analysis as well as by documentary sources; streams that had been straightened and embanked are allowed to meander and flood their plains. The term also embraces the reintroduction of regionally extinct animals, with predatory birds a favourite group. In Scotland, the late twentieth century saw the beginnings of a debate over the reintroduction of the beaver (*Castor fiber*). It would need a broad riparian habitat with aspen, willow and perhaps birch, and this would create habitat for waterfowl and otter as well.[54] But other interests, such as timber growing, aquaculture and fish migration, might be adversely affected which was not the case with the reintroduction of the red kite (*Milvus milvus*) into England and Scotland, where it had been extirpated in the eighteenth century.

We might think, though, that 'nature' as a whole has value outside those areas which have been designated as 'natural' or something similar. One calculation places a dollar value on nature's 'services' (processing wastes, supplying energy, regulating climate and many more) and notes that it exceeds the human economy. The data worked out suggest that the world's gross national product is about US$18 trillion ($10^{12}$) per year but that the services of ecological systems and the natural capital stocks are in the range of US$16–54 trillion per year with an average of US$33 trillion per year, so the value of the natural world is at least 1.8 times that of the human-directed economy. Nearly all the natural services are outside the market yet are critical to the functioning of the Earth's life-support system.[55] Some interpret that as meaning its value must be preserved, others that it is acceptable to trade some of it for greater immediate comfort. What is certain is that the hunter-gatherer's 'felt flow' of nature ('moose are good to eat and good to think') scarcely exists in the PIE era.

Resources and structures: minerals, cities and water

Few human activities come with as much attitudinal baggage as mineral extraction. As old as tool-making itself, it was a small-scale activity until the eighteenth century except for a few coal-producing regions. But its explosion on a worldwide basis that has altered the face of the Earth permanently in many places, and the often obvious concentration of production in poorer countries, have left a legacy of distrust and abhorrence. Although many metals can be recycled, industrial minerals, such as cement and salt cannot, and energy minerals such, as coal and oil, clearly have a one-way passage through the economic system. Though small-scale mining persisted (there were still gold rushes in Amazonia, for instance), the large corporations mining on a large scale dominated environmental considerations. In a typical PIE year, over 1 million tonnes per year of nickel, 12 million tonnes of copper, 992 million tonnes of iron ore, 1,560 million tonnes of cement, and 4,655 million tonnes of coal were produced. Their environmental impacts constitute a long list, especially where open-cast mining is the adopted technology. The previous land cover is eradicated completely at the mining site and at the locale of any waste disposal, and emissions will extend this zone. The land may also be changed in its actual contours and become unstable, and any inadequate groundwork or dams possess the potential for land-form change as well as the destruction of human-made structures. The lifetime of a mine is always limited and not all companies engaged in land rehabilitation after closure, with abandoned pits and plant an obvious reminder of the effects of the mine. Those effects have been carried on for many years, especially in the alterations of the local water table and in the quantity and quality of water flowing from the mining area. Mine drainage is usually acid and its quantity raised by being pumped from underground workings. Tailings yield contaminants to soil and water through leaching and may also contribute dust to an atmosphere already loaded with emissions from mineral processing; mines may also release methane to the local atmosphere. Mining attracts workers and so the building of settlements and communications also brings about change, especially where the quality of housing, water supply and sewage disposal is poor.[56]

The attitudes taken to large-scale mining were, in very general terms, a cascade of concern from citizen groups and non-governmental organisations through governments to the large corporations. Only when forced to take social and environmental audits and to internalise some of those costs have companies take action, during most of the post-1950 years. None of them, however, challenges the predominant paradigm that extraction of minerals on a large scale (even in the face of recycling possibilities in the case of metals) is a source of income to some of the poor as well as of material goods to the richer inhabitants of the world. There are concerns that the lifetime of some mineral reserves is distinctly limited but counter-arguments based on substitution and technological progress are also deployed. In general, there is still frontier feeling about many mineral operations even if the very large companies (Anglo-American, Rio Tinto, and Broken Hill, for example) have espoused more responsible policies in our period of interest.

One of the main destinations of minerals is the world's cities. The growth of these agglomerations has been an outstanding phenomenon of the last 100 years and has accelerated in the last fifty. The use of steel framing for buildings and of concrete walls and other urban surfaces took up large quantities of iron ore and cement, and the growth of computers and their wiring has increased the demand for copper. The electronics industry has added new materials and forms to the waste streams of cities: electronic wastes comprise from 2 to 5 per cent of the municipal solid waste stream. Rapid technological advances and lower product prices for more powerful machines contributed to shorter product life-spans and frequent replacement. Yet, consumers are likely to store their old electronics: three-quarters of all of the computers sold in the United States remain stockpiled in a garage, cupboard, or storage space. In 2002, the European Union issued directives which forbid certain chemical components of electrical and electronic machines along with a waste-disposal regime which encourages manufacturers to recycle all the component materials.[57]

Yet in many parts of the world, the metabolic problems of cities were still those prevalent in the industrial world in the nineteenth and early twentieth centuries except they were made worse by the rapidity of growth and by the higher material expectations of many migrants to them. Cities are home to more people than ever before. In 1900, only 160 million people, one-tenth of the world's population, were city dwellers. But in 2005 it was announced that half the world's people lived in urban areas, which was a twenty-fold increase in numbers.[58] The metabolism of cities is an emblem of their environmental relationships. In the Baltic cities, for example, 26 per cent of the population of the region (22 million people) need an area some 200 times the area of the cities for wood, papers, fibre and food. In terms of land use, this requires eighteen times the area of forests, fifty times for agricultural land and 133 times for marine ecosystems. The calculation for the assimilation of wastes provides for a wide range of values but for carbon dioxide these amount to 390 to 975 times the area. Hence the total terrestrial and marine appropriation is 565 to 1,130 times the total area of the cities or about 60,000 to 115,000 square metres for the average citizen, amounting to 75–150 per cent of the whole Baltic basin. If this approach is uprated to the 744 cities in the world which, in 1996, had over 250,000 inhabitants, then they produced, for example, 2,099 million tonnes of carbon dioxide which would need 46.45 million square kilometres to sequester the carbon, with about 41.46 million square kilometres available. All these processes are 'hidden' with no price in the economy and are seldom perceived by policy, real though they are.[59] In total, urban areas take up just 2 per cent of the world's surface but consume the bulk of changes in six areas (water, waste, food, energy, transport, and land use); they are therefore major contributors to environmental processes. Not least, they impel the use of the packaging of food which in the United Kingdom, contributed about 11 per cent of the energy use in the food-supply system.[60] The example of Hong Kong provides further local insight into world-scale data. In studies carried out from a 1971 baseline, jumps in affluence are shown by, for example, an increase of 400 per cent in the consumption of plastics and 530 per cent more putrescible materials. One million styrofoam lunchboxes per day add up to 120 tonnes per day of landfill. Each person (in 1997) generated 1.3 kilograms per day of solid waste, which was an 80 per cent increase since 1971. All this challenged the normal Hong Kong method of dealing with wastes by dilution, and carries lessons for the future urban development of the rest of China.[61] In more general terms, landfill has created worldwide another type of environment, this time underground, though probably interactive with the atmosphere because methane is often produced.

Different actions are needed to make cities and the vast areas they affect better for people and for the planet. Cities can align their consumption with realistic needs, produce more of their own food and energy, and put much more of their waste to use, in programmes which are usually labelled 'sustainability'. In such a setting, one of the fascinating things about cities is the generally negative cultural image they have had for so long. They have been the fount of problems in contrast to the 'purity' of rural areas in the eyes of many commentators and professionals. Apart from the pleasure they can give to those who like to live in them, there seems little doubt that they are often the generators of new thinking and practices, in the environment sphere no less than others. Cities can but express the dominant culture, however, and, though a few urban centres have expressed 'green' ambitions, by the year 2000 no truly 'environmental' city had emerged.[62] Some have little choice, having undergone rapid growth in areas of high environmental risk, exacerbated by, for example, water extraction leading to land subsidence. Bangkok sits on 4,550 square kilometres of land that has to be pumped during the monsoon when 150 centimetres of rain falls. Yet, locally, the land is subsiding by 10 centimetres a year, with some zones totalling 20 to 160 centimetres below the 'natural' surfaces of 0.5 to 2.0 metres above sea-level. In Kingston (Jamaica), peri-urban deforestation adds to the toll of earthquakes, landslides and liquefaction of soils in maintaining the hazards underlain by high-intensity rainfall and 'natural' slope failure.[63]

One outstanding problem of poorer cities is water supply, where the needs of the individual and of the urban metabolism cannot be met with high-quality supplies. In some ways this is an early indication of what is increasingly seen as a problem of shortage in many parts of the world, both in cities and in rural areas.[64] The amount of technology brought to bear on water supply has been very great, though it has not changed much since the nineteenth century: the dam, the pump and the pipe are still the basic ingredients. (The number of large dams climbed from 5,000 in 1950 to more than 45,000 in recent years: an average construction rate of two large dams a day for fifty years.) Re-use and greater water efficiency generally are minor considerations and usually prompted by local prices rather than by broader environmental considerations, these in turn usually being externalised costs.[65] In fact, cities are responsible for only about 10 per cent of worldwide water withdrawals but the concentrated demand means that complex and capital-intensive infrastructures are needed, and these are often the cause of unplanned environmental impact: downstream from a large dam almost every aspect of a river's biology and chemistry is changed. There is exacerbation from water losses in the urban system, sometimes as high as 40 per cent even in the Organization for Economic Co-operation and Development (OECD) countries. One major problem with water is the accessibility of the resource which is largely confined to fresh water flowing through the solar-powered hydrological cycle.[66] One estimate suggests that humans have now appropriated 26 per cent of the land-based evapo-transpiration and 54 per cent of the equivalent land-based liquid resource: about 30 per cent, therefore, of the resource is in human hands. The 54 per cent figure is withdrawals of all kinds and about 18 per cent is 'consumption', that is, diverted to vapour. The spatial distribution and effects of water use are highly uneven: agriculture worldwide takes about 70 per cent, with the figure being 90 per cent in some low-income economies.[67] Here lie some of the widespread effects of, for example, the salinisation of soils that affects some 20 per cent of the world's irrigated lands and which is increasing at about 2 million hectares a year. One unseen effect is the drawdown of groundwater and aquifers, with the worst areas affected being the northern plain of China, the Punjab, parts of south-east Asia, North Africa and the Middle East, and the western United States. Many rivers now release no water to the sea in the dry season: the Huang He (Yellow) river of China, the Colorado river of the south-west United States and Thailand's Chao Phraya join the rivers of the (former) Aral Sea in this respect. If, indeed, the food demands of some nations are to be met by a grain trade then, with one tonne of grain needing 1,000 tonnes of water, the rain-fed grain-growing zones of the globe assume a greater strategic importance.[68] The consumption of a meat-rich diet may require twice the water input of a vegetarian diet: rice production needs twenty times less water for the same calorific value as that of beef. Industry consumes about 22 per cent of water withdrawals (59 per cent in high-income and 10 per cent in low-income economies) but can be subjected to efficiency gains. Unilever, for example, reduced its water use to 4.3 cubic metres per tonne of product from 6.5 cubic metres between 1998 and 2004. Yet, while western industries in the 1980s used 50 tonnes of water to make a tonne of finished paper, China used 100 to 300 tonnes.[69]

Work in the 1950–2000 years has removed any doubt that human activity can change other facets of the hydrological cycle: land use and cover change can alter local precipitation. The heterogeneity of land cover (that is, its fragmentation) increases turbulence in the morning, leading to mesoscale circulation and the incidence of precipitation in the afternoons.[70] But nobody knows the overall effects of the loss of 110,000 square kilometres of forest every year or of the 700 per cent increase since 1950 in the stock of impounded water, with a residence time now averaging forty-five days. Climate change will no doubt affect water availability but the pace of the last fifty years, if maintained, will ensure that economic development and population growth are more significant influences.[71] One finding of all these data is the knowledge that just as six countries contain about half the water resource, several are in a condition of acute or potential shortage: Egypt, Yemen, the United Arab Emirates and Saudi Arabia all use 100 per cent of their water income.[72] No better example from the Middle East can be seen than the Yarquon-Taminin aquifer. Eighty per cent of this is tapped by Israel from its West Bank acquisitions but would come under Palestinian control if an independent state is created. More widely, taxonomies of political action to secure water can be devised and all of them end in struggle; some opinions have it that inter-state conflict over water resources is more likely than over any other resource. What is likely as well is that water becomes not a locally or regionally distributed resource but enters worldwide trade. Towing icebergs is perhaps fanciful but a Norwegian company has a contract to deliver seven million cubic metres per year in bags to Cyprus. Similar towing of water bags delivers water to some Greek islands. In 1960s North America, vast, continental-scale canal–tunnel–pipeline projects which basically take northern Canadian water and deliver it as far south as Mexico, came up from time to time. Most were resisted on grounds of cost, national politics or environmental impact, but they are not totally outside technological possibility.

Water is so basic to life that it is an integral part of daily patterns, of everyday language and imagination. It carries cultural codes about community, health and wealth.[73] Thus, the way in which it is moving from control by local communities into the portfolios of multinational companies (effectively a form of dematerialisation)[74] is at odds with the the cultural codings and, at the same time, part of the separation of most people from a part of nature that sustains them. This is especially so in remote places such as the James Bay area of Quebec where an area the size of France was to be flooded for hydroelectric power. This meant the displacement of 15,000 Cree and Innuit native people, judged to be of much less importance than cheap electricity for southern Canada.[75] Nevertheless, the protracted land battle of the 1980s ensured that the autonomy of the Cree and Inuit communities was enhanced by the strengthening of the hunting economy and society, and by their legal standing in the initiation of political and administrative action.[76]

Marine complexities: no blue-green peace

The seas and open oceans are a route for trade and a source of a few minerals but the period since 1950 was dominated by two processes: what organic materials are taken out and what largely inorganic materials are put in: fisheries and pollution. The areas most likely to have experienced heavy impact were those near the shorelines, and especially the estuaries and reefs which are the most biologically productive zones of the marine world. The emblematic event for marine contamination is that of the emissions of mercury into Minamata Bay in Japan. The poisoning of humans was first detected in 1956 but the effects were still present at the turn of the century due in part to the refusal of both the Chisso Chemical company and the Japanese government to acknowledge what had happened.

Fishing (in which organisms such as squid can be included but not whales) was still both artisanal and industrial but both groups were the beneficiaries of World War II technology in the form of radar, sonar, and nylon for nets. Thus, the impact of a given amount of fishing effort has been increased by many times. Most artisanal fisheries produce human food whereas about one-third of the industrial fishery catch is converted into fishmeal to be fed to farmed fish or to cattle. The basic data for salt-water fishing in this era start with a world catch of 21 million tonnes in 1950 rising to 116 million tonnes in 1996 and then falling to 94.8 million tonnes in 2000. It is said that under-reporting by China was significant and that, if the by-catch and the unre-ported catches are added in, then the 2000 figure may be more like 130 million tonnes. In the last fifty-odd years the industry has fished down the food web to lower trophic levels and has also simplified food webs by extracting so many fish that variability of populations is high and predictability of presence and catch is low. Thus, human-induced impacts can be confused with environ-mental changes. The story of the Peruvian achoveta exemplifies this: it was scarcely fished at all in the 1950s, yielded 13 million tonnes in 1970 but only 2 million tonnes in 1974 and 0.8 million tonnes in 1984 but, by 1995, was back to two-thirds of its 1950s level. This is usually presented as a sequence caused by El Niño but high-impact fishing may also be implicated.[77] The number of fish stocks that are no longer commercially viable has increased greatly, espe-cially where carnivorous fishes are the targets, as with cod, whose presence in the Atlantic and North Sea has plummeted. Off Greenland, falling sea tem-peratures after 1960 and in 1982–4 meant fewer cod at a time also of intensive fishing, so that there was a total collapse and the industry shifted to shrimp and halibut. The available technology has ensured that all of these are capable of being over-fished quite easily.[78]

In the face of these data, the great growth has been in aquaculture. In 1950, aquaculture comprised 1 per cent of fish supply but, by the end of the century, it topped 25 per cent. It was supposed to relieve pressure on open-water fish-eries but, if (as in the case of the popular salmon and shrimp farms) the animals are fed fishmeal and fish oil, then it typically took 5 kilograms of wild fish to produce 1 kilogram of farmed fish. Only if herbivorous fish were farmed is the equation more favourable. The world leader in aquaculture is China, with about 60 per cent of world output. The western world is much given to salmon farming, where the top producers were Norway, Britain, Canada, the United States and Chile. The marine-support area for salmon is

400–500,000 times the area of the farm, and the Nordic industry contributes as much nitrogen as the untreated sewage of 4 million people as well as antibiotics and pesticides. In the United States, aquaculture is a net consumer of energy rather than a producer. Shrimp farming also relies on wild fish for food and is heavily implicated in the loss of half the world's mangroves, opening coasts to greater vulnerability to storms as well as reducing the breeding grounds for finfish and for the shrimp species themselves.[79] Retaining some mangrove in an area of coastal shrimp farming seems to improve the productivity of the target species.

The long-term lesson of open-sea fishing is that there has never been a conservative set of practices. As one fish species is consumed, then there is a move onwards to another. Technology, increased outreach in distance and depth, the revaluing of spurned species and the masking effect created by aquaculture have all brought about the depletion of a valuable resource. Only now are there a few examples of carefully managed stocks in a world which seems reluctant to follow their example. In some ways, fisheries are the last great hunting economies. But hunting cultures never seemed to run down their food sources in this way, for obvious reasons. The great sticking point seems to be the livelihoods of the fishermen who claim that they cannot do anything else, as do, interestingly, many farmers. They manage to convince governments and other bodies that the scientific advice is either wrong or need not be implemented yet, which is an interesting comment on the normally hegemonic role of science. That science was highly influential in whale management, where for many years it hammered home the decline in whale numbers, and the International Whaling Commission set in place a moratorium on whaling. Just after 2000, this was broken by Norway and Japan with their 'need' to kill whales 'for scientific purposes' along with, as the Japanese put it, their 'right' to eat whale. Though culturally interesting, this is just nationalist nonsense. In fact, whale-watching is a fast-growing activity in Japan, one which has grown much faster than the average world rate throughout the 1990s. Between 1994 and 1998, it grew by 16.8 per cent per year; from 1991 to 1998, the average increase was 37.6 per cent per year. In 1998, some 102,785 people went whale- and dolphin-watching in Japan. The most commonly watched cetaceans are humpback, Bryde's, minke, and sperm whales, as well as bottlenose and other dolphins. Three of these, minke, Bryde's and sperm whales are currently being targeted by the Japanese whaling industry. Norway has experienced growth in whale-watching at 18.8 per cent a year since 1994 at places like Andenes in northern Norway, which features sperm and other whales, and at the Tysfjord area where, in autumn, orcas come in close to feed on herring and are watched from the land. In some other communities there has been a conflict between old and new attitudes to whales.[80]

The natural sciences also contributed the bulk of our knowledge about the contamination of the oceans. The inputs from the land constitute a long list in which the dominant elements are trace metals, synthetic organic compounds [for example, PCBs (polychlorinated biphenyls) and pesticides], hydrocarbons (adding to some natural leakage from ocean floors), carbon dioxide, radioactivity, litter, phosphorus, nitrogen and silicon. To highlight just one flow: some 6 million tonnes of litter were deposited by commercial ships every year. There is an overall difference in impact between the open oceans, where dilution is still effective, and the regional seas where there enclosure makes concentration significant. In the open oceans, however, synthetic organics and radioactivity are detectable because they decay only very slowly. In the regional seas there are abiotic zones caused by eutrophication (as in the Gulf of Mexico and the Baltic), and poisons can build up in food chains: mercury is a recurring example. But control is more possible here and restoration is more feasible, albeit with time lags of a decade or more. Oil spills are another offshore and coastal hazard though, unless they happen in very cold waters, the long-term effect is less than often feared.[81] Mineral extraction from the seas concentrates on seabed minerals together with salt and magnesium from seawater, with consequent effects on turbidity, the resuspension of pollutants and the burial of habitat.[82]

The human impact on the planet's defining characteristic is therefore far from negligible. It is mainly on the continental shelves and enclosed regional seas, but measurable in the open oceans as well. Sediment, other particulate matters (especially plastics) and chemicals of a persistent nature are up in concentrations, whereas the quantity and diversity of fish, whales, seagrass beds and coral reefs are down, with the last being always a topic of dispute over causes. The volume of science is an apt tribute to the volume of water; the quantity of care that is needed to keep the oceans as a predictable and stable element of our environment seems lacking. This is put into an ambiguous perspective by the work of Hoffmann (cited in chapter 3) and by other recent work which argues that overfishing (interpreted as over-use of any marine taxon) is a long-standing historical process which has led especially to the collapse of coastal ecosystems. It has happened in the past to kelp forests, cod, abalone, sea otters, coral reefs, turtles, seagrass beds, oyster reefs and has produced anoxia and eutrophication in many localities worldwide. The concern for the last fifty years is increasingly that such processes are no longer confined to the shallow offshore zones but can affect the deeper oceans as well. Inevitably, shifts in ocean temperatures connected to global warming form part of the complexities involved in understanding marine populations and their reactions to heavy levels of extraction.[83]

INSIDER KNOWLEDGE

The period 1950 to 2000 was characterised overall by an intensified search for, and use of, resources. Areas hitherto 'unexplored' came into the fold because of better technology. Oil from Arctic ocean floors, leaner ores for minerals, clearing of old-growth tropical forests and evaluation of mineral recovery from the beds of deep oceans are examples of a reach hitherto undreamed of. More energy was available to contact and to change the world, not least in the form of rockets putting observation satellites into orbit. All these processes have impacts. Some can be ameliorated by careful management, some are just left in the hope that they will cure themselves, and some are prone to cause instabilities on a variety of scales. If the ocean beds are mined then nobody expects the kind of attention to restoration that characterises the British open-cast coal industry, for instance. When rainforests are converted to grow soya for animal feed but the soils melt into a solid mass, then poor-quality grass-land is left for settlers with no choices. Yet Arctic pipelines are scrutinised daily from satellites so that any leaks are quickly known. Add in the technological ability not only to find, but also to characterise and tag: note that each apple at a supermarket has its little label for variety and source. How long before each apple has its own number, as do cattle in BSE-free (more or less) coun-tries? These actions have a context which is different from any previous era and which can be analysed into a number of components, though this eludes their essential character of all taking place at once.

Increased population, higher consumption

Though 80 per cent of the world's population is in low-income economies, there are enough of the high- and middle-income groups to build a demand for all kinds of products, nearly all of which demand energy inputs from extra-solar sources. The emergence since the 1990s of China as a source of demand will, given its population (1,294 million in 2002) alter many environmental interactions, not least in the supply of energy and the resulting emissions. Yet there are unknown instabilities in human populations: the effect of HIV/AIDS is one such, depending most likely on the time-lag before a vaccine becomes available. Another is the possibility of worldwide pandemics of infectious disease akin to the 'Spanish flu' of 1919. The growth of population has meant that an average per caput allocation of arable land of 0.44 hectares in 1960 had fallen to 0.27 hectares by 1990. One consequence is that nine countries with little potential to expand crop land account for over 50 per cent of world pop-ulation growth.[84] Intensification to an industrial style of food production is their only course. Though much of the world's population is 'poor' in the sense of low incomes, they have aspirations which demand high levels of resource use: the arguments over the Kyoto Protocol contain the major theme of the future contribution of higher per caput energy consumption by low-income economies. Examination of the ways in which, for example, Thailand has lifted itself out of general rural poverty reveals that road construction and the access to motor transport are key elements. In addition, there seems no upper asymptote to the capacity of the rich to suck up resources, whether in the shape of immense private vehicles, long-distance air travel for people and luxury foods, or ever-bigger portions of meat. The possession of goods and access to services seem to fit the economists' class of 'positional goods' which are demanded so as to be richer (and be seen to be richer) than other people; the price in the end is environmental change. The influences on 'western' consumption patterns are so diverse (government, education, media and entertainment industry do not exhaust the list) that a collective 'leave it to the market' has been taken to be the only practical response to the developments of the post-industrial era.[85]

Such change may also have had an explicit ideological content. In the Czech Republic, for example, there were specific Party measures after 1948 which resulted in the loss of agricultural land to, for example, the open-cast coal-mines of northern Bohemia and to other state industrial enterprises. After 1990, private property was restituted, there was partial privatisation of state holdings and access to markets was restored, leading to a decrease in arable land, an increase in pastures and meadows and more urban expansion.[86] Overall, it seems as though the impact of 'natural' conditions, such as soils and climate, is more important under the free capitalism of recent times than under state socialism, a feature seen in the 'conquest of nature' campaigns of the Soviet Union and China at various times in the twentieth century.

Technology and 'progress'

Nobody can sit writing at a computer and complain about technology. The benefits it has brought to some humans, and is capable of delivering to most, are immense. More broadly, a number of general trends in technologies seemed to be emerging in the late twentieth century:

1   The rate of change was more rapid. One invention is rapidly rendered obsolescent by the next in the a form of 'future shock'. Although this phenomenon can be exaggerated because it is confined mostly to the high-income economies, some of the effects spread out to other regions especially in the form of demands created by the ubiquity of media such as television controlled from a high-income economy; the spread of hamburger bars and ubiquitous soft drinks is accompanied by the consumption of resources and subsequent emissions to air and water. This accelerating rate of change was underlain by the cultural attitude that science will find 'magic bullets' to overcome any problems.

2   Change was often predicated on the availability of cheap electricity. Much of this is generated either by the inefficient combustion of fossil fuels (in the sense that the conversion efficiencies are usually low) or from large dams with their consequent problems, and then there are transmission losses. Though there are still huge reserves of coal and natural gas, it may be that the years around AD 2000 saw the peak of oil production (Fig. 6.1). Hitherto, energy content in goods and services was scarcely worth mentioning but increasingly it became noticed, though not yet fulfilling the prophecy of H. G. Wells's *The World Set Free* of 1914 in which the value of coins was set in terms of a number of units of energy.[87]

3   Technology became miniaturised. Many devices are now so small that they can be owned and operated by one person: the tape-playing Walkman, followed by the even smaller (but with a higher capacity) MP3 player is a an obvious sequence. The laboratory and the manufacturing plant can go well beyond these toys and into the heart of things: the structure of the atom, stem-cell research, nanotechnology, cell-target pharmaceuticals, all rely on an ability to manipulate the smallest components of the living world.

4   As many environmental issues became better understood in the period 1950–2000, they were highlighted more and more by non-local bodies. One set of agencies were the international non-governmental organisations with an environmental focus, such as Greenpeace and Friends of the Earth, along with groups devoted to wildlife generally, or mammals, or mountains. The other comprises the world-level official bodies such as the United Nations Organization, with its Environment Programme as the spearhead but with many others (FAO, WHO, WMO, for example) carrying a worldwide or, indeed, global concern. All of these produce action plans and many publications but usually rely on others (typically the agencies of the nation state) to carry out the objectives, with varying degrees of commitment. The same is true of multination blocs, such as the European Union. There are also international bodies, such as the International Atomic Energy Agency (IAEA), whose attempts since 1957 at regulation of atomic energy (with linkages to military developments) have secondary environmental effects at the very least.

There are always consequences for nature. More and more human activities impinge in some fashion, with technology increasing the penetration for knowledge as with satellite monitoring on the one hand and the voyeuristic intrusion into the life of wild creatures presented on television, made possible by mini-cameras and special lenses. There is also more opportunity for action, as with contemplating recovering minerals from the beds of the deep oceans. If the moon had a use other than as a dump for toxic wastes (politically unacceptable at present) then a multinational company would be exploiting it by now. Increasingly, there is competition for resources which may lead to conflict between different groups of people (this is not, of course, new) and in which water looks as if it may be the commonest cause, followed by environmental carelessness in extracting a resource at lowest cost, as with Nigerian oil. Getting access to oil reserves in which their size is diminishing at the world scale may well be a political priority for those nations with the greatest per caput demands. Such scrambles are unlikely to be accompanied by an enhanced environmental concern and more likely by inter-state conflict.

It is easy to purvey a sense of doom about the 1950–2000 period. There is no doubt that technology was a boon to many and could well have been so to many more. It could also have been used to protect the environment in the sense of reducing the chances of unpredictable and damaging fluctuations, and in keeping a diversity of life which is a source of delight and wonder as well as of genetic variation. Yet there is a widespread feeling that it became self-fuelling and that no amount of social control will now keep any check on what is feasible, and that what is realisable will, in fact, be used. That is an assent to technological determinism and a big step in the whole history of human–nature relationships.

SUPERPOWER: COALESCENCE AFTER 1950

A key concept of the 1990s was 'globalisation'. In its material effects, this meant the human ability to transport and to communicate worldwide along with the added capacity to 'use' the atmosphere and space. Thus, we are in the realm of the truly global, beyond that delineated so far as worldwide. The development of satellites, rocket propulsion and instant electronic communication by the post-industrial economies has produced a dominant technologically based culture in which other places share to varying degrees, with few wanting to opt out. North Korea might be seen as one of the latter but its inability to feed its population means an eventual shift towards the dominant modes of society and production. A major cultural feature is commercial penetration on the back of cheap transport so that the same brand names are seen worldwide with only minor regional variations: Toyota and CNN are the obvious examples, though the incidence of the pizza seems to be even higher. The multinational company (MNC) is a bedrock of manufacturing and now most outsource manufacturing so that it takes place in cheap-labour countries; the same is true of food production when baby sweetcorn is flown thousands of kilometres to western supermarkets, with minimal returns to the producers. The criss-crossing of the planet entrains all kinds of organisms and materials, not least of which is the spread of disease. Apart from the obvious examples, Britain imports 5 million vehicle tyre carcasses each year for recycling and these may form a breeding ground for the Asian mosquito *Stegomyia albopicta* which has been associated with twenty-three diseases including dengue fever and West Nile virus, both of which might now get a hold in Britain.

All this commerce requires energy and the superpower is the biggest consumer of all though the energy needs of countries emerging into industrialisation (notably India and China) are immense. But the desire to maintain levels of access to power in all senses of that word leads to far-reaching political action: the United States's impulses to maintain stability in the Middle East after World War II at the expense of democracy can be interpreted as providing secure oil supplies.[88] As more oil is secured, it seems as if more has to be devoted to military purposes to secure those supplies and the next tranche as well. There has been some rejection of this pattern of a unipolar world, and some dissenters mark their protest at G8 meetings, whose cost and energy use would support a small African country for a year. Political power goes along with high per caput energy use, at levels which bring in the whole globe because the levels of emissions are so high. Superpower means super impact on the environment. There is more dust, for example, so that geomorphologists can talk of a 'dimming world'. This term is derived from measurements of aerosol particulates (mostly sulphates, soot and organic matter) which both reflect away solar radiation and absorb it. There was a solar dimming of 1.3 per cent per decade over land in 1961–90 but this was outpaced by an increase of solar radiation after 1983; so that any cooling effect was outpaced by the emission of greenhouse gases. Sulphur emissions declined by about 22 per cent after 1990: such trends ameliorate local pollution problems but take away a counter-influence to global warming.[89]

Dimming seems less important than the great encroachment upon the composition of the atmosphere, where various gases have enhanced the capacity of the atmosphere to retain heat and thus produce what is known as the 'greenhouse effect'. The late-twentieth-century emphasis on methane and carbon dioxide was preceded by concern about chlorofluorocarbons (CFCs) and their production of 'ozone holes' over the poles. The Montreal Protocol of 1987 (amended in 1990 and 1992) was supposed to alleviate that process though the ozone layer still shows an alarming thinning in some years. The nature of a warmer globe is assessed in two ways. Firstly, any current trends which seem to fit the idea are assigned to that cause. Thus the rise in global temperature at the end of the twentieth century is seen as one result, as are species shifts which see the retreat of the cold-tolerant and the advance of the warmth-seekers. Second, complex models predict likely effects and their regional variations, including the complex feedback effects of such elements as cloudiness or the breakdown of tundra peats giving off methane. This procedure has been the subject of immense amounts of science and is constantly reviewed and presented with appropriate caveats. Thus, it represents the best that science can contribute to the question of the global environment. The ways in which a wriggling government in the grip of multinationals can then avoid action ('we need to understand the problem better') would invoke farcical laughter if the subject were (a) not so serious and (b) not an example of a more widespread phenomenon. Fishing and whaling in the PIE era suffered similar dissonances.

Although there is now one globe as never before, it is a globe that is produced by superpower in the sense of access to energy applied through technology and driven by the levels of consumption and the values of the United States. That nation has provided the beef and the vehicles in the period 1950–2000. The origins of this ascendancy lie before 1950, most clearly in the entry of the United States into two world wars and, in particular, the breakdown of isolationism after 1941 and the perceived need to contain communism during the Cold War. The impact has been out of all proportion to the 5 per cent of the world's population who live in the United States, and can be characterised as the outreach of massive energy consumption mediated via technology. There is a basis of mass production and mass communication conducted by companies who make possible the other developments. These centre around (a) the possibility of mass destruction, since there is a capacity to deliver immense amounts of weaponry anywhere in the world, rather dwarfing the concept of the superpower when the word was first used in 1944; (b) mass consumption, pioneered by the United States as in the late 1950s when it had two-thirds of the world's television sets, and (c) the promotion of its culture to be consumed elsewhere, with the primacy of the English language as a carrier.[90]

The environmental consequences again focus on the global reach, as discussed above, but we can note the United States as being in the vanguard of levels of material use and acquisition, including travel and therefore in per caput use of energy and hence of impact in the course of resource use, processing and the emission of wastes. The United States has also exported attitudes towards the environment: the idea of wilderness as an untouched nature apart from human use except for very light-touch recreation has been implanted in a number of other countries, though revisionist views about the manipulation of US wildernesses by native Americans are undermining the concept in its heartland. The type of mass destruction feared by the authors of the 'nuclear winter' scenarios never came about but environmental destruction is forever part of the history of the war in Vietnam. But, of course, if several other nations could have achieved those levels of power then they would have done so.

There have been at least three waves of integration of the world, starting with migration and trade in the period before about 1500 and then being catapulted to a new level by access to fossil fuels and the accompanying technologies.[91] All are outclassed by the post-World War II integrations led by the United States and which in the PIE era qualify unreservedly for the first time for the label 'global'.

NO POWER HERE

This phrase comes from a modern (1965) Japanese *haiku* by Seishi Yamaguchi[92]

Umi no ue	Flight over water –
tobu setsurei no	the gods of snowy mountains
kago mo naku	have no power here

It conveys poetically the idea of separation of function and of powers which the section on coalescence has tried to refute. Yet the PIE world has been one of many fragmentations and separations. Two linked processes have dominated the environmental history and its potentials at the end of the twentieth century: miniaturisation and advanced biotechnology. The first is the latest stage in a long process of stratification and segmentation that satisfies the wishes of individuals. When the technology is personal then the immediate demands of the individual can be addressed. 'My music' on an MP3 player replaces the shared experience of a concert and avoids anything that might be unexpected or disturbing. The mobile phone replaces the community's box; medication is formulated to a particular physiology and 'designer babies' will make it further into the world than the tabloid press. In such a setting, the intellectual rise of postmodernism with its emphasis on avoiding the ubiquitous and the absolute in favour of the local and the relative is not surprising. But there was scope for debate over the location of power: if it was indeed becoming radically decentred than the influence of the multinationals seemed to constitute a new class of vigorous controllers of lifestyle. If we all subscribe to 'my news', then what information do we have in common?

Being able to manipulate the basic material of the living cell has opened new vistas of tailoring. The main thrusts have been in matters of human health, such as replacement parts grown from embryonic stem cells, and the so-called GM (genetically modified) crops in which resistance to disease or to a herbicide is implanted in the crop's genetic material. Neither technique was, in 2000, well advanced and it is clear that there is much more to be developed. In both cases, technological advances were developed in their own subcultural worlds though there are more binding medical ethics committees than planetary health committees with compulsory powers. The GM advocates concentrate on higher yields without worrying about the social implications of the cost and supply structure that go with them. Beyond both, there is the fear that uncontrollable harmful organisms will be released and that insufficient international protocols will ever be in place to ensure that the 'Frankenstein Effect' is simply a story.[93] In the immediate future, the development of a new species by genetic modification means that it can be patented, which is a fresh development in the history of humans' relationship with the non-human.

The environmental effects of these technologies were only just above the horizon in 2000 and so seem speculative in this context. Communicability of an individual kind might mean less need for travel (and so less consumption of fossil fuels) though the habit of using travel as a symbol of status is probably deeply entrenched. Biotechnology at the field level might mean coalescence as well because many fewer crop varieties might dominate a region if a broad-spectrum tolerance is implanted. The absence of weeds may diminish the populations of all other species if a crop monoculture of great genetic purity is the only life in the field. A segmented market of individual choice militates for an analogous land-use pattern, with piecemeal conversions, for example, to supply seafoods from shrimp farms at the expense of mangroves, soy-bean farms instead of tropical forest and golf courses to replace agricultural land or pasture. On some coasts the makeover is virtually complete with artificial islands extending the pleasure zone, as in Dubai. Thus, 'nature' becomes a separate category with fences (often literally so) around 'reserves'; the non-human world becomes in some cases (especially so in National Parks) something recreational, to be entered or observed largely for pleasure: a reserve for tourism, in fact. Tiny cameras mean that no part of the life of a wild bird need be unobserved, on a television screen.

There is a paradox in the sense that tiny cameras and other micro-technology mean a heavier footprint. There has been little sense of decoupling the processes of the planet's ecology from those of human economies, still less of the somewhat vague notions of dematerialising the economies of the PIE world.[94] At the end of the millennium, there was a growing, but still politically ineffective, questioning of whether all these brave new developments were part of a Promethean mythology in its widest sense.

SCREENING THE WORLD?

Max Frisch (1911–91, a Swiss architect and writer) famously said that 'Technology [was] the knack of so arranging the world that we don't have to experience it.'[95] His meaning was clear, though we might want now to modify it when we think of the multitude of different ways in which information about the world is made available. Many of these, however, are heavily medi-ated via technological devices. Thinking about the representations of human–environment relationships in the second half of the twentieth century became even more difficult than for earlier times, not least because of the diversity of it. To make a provocative generalisation: representations are engaging in a polar fashion with the very small and with the global/long term, whereas politics deals with immediate crises and the middle term but turns away from the longer and wider views. The aesthetics of nature are akin to music: they present us with an exposed mathematical order in which the information content becomes meaning. We then transmit much of that meaning with a vocabulary of symbolisms from our culture's life. We might wonder whether aleatory music has a parallel in the close-ups of birds' nests that use miniature cameras and so frame only a tiny piece of the entire ecosys-tem of which they are a part; possibly more persuasive is the parallel between human rapaciousness and the attraction of large mammalian predators, including the fascination with sharks. Equally fascinating are films about films where the sight of a lion eating or sleeping (or indeed copulating) is witnessed by about a dozen safari vehicles parked in a circle around the animal. What sort of experience are we getting?

Some thoughtful interpreters try to connect Earth and mind more directly.[96] The development of 'land art' can be seen in this light, with its explicit origins in the 1970s with works such as Robert Smithson's *Spiral Jetty* (1970) and Walter de Maria's *Lightning Field* from the 1970s;[97] Michael Heizer's *Double Negative*, also from the 1970s, consists of two trenches cut into the eastern edge of the Mormon Mesa, north-west of Overton, Nevada. The trenches line up across a large gap formed by the natural shape of the mesa edge. Including this open area across the gap, the trenches together measure 457 metres long, 15.2 metres deep, and 9.1 metres wide. Some 240,000 tons (218,000 tonnes) of rock, mostly rhyolite and sandstone, were displaced in the construction of the trenches: 'there is nothing there, yet it is still a sculpture' said the artist.[98] Compared with the great scale of these examples from the United States, the work of, for example, Andy Goldsworthy and Richard Long in (mostly) the United Kingdom is more domestic and often intended to be temporary. Whereas the large-scale works have been criticised as ecological vandalism, Goldsworthy, for example, tends to construct 'monuments' to the local materials, be they wood or stone or even leaves. Some metal objects, however, might reflect what lies beneath, as in Tyneside circles of scrap steel that allude to the de-industrialisation of their site.[99] (In San Francisco, not surprisingly, a piece for a gallery that sits near the San Andreas fault will challenge 'the viewer's notion of what constitutes a work of art by blurring the distinction between the natural and the man-made, while also drawing attention to nature's potential to undermine or destroy the works created by humans'.)[100] More puzzling is the well-known pattern of 'wrapping' nature (as well as buildings) as practised by Christo (1935–): his *Running Fence* in Marin and Sonoma Counties, California, and *Valley Curtain* in Colorado (both in the 1970s) shroud and divide the landscape and its views rather like his enclosure of the Berlin Reichstag in fabric. Perhaps we can take this as pointing to our ability to be distanced from the land and its structures.

For more slightly more traditional art, though, there was the example of the use of 'materials with the energy of the world in them' in the output of Joseph Beuys (1921–86). A citizen of East Germany, he often used felt and fat in constructions because they were full of that energy. They also looked thrifty, a necessary ingredient of life east of The Wall. One of his major projects was at Kassel in Germany, the planting of 7,000 oaks from 1982 onwards, as an engagement with environmental degradation. This led to *The End of the Twentieth Century* of 1982–3, (part of which is reproduced at the head of this chapter) in which basalt blocks look like the fallen tombs of a former age. The hole creates a metaphorical wound which is soothed with clay and felt and refilled. One commentator thought that Beuys 'shifted the optical and semiotic field of art towards a new and potentially troubling theatricality' and the exhibition in London in 2005 certainly produced a thoughtfulness among its viewers.[101] There is also the question of 'green' cinema where explicitly environmentalist films usually deal with toxic emissions such as pesticides or the threat of nuclear-plant meltdown or even futuristic scenarios of dystopias. These are scarcely significant beside the vast output of implicit attitudes when scenery is placed before us along with uplifting music, and we are invited to admire it no matter what its history has been. The moorlands of the Welsh Marches and the open grasslands of the High Plains alike are overgrazed and have seen much rural depopulation owing to lack of thought for the natural conditions, but nevertheless are presented as likely parts of idylls. This tendency to regress is shown most vividly in Akira Kurosawa's *Dreams* (1990), where the last episode (the film is a series of short stories with an emphasis on death and destruction) shows the funeral of an old woman whose coffin is preceded by dancing children, moving alongside a river harnessed by only the most gentle of watermills.[102] There is no electricity and no tractors and the 103-year old man quizzed by the narrator character says that scientists fail to see to the heart of things. It was nostalgia of the most appealing kind. There has been no agreement as to how television and other media had altered perceptions of the world compared with, say, the mid-nineteenth century. Nor was class difference explored: even those who had access to 'nature' very likely had parents, school expeditions and, of course, books to tell them what to expect.

Words were still a major communication route for environmental representation, from novels to poetry and journalism, and the production of web pages by the million. A kind of meta-verbal layer was the development of explicitly environmental philosophy and ethics, with the full panoply of academia deployed, as books, lecture series, jobs, journals and learned societies. This is not the place for a review of the field but it may suffice to say that a number of different and often irreconcilable positions evolved, including the text-based exegeses at one extreme and the post-modern relativist at the other. In the year 2000, it looked very much as if the dominant world view was still that of humans who were largely outside the ecology of a planet whose systems could be manipulated without limit provided the right knowledge and the right social systems were in place. The unbelievers were the minorities who had to make the running if change was to happen.

TENSIONS

Despite the greater abundance of goods and services that some people enjoy, the outcome of the interplay of coalescence and fragmentation has been one of strain and of tension. There are more people with higher incomes and better health than ever before (and, cynics might say, with more time to worry) yet the future seems uncertain. The present work is one of history and not futurology but some of the anxieties are rooted in the last fifty years, if, indeed, not before. The trends in loss of total life on the planet, with about 5 per cent of vertebrates being threatened species and biotic functions altogether eliminated from 9 million hectares and significantly diminished in a further 300 million hectares, are plain to see.[103]

The domestication of the planet might seem a non-issue because it is clearly a long-term activity of humanity. In an apparent continuation of the process of naming, it seems as if science and technology have allowed the labelling of almost everything, currently producing the stickers on apples and the ear-tags on cows. With domestication comes the loss of the wild both in the sense of habitats and in the diminution of the genetic variety inherent in wild species and in localised domesticates. What started with Neolithic taming of cattle was in 2000 at the stage of genetically modified varieties of soy beans and a cloned sheep. What was different was the accuracy with which the breeding was done and the rapidity with which it was accomplished. The vehicle, as might be expected, was access to plentiful energy in the form of well-equipped laboratories and lecture rooms, and as knowledge.

The carbon-based economy of the world was expected to go on as long as coal and oil lasted. The promises by industry of resource recovery rates being vastly improved added to a sense of security as did a politically stable world in which friends and enemies could be identified and kept in their places. The changes of the 1980s into first a multi-polar world and then into a single-power world changed many views of the human environment. Freedom to consume goods and services that followed the breakdown of Communism at its strictest in Russia and China (and their satellites) plus the burgeoning wealth of India mean that any tendencies to instability caused by resource use and emission levels are made worse. The obvious manifestation has been climate change, now accepted as a fact by all except the most denial-prone governments and multinationals. In addition, pressures to produce more fish and aquaculture products, to cater for beach tourism and simply to find somewhere to live have forced communities in low-income economies to occupy and transform zones along coasts and on steep slopes which are vulnerable to geophysical hazards. While no human intervention is likely to ameliorate the magnitude of earthquakes and tsunamis, for example, more heat retained in the atmosphere may well exacerbate the frequency and intensity of tropical storms and thus damage to coastlines as well as inducing landslides. It looks as if many geophysical hazards (drought and fire included) may get worse in a warmer world.[104] The outcome, therefore, is of an environmental instability with unpredictable amplitudes of fluctuation.

Instability itself became better understood in the late twentieth century with the development of ideas such as chaos theory. The attractive image of a butterfly beating its wings in Tokyo and causing a storm in New York has been much refined. Nevertheless, the basic idea that small initial differences can lead to very large-scale consequences and, moreover, that the outcomes are not likely to be the same twice, has had considerable resonance in climatic studies and has been applied eagerly to discuss possible environmental outcomes. It can be argued that much of the international action taken in the second half of the twentieth century, which had direct or indirect environmental implications, was aimed at reducing instability and unpredictability. The realisation that everything is connected to everything else and that the mediation was via the atmosphere was a sharp jolt to those who suddenly realised that nobody owned it. The emissions trading at the centre of the Kyoto process is a first step in the application of conventional economics to what had hitherto been a common resource. Environmentalists lobbied hard in the 1960–90 period for an equilibrium world as distinct from the growth model fostered by neoclassical economics; both parties, however, rather turned their backs on 10,000 years of history in which equilibrium was characteristic of neither the natural nor the human-made world, let alone the hybrids of both. The legacy of the nineteenth century has been to expect growth for ever. Yet again, let it be emphasised, nobody has disproved that there is not an upper carrying capacity level for the species *Homo sapiens*, depending on the support level of other parts of the global ecosystem.

Hence, the existence of international bodies concerned directly with the environment and the attention given to it by the more powerful national governments in the 1990s have confirmed one thing. V. I. Vernadsky's conception of a noösphere in which the physical -spheres (atmos-, litho-, bio-) are brought under the network of human thought and communication is more or less accomplished, though some people are more connected than others, not surprisingly. Vernadsky (1863–1945) and his interpreters have thought that this implosion would lead to a new human culture: a coalescence of all that is good about interdependence.[105] This brings us to the question of authority. In the 1950–2000 period there was a shift from the printed word (of books and newspapers especially) to visual expressions of cultural influence. 'I saw it on TV' replaces 'I read it in the papers'. The role of television and the computer is central: many advertisements on the small screen of television actually depict a computer screen as part of the sequence as a sign of authority, with the other cultural dominant of science manifested as a white coat whenever relevant. So environmental knowledge is probably more widespread than ever before but it is at a safe distance and there is an 'off' switch. Cultural attitudes can emerge which are detached from their ecology: fire in the home, for example, is tamed as candles, log fires and cigarette lighters whereas the media pictures from the wild are of war and disaster. No wonder that controlled fire in sclerophyll woodlands is outlawed and fuels build up, as in Australia and the Mediterranean.

The setting for many of these issues is that of consumerism. People formulate their goals through acquiring goods and services far removed from subsistence needs or even those of elegant sufficiency. Shopping seems to give an identity now available to many more individuals than the differentiated access of earlier rich people in, say, the Bronze Age or even the nineteenth century. Former luxuries are now commonplace and so a positional good has to be sought which very often involves further environmental impact. The whole history of sugar (as discussed in chapter 3) is a case in point.[106] The ability of the social sciences to understand all these interactions is in flux but has changed markedly since 1895 when Durkheim urged sociologists to ignore human biology and geography in their search for explanation. Globalisation has brought the limits to nature back into the frame whereas, in much of the twentieth century, 'environment' was simply a barrier between humanity and the modern world. The thinking may have changed but many of the forces associated with it (especially in terms of technology) continue to expand.[107]

The 1990s ended with a resurgence of a question which had seemed settled perhaps twenty years earlier. The rising price of oil and rumblings about the longevity of the reserves, the future attitudes of oil-producing nations that were becoming Islamist in polity, and the refusal of the United States (especially but not solely) to join any form of 'use-less' movement, all fused in a revaluation of the virtues of nuclear power. In many countries the fission power station has been a symbol of disaster (actual or potential) and an expensive source of electricity. But all such objections pale beside an alternative in which power supplies are subject to reductions and cuts. No juice, no shopping: end of the western world and its consumption patterns (Table 5.2). Even nuclear fusion (fifty years of development and perhaps never able to produce more than it consumes) will become the site of massive international co-operation, which is a real sign of desperation.

The PIE world then exhibits many things at once, but the tension at its heart between rich and poor has many environmental consequences and all lead to an increased chance of instabilities of many kinds. These are likely to show greater amplitudes of fluctuation than has been the case in the past and probably (in spite of the remarkable successes of modelling) predictable

TABLE 5.2 Levels of consumption

World totals (million tonnes):

Product	1960	1997
Wood	2,000	3,250
Meat	75	220
Grain	900	2,100
Fish (excl. aquaculture)	79	82

These numbers are estimated visually from graphs and so should not be taken as precise.

Annual per caput levels, late 1990s:

Country	Meat (kg)	Fossil fuels (kg of oil equivalent)	Passenger cars per 1,000 people
USA	122.0	6,902	489
Japan	42.0	3,277	373
Poland	73.0	2,585	209
Indonesia	9.0	450	12
China	47.0	700	3.2
India	3.4	268	4.4
Nigeria	12.0	186	6.7
Zambia	12.0	77	17

Source: *World Resources 2000–2001*, Washington DC: World Resources Institute, 2000, Box 1.11, pp. 26–7. Some of the data are the WRI's own, some from the World Bank and some from FAO.

only imprecisely. History does not equip us all that well to deal with such a world.

A HASTE LAND?

If we examine the 'fundamental factors' in the metamorphoses of fifty years of post-industrial development, then we find significant changes in all of them:

1. The increasing capacity for communal action by states and corporations has been enhanced by the arrival of the United Nations and its component agencies. They have been able to direct action by pointing out the international nature of problems but have probably been less responsible for changes in the field than the World Bank, whose fiscal policies have had major environmental repercussions in developing countries. The effects of super-powers, such as the Soviet Union and the United States are also clear, as are the impacts due to the multinationals, especially those in the fields of energy access and of agricultural produce.

2. Colonial regimes have become politically unacceptable and many empires have broken apart. Their legacy is in the terms of trade between the post-industrial economies and their poorer suppliers in which subsidies to, for example, developed-world agriculture affect the environments of both suppliers and consumers.

3. The world market for both cheap and precious goods has extended to unpredictable items as well as to those which were foreseeable in industrial times. The selling of wild places for tourism is an example of the latter, but the lengths to which water is now being traded (and perhaps will be fought for) are an emergent phenomenon.

4. Population growth had a lower profile in the 1990s than in, say, the 1960s, and there are now some localised worries about replacement rates and 'greying'. But, in most of the world, the numbers are still rising and adding to the demands on resources and waste-processing capacities. A big unknown is the long-term effect of HIV/AIDS: will its late-twentieth-century effects be seen eventually as a 'blip' or as the beginning of quasi-permanent effects?

5. The accessibility of technology has changed with the movement of some highly visible items into private use and ownership. This has increased expectations, especially because the items are often so small. Many consumer electronics have a short lifespan, are heavy consumers of resources and are not easily recycled because they do not get in the way as an old motor car or refrigerator does.

6. The complexities and impacts of all these changes have resulted in a burgeoning of environmental law, led by the most highly industrialised regions but with some attempts (not always enforced) by developing

economies. These last often perceive legal restrictions as inimical to 'progress' but it is worth noting that loud voices in the United States have claimed that restricting carbon output to the atmosphere would wreck the economy.

Few of these abstractions in themselves convey the dominance achieved during 1950–2000 of two great planetary manipulators:

1 Demand for meat and fish. Animal protein has always been a sign of status and the continuity of this trait seems to have diminished very little. Such is its standing that low-income people in high-income economies will eat poor-quality meat rather than better-quality fruit and vegetables; plant- and fungus-derived foods are marketed as disguised meat. Marketable animals (especially beef cattle) consume immense amounts of plant material, which would otherwise be direct human food, as well as large quantities of fish meal and the remains of other animals, as was shown in the British BSE epidemic which peaked around 1992. Thus, the consumption of meat and its effects on forests, grasslands, the sea and energy consumption are primary environmental manipulators.

2 There are many kinds of motors in the world but most are powered directly or indirectly by the combustion of hydrocarbons. The getting of the fuels and the necessary refining, the land-use changes that are brought about by roads and vehicles, and the effects of emissions both local and global place this addition to the somatic repertoire of humans very high on the list of environmental management and impact sources. While there is no denying the benefits brought to many by all kinds of motors (with the possible exception of blinged-up vehicles with 1,000-decibel sound systems), the external costs both directly to humans and indirectly via the environment are very rarely added up.

The common factor in many of these changes is the speed with which they occur and the rapid metamorphoses as a process 'matures': this is as true of new agricultural chemicals as it is of computers, and it certainly seems to be true of climatic change, if the models of positive feedback loops are correct. The way in which a luxury item, such as meat, became commonplace in the West after 1950 (even in Japan, for example) has amazed most observers. The desire for speed, which produced Concorde (and its load of emissions), propels many other processes, not all of which seem to produce much benefit except that of being able to say that the person is living 'in the fast lane'. In the case of food, immense amounts of energy are spent transporting it across continents. The Italian-inspired 'slow food' movement founded in 1986 constitutes one response, though how well rooted it became is difficult to assess. In 2000 the parent group instituted the Slow Food Award for the Defense of Biodiversity with the goals of publicising and rewarding activities of research, production, marketing, popularisation and documentation that benefit biodiversity in the agricultural and gastronomic fields. Yet none of these interactions seems to have produced any sense of security: there were instabilities in the population-resources-environment system, a floundering over the mitigation of, or compensation for, environmental problems caused by the North but inflicted on the South. Insecure nation states and insecure people seemed unlikely to take the social and political hazards incurred in tackling environmental risks.[105]

It seems a cliché to say that the world in 1950–2000 was poised to choose one of several different paths and, indeed, the metaphor is too rigid. Within the years after 1945, shuffle and muddle were as common as were the ringing declarations of those who wanted to 'end poverty' or echoed the Morgenthau plan for Germany in 1945 which was to make it a pastoral and de-industrialised nation, or even contemplated a population level that could be supported by foraging. Somewhere in that fifty years, however, the idea of keeping a region 'down' disappeared.

NOTES

1. A. P. Chester, 'Globalisation and the new technologies of knowing', in M. Strathern (ed.) *Shifting Contexts. Transformations in Anthropological Knowledge*, London and New York: Routledge, 1995, 117–30.
2. D. Lowenthal, *The Past is a Foreign Country*, Cambridge: Cambridge University Press, 1985.
3. V. Smil, *General Energetics. Energy in the Biosphere and Civilization*, New York: Wiley Interscience, 1991, ch. 14.
4. So probably the total area devoted to motor vehicles is approaching 3 or 4 per cent? The road area comes from a contribution to M. O'Hare (ed.) *Does Anything Eat Wasps?*, London: Profile Books, 2005, p. 145.
5. Though of course they were: international aid helped them avoid total starvation.
6. It is difficult to judge whether those few nations whose governments actively eschew western ways (such as North Korea, Iran, Bhutan and until recently China) reflect the wishes of their populations. When given the chance to change, they usually take it.
7. In the United States, substitution by electricity reached its maximum penetration by about 1930, three decades after it had begun. In the low-income economies this process is still far from complete.
8. F. Fukuyama, *The Great Disruption. Human Nature and the Reconstitution of Social Order*, New York: Free Press, 1999. A disruption, of course, has another side beyond the rupture. Environmentalists sometimes talk of the twentieth century as 'the great transition'.
9. There are always good summaries in the annual publication of the Worldwatch Institute, *Vital Signs*. These data are from *Vital Signs 1999–2000*, ed. L. Starke, London: Earthscan, 1999, p. 97 et seq.
10. The most exposed neo-Malthusian has been the biologist P. R. Ehrlich. The first high-profile book was *The Population Bomb*, New York: Sierra Club/Ballantine Books, 1968, followed by, inter alia, *The Population Explosion* (with A. H. Ehrlich),

New York: Simon & Schuster, 1990; P. R. Ehrlich, A. H. Ehrlich and G. C. Daily, *The Stork and the Plow: the Equity Answer to the Human Dilemma*, New York: Putnam, 1995. I have written a short account of his work: 'Paul Ehrlich', in J. A. Palmer (ed.) *Fifty Key Thinkers on the Environment*, London and New York: Routledge, 2001, 252–60.

11. R. Clarke (ed.) *Global Environment Outlook 2000*, London: Earthscan/Nairobi: UNEP, 1999, ch. 4.

12. The Third Assessment Report of the IPCC (Intergovernmental Panel on Climatic Change), 2001, p. 2. Accessed in March 2005 at www.ipcc.ch/pub/spm22-01.pdf.

13. IPCC, *Climate Change 2001: Impact, Adaptation, and Vulnerability*, p. 3, accessed in March 2005 at www.ipcc.ch/pub/wg2SPMfinal.pdf. The relevant studies are mostly in north and north-western Europe and North America.

14. For example, A. Pitman, R. Pielke, R. Avissar, M. Clausen, J. Gash and H. Dolman, 'The role of the land surface in weather and climate: does the land surface matter?', *IGBP Newsletter* **39**, 1999, 4–9; J. Fuhrer, 'Agroecosystem responses to combinations of elevated CO_2, ozone and global climate change', *Agriculture, Ecosystems and Environment* **97**, 2003, 1–20.

15. S. Lavorel, E. F. Lambin, M. Flannigan and M. Scholes, 'Fires in the Earth System: the need for integrated research', *Global Change Newsletter* **48**, 2001, 7–10.

16. V. Smil, *General Energetics. Energy in the Biosphere and Civilization*, New York: Wiley Interscience, 1991, ch. 10; *idem, Energies*, Cambridge MA: MIT Press, 1999, ch. 5. There are no better places in which to read about energy flows of all kinds on the planet than in Smil's books.

17. V. Smil op. cit. 1991; L. Starke op. cit. 1999, pp. 48–55.

18. R. Clarke op. cit. 1999, p. 336.

19. D. Cosgrove, 'Contested Global Visions: One-World, Whole-Earth, and the Apollo space photographs', *Annals of the Association of American Geographers* **84**, 1994, 270–94, thought that the photo showed the Earth as an ownable commodity. The outcome awaits the verdict of time.

20. M. Degg, 'The 1992 "Cairo earthquake": causes, effects and response', *Disasters* **17**, 1993, 226–38; Y. Q. Zong and X. Q. Chen, 'The 1998 flood on the Yangtse, China', *Natural Hazards* **22**, 2000, 165–84.

21. For a comprehensive overview see A. M. Mannion, *Agriculture and Environmental Change. Temporal and Spatial Dimensions*, Chichester: Wiley, 1995.

22. D. Goodman and M. Redclift, *Refashioning Nature. Food, Ecology and Culture*, London and New York: Routledge, 1991.

23. Goodman and Redclift op. cit. 1991.

24. http://www.defra.gov.uk/news/latest/2005/food-0715.htm

25. M. Williams, *Deforesting the Earth*, Chicago: Chicago University Press, 2003, ch. 14.

26. Note that an error term is now possible. F. Achard et al., 'Determination of deforestation rates of the world's humid tropical forests', *Science* **297**, 2002, 999–1002.

27. See the account in M. Williams op. cit. 2003, ch. 14.

28. E. F. Lambin and H. J. Geist, 'Regional differences in tropical deforestation', *Environment* **45**, 2003, 22–36.

29. M. Gadgil and R. Guha, *This Fissured Land. An Ecological History of India*, Delhi: Oxford University Press, 1993, ch. 6.

30. See for example, R. Walker, 'Theorizing land-cover and land-use change: the case of tropical deforestation', *International Regional Science Review* **27**, 2004, 247–70; H. Geist and E. Lambin, 'Is poverty the cause of tropical deforestation?', *International Forestry Review* **5**, 2003, 64–6; R. Bonnie, S. Schwartzman, S. C. Stier and S. F. Siebert, 'Tropical reforestation and deforestation and the Kyoto

protocol', *Conservation Biology* **17**, 2003, 4–5; E. B. Barbier and J. C. Burgess, 'Tropical deforestation, tenure insecurity and unsustainability', *Forest Science* **47**, 2001, 497–509.

31. Y. Malhi, P. Meir and S. Brown, 'Forests, carbon and global climate', *Phil. Trans. R. Soc. Lond.* **A 360**, 2002, 1567–91.

32. E. G. A. Olsson, G. Austrheim and S. N. Greene, 'Landscape change patterns in mountains, land use and environmental diversity, Mid-Norway 1960–1993', *Landscape Ecology* **15**, 2000, 155–70.

33. O. Kandler, 'Historical declines and diebacks of central European forests and present conditions', *Environmental Toxicology and Chemistry* **11**, 1992, 1077–93.

34. This was in 1966. The actual quote was probably more like 'A tree is a tree. How many more do you have to look at?'

35. See both the W. L. Thomas and the B. L. Turner edited volumes frequently referred to above: neither takes recreation seriously. J. R. McNeill does not mention it as a twentieth-century phenomenon. A. M. Mannion (op. cit. 1997, 2nd edn) certainly does.

36. See, for example, M. Clawson and J. L. Knetsch, *Economics of Outdoor Recreation*, Baltimore: Johns Hopkins Press for Resources for the Future, 1966, for a pioneering volume. A concise overview is in J. Pigram, *Outdoor Recreation and Resource Management*, Beckenham: Croom Helm, 1983.

37. I. G. Simmons, *Rural Recreation in the Industrial World*, London: Edward Arnold, 1975. Cities are, unsurprisingly, not covered.

38. A BBC News story ('Piste pressure on alpine plants') accessed on 18 April 2005 at http://news.bbc.co.uk/go/pr/fr/-/1/hi/sci/tech/4449637.stm. Also, C. Rixen, V. Stoeckli and W. Ammann, 'Does artificial snow production affect soil and vegetation of ski pistes? A review', *Perspectives in Plant Ecology, Evolution and Systematics* **5**, 2003, 219–30.

39. A. C. Gange, D. E. Lindsay and J. M. Schofield, 'The ecology of golf courses', *Biologist* **50**, 2002, 63–8.

40. A. T. Grove and O. Rackham, *The Nature of Mediterranean Europe. An Ecological History*. New Haven and London: Yale University Press, 2001.

41. In the early twenty-first century, salmon fishing in northern Scotland's wild areas of Caithness would cost at least £1,000 per week per person, and a day's pheasant shooting in Somerset some £500. Scottish deer stalking was advertised at £250 per stag, though it was unclear if that depended upon results.

42. K. R. Jones and J. Wills, *The Invention of the Park. From the Garden of Eden to Disney's Magic Kingdom*, Cambridge: Polity Press, 2005.

43. F. C. Coleman, W. F. Figuueira, J. S. Ueland and L. B. Crowder, 'The impact of United States recreational fisheries on marine fish populations', *Science* **305**, 2004, 1958–60.

44. D. Biggs, 'Managing a rebel landscape: conservation, pioneers, and the revolutionary past in the U Minh forest, Vietnam', *Environmental History* **10**, 2005, 448–76.

45. National Research Council [of the USA] Committee on the Effects of Herbicides in Vietnam, *The Effects of Herbicides in Vietnam. Part A. Summary and Conclusions*, Washington DC: National Academy of Sciences, 1974. (Part B contains the detailed working papers.) It is interesting that the calculations on tree growth were framed in terms of 'merchantable timber'. Nobody should forget the seminal work by A. H. Westing in this area, as in, for example, his edited collections, *Herbicides in War. The Long-term Ecological and Human Consequences*, Stockholm: SIPRI, 1984, and *Environmental Hazards of War: Releasing Dangerous Forces in an Industrialized World*, London: Sage, 1990.

46. G. J. De Graaf and T. T. Xuan, 'Extensive shrimp farming, mangrove clearance and marine fisheries in the southern provinces of Vietnam', *Mangroves and Salt Marshes* **2**, 1998, 159–68.

47. J. W. Readman, B. Oregioni, C. Cattini, J. P. Villeneuve, S. W. Fowler and L. D. Mee, 'Oil and combustion-product contamination of the Gulf marine environment following the war', *Nature* **358**, 1992, 662–5. There is a popular treatment of the immediate effects on wildlife in M. McKinnon and P. Vine, *Tides of War*, London: Boxtree, 1991.

48. See BBC News reports from various sources in the first half of 2005.

49. An immense literature was generated. See P. Ehrlich et al., 'Long-term biological consequences of nuclear war', *Science* **222**, 1983, 1293–1300; L. Dotto, *Planet Earth in Jeopardy. Environmental Consequences of Nuclear War*, Chichester: Wiley, 1986. The leaders at the time were Ronald Reagan and Mikhail Gorbachev.

50. It was his farewell address to the nation before leaving the Presidency. The full text is available on numerous websites.

51. The 'official' data from IUCN's Red Books are available on www.iucnredlist.org, and there is historical information from the Committee on Recently Extinct Organisms on creo.amnh.org, and a splendid individual's website www.extinct. petermans.nl/extinct. All accessed in July 2006.

52. The literature is enormous and the general media attention being high, there are many other sources of information from, for example, television and the Web. There is everything from mysticism to mountainous government reports. If the printed source is taken as more permanent, and something more than local advocacy is required, then possible material includes: W. M. Adams, *Future Nature: a Vision for Conservation*, London: Earthscan, 2003; *idem, Against Extinction: the Story of Conservation*, London: Earthscan, 2004; B. Green, *Threatened Landscapes: Conserving Cultural Environments*, London: Spon, 1991; J. A. McNeely, *Conservation and the Future: Trends and Options towards the year 2025*, Gland: IUCN 1997; J. G. Nelson and R. Serafin (eds) *National Parks and Protected Areas: Keystones to Conservation and Sustainable Development*, Berlin, Springer 1997; P. Marren, *Nature Conservation*, London: Collins New Naturalist 91, 2002; J. A. Burton, *The Atlas of Endangered Species*, London: Apple, 2000; R. W. Sellars, *Preserving Nature in the National Parks: a History*, New Haven and London: Yale University Press, 1997; M. W. Schwartz, *Conservation in Highly Fragmented Landscapes*, London and New York: Chapman & Hall, 1997; D. Ingram, *Green Screen: Environmentalism and Hollywood Cinema*, Exeter: University of Exeter Press, 2004.

53. The *OED* gives a first example from 1966.

54. J. van Andel and J. Aronson, *Restoration Ecology*, Oxford: Blackwell, 2005; R. Ferris-Kaan (ed.) *The Ecology of Woodland Creation*, Chichester: Wiley, 1995; P. Collen, 'The reintroduction of beaver (*Castor fiber* L.) to Scotland: an opportunity to promote the development of suitable habitats', *Scottish Forestry* **49**, 1995, 206–16.

55. R. Constanza et al., 'The value of the world's ecosystem services and natural capital', *Nature* **387**, 1997, 253–60.

56. Much of this material comes from a special issue of *Industry and Environment* (Vol. **23**, 2000), a UNEP publication.

57. Directives 2002/95/EC on the restriction of the use of certain hazardous substances in electrical and electronic equipment and 2002/96/EC on waste electrical and electronic equipment.

58. M. O'Meara, *Reinventing Cities for People and the Planet*, Washington DC: Worldwatch Paper #147, 1999.

59. C. Folke, A. Jansson, J. Larsson and R. Constanza, 'Ecosystem appropriation by cities', *Ambio* **26**, 1997, 167–72.

60. J. N. Pretty, A. S. Ball, T. Lang and J. I. L. Morison, 'Farm costs and food miles: an assessment of the full cost of the UK weekly food basket', *Food Policy* **30**, 2005, 1–19.

61. K. Warren-Rhodes and A. Koenig, 'Escalating trends in the urban metabolism of Hong Kong: 1971–1997', *Ambio* **30**, 2001, 429–38. This builds upon the pioneering paper of K. Newcombe, J. Kalma and A. Aston, 'The metabolism of a city: the case of Hong Kong', *Ambio* **7**, 1978, 3–15.

62. P. Hall, *Cities in Civilization*, London: Weidenfeld & Nicolson, 1998 is a rich source of information. The judgement about 'environmental' cities is mine, not his.

63. A. Gupta and R. Ahmad, 'Geomorphology and the urban tropics: building an interface between research and usage', *Geomorphology* **31**, 1999, 133–49.

64. The world's irrigated area grew at 2.3 per cent per year in the 1970s but dropped back to 1.4 per cent in the 1990s. Most of the recent expansion has been in Asia, with other continents being static or even declining a little. See L. Starke (ed.) *Vital Signs 1999–2000*, London: Earthscan, 1999, 44–5.

65. P. E. O'Sullivan and C. S. Reynolds (eds) *The Lakes Handbook*, Oxford: Blackwell, 2004–5, 2 vols.

66. There is also desalination, mostly confined to small islands such as Malta and to nations with cheap energy supplies (such as Saudi Arabia); plans to tow icebergs from Antarctica to, for example, Western Australia arise from time to time.

67. S. L. Postel, G. C. Daily and P. R. Ehrlich, 'Human appropriation of renewable fresh water', *Science* **271**, 1996, 785–8.

68. S. L. Postel, 'Water for food production: will there be enough in 2025?', *Bioscience* **48**, 1998, 629–37.

69. V. Smil, *Global Ecology. Environmental Change and Social Flexibility*, London and New York: Routledge, 1993.

70. R. Avissar and Y. Liu, 'Three-dimensional numerical study of shallow convective clouds and precipitation induced by land surface forcing', *Journal of Geophysical Research* **D 101**, 1996, 7499–518.

71. C. J. Vörösmarty, P. Green, J. Salisbury and R. B. Lammers, Global water resources: vulnerability from climate change and population growth', *Science* **289**, 2000, 284–88; C. J. Vörösmarty, 'Global change, the water cycle, and our search for Mauna Loa', *Hydrological Processes* **16**, 2002, 135–9.

72. Brazil, Russia, Canada, Indonesia, China and Colombia. Not all this is accessible as discussed in the preceding paragraphs.

73. V. Strang, *The Meaning of Water*, Oxford and New York: Berg, 2004.

74. As used in J. Blatter and H. Ingram (eds) *Reflections on Water: New Approaches to Transboundary Conflicts and Cooperation*, London and Cambridge MA: MIT Press, 2001.

75. B. Cohen, 'Technological colonialism and the politics of water', *Cultural Studies* **8**, 1994, 32–55.

76. R. Niezen, *Defending the Land: Sovereignty and Forest Life in James Bay Cree Society*, Boston MA: Allyn & Bacon, 1998.

77. D. Pauly et al., 'Fisheries impact on ecosystems and biodiversity', *Nature* **418**, 2002, 689–95.

78. P. Lysten and O. Otterstad, 'Social change, ecology and climate in 20th-century Greenland', *Climate Change* **47**, 2000, 193–211.

79. R. L. Naylor et al., 'Nature's subsidies to shrimp and salmon farming', *Science* **282**, 1998, 883–4.

80. Data from International Federation for Animal Welfare accessed on 28 May 2005 at www.ifaw.org/ifaw/dfiles/file_106.pdf, a document of 2001. Iceland also catches some whales and in 2005, South Korea wanted to join the outcasts. A fraught meeting in 2006 reflected the buying of votes by Japan in the form of development aid.

81. T. D. Jickells, R. Carpenter and P. S. Liss, 'Marine environment', in B. L. Turner et al. (eds) *The Earth as Transformed by Human Action*, Cambridge: Cambridge University Press, 1990, 313–34.

82. R. H. Charlier, 'Impact on the coastal environment of marine aggregate mining', *International Journal of Environmental Studies* **59**, 2002, 297–322.

83. J. B. C. Jackson et al., 'Historical overfishing and the recent collapse of coastal ecosystems', *Science* **293**, 2001, 629–37; M. Barange, F. Werner, I. Perry and M. Fogarty, 'The tangled web: global fishing, global climate, and fish stock populations', *Global Change Newsletter* **56**, 2003, 24–7.

84. Namely, India, China, Pakistan, Bangladesh, Mexico, Iran, Vietnam, Ethiopia and Egypt. The Philippines and Costa Rica come close.

85. It is interesting (and indeed encouraging) that a 'Commission on Sustainable Consumption' was chaired by a former minister in one of Britain's more explicitly right-wing governments. J. Gummer (Chair) Oxford Commission on Sustainable Consumption, November 2004. Available in 2005 on www.environmentdaily.com/docs/41105b.doc

86. I. Bičik, L. Jeleček and V. Štěpánek, 'Land use changes and their social driving forces in Czechia in 19th and 20th centuries', *Land Use Policy* **18**, 2001, 65–73.

87. H. G. Wells, *The World Set Free*, London, Macmillan 1914; the energy value prophecy may be as yet unfulfilled but note the following from the same book:

> Certainly it seems now that nothing could have been more obvious to the people of the earlier twentieth century than the rapidity with which war was becoming impossible. And as certainly they did not see it. They did not see it until the atomic bombs burst in their fumbling hands . . . All through the nineteenth and twentieth centuries the amount of energy that men were able to command was continually increasing. Applied to warfare that meant that the power to inflict a blow, the power to destroy, was continually increasing. There was no increase whatever in the ability to escape . . . Destruction was becoming so facile that any little body of malcontents could use it . . . Before the last war began it was a matter of common knowledge that a man could carry about in a handbag an amount of latent energy sufficient to wreck half a city.

88. R. Robertson, *The Three Waves of Globalization*, London and New York: Zed Books, 2003.

89. G. Stanhill and S. Cohen, 'Global dimming: a review of the evidence for a widespread and significant reduction in global radiation with discussion of its probable causes and possible agricultural effects', *Agricultural and Forest Meteorology* **107**, 2001, 255–78; U. Lohmann and M. Wild, 'Solar dimming', *Global Change Newsletter* **63**, 2005, 21–2; D. I. Stern, 'Reversal of the trend in global anthropogenic sulfur emissions', *Global Environmental Change* **16**, 2006, 207–20.

90. D. Reynolds, 'American globalism: mass, motion and the multiplier effect', in A. G. Hopkins (ed.) *Globalization in World History*, London: Pimlico, 2003, 243–60.

91. R. Robertson op. cit. 2003.

92. Seishi Yamaguchi (trans. Takashi Kodaira and A. H. Marks) *The Essence of Modern Haiku*, Atlanta: Mangain Inc., 1993, 198.

93. There is an extended discussion of pros and cons in A. M. Mannion, *Agriculture and Environmental Change*, Chichester: Wiley, 1995, ch. 10.

94. All this sounds suspicious to the low-income economies whose people would like better material standards and who fear that emissions-trading, for example (as laid out in the Kyoto process), will be like other forms of trade and disadvantage them. Talk of them bypassing the carbon economy in favour of some later stage sounds to them very like us pulling up the ladder.

95. This quotation comes from a novel, *Homo Faber*, of 1957.

96. See J. Gold and G. F. Revill, *Representing the Environment*, London and New York: Routledge, 2004.

97. Alas, too many to reproduce here. But all easily available on the Internet.

98. See http://doublenegative.tarasen.net/double_negative.html (accessed January 2006).

99. D. Matless and G. Revill, 'A solo ecology: the erratic art of Andy Goldsworthy', *Ecumene* **2**, 1995, 423–48.

100. Seen at http://www.thinker.org/fam/press/press.asp?presskey=171 in January 2006.

101. Quotation from P. Wollen, *Raiding the Icebox: Reflections on Twentieth Century Culture*, Bloomington: Indiana University Press, 1993, p. 158; see also C. Tisdall, *Joseph Beuys*, London: Thames & Hudson, 1979. *The End of the Twentieth Century* is in the Bayerische Staatsgemäldegalerien, Pinakothek der Moderne, in Munich, Germany.

102. Warner Home Video SO11911, c.1993, in VHS format. (PG certificate); in DVD, a region 1 disk ASIN B0007G1ZC, 2003.

103. D. W. Morris, 'Earth's peeling veneer of life', *Nature* **373**, 1995, 25.

104. S. Lavorel, 'Global change, fire, society and the planet', *Global Change Newsletter* **53**, 2003, 2–6.

105. P. R. Samson and D. Pitt (eds) *The Biosphere and Noosphere Reader: Global Environment, Society and Change*, London and New York: Routledge, 1999; Vernadsky mostly published in Russian but a key paper, 'The biosphere and the noösphere' appeared in *American Scientist* in 1945.

106. P. N. Stearns, *Consumerism in World History. The Global Transformation of Desire*, London and New York: Routledge, 2001.

107. M. Albrow, *The Global Age. State and Society beyond Modernity*, Stanford: Stanford University Press, 1997.

108. R. E. Kasperson, J. X. Kasperson and K. Dow, 'Global environmental risk and society', in J. X. Kasperson and R. E. Kasperson (eds) *Global Environmental Risk*, Tokyo: UNU Press/London: Earthscan, 2001, 1–48.

CHAPTER SIX

Emerging themes

FIGURE 6.1 *Garden of the Ryoanji temple in Kyoto.*
Photograph by I. G. Simmons.

The garden of the Ryoanji temple in Kyoto, Japan. A fire in the temple grounds allowed the construction of this garden to begin in 1488 though, following Japanese practice, some parts have been rebuilt from time to time within the 30 by 10 metre frame. Like most of the medieval cathedrals of Europe, the actual designers are anonymous, though the names of two labourers of the *sensui kawaramono* ('river-bank workers acting as gardeners') class are found on the back of one of the rocks.

In the Muromachi era (1338–1573) simplicity, honesty and under-statement were prized in most types of design, including architecture. This quality (*wabi*) was especially prevalent in Zen Buddhist establish-ments where meditation was aided by concentration on the blanks as much as on any objects. Hence this strong horizontal design dominated by raked sand. In the sand are fifteen 'islands' of rock, all of which except one seem to be flowing from left to right which is, of course, against the current of written Japanese, which is top to bottom, right to left. There are no trees. There is a legend that no matter how often visitors count the rocks they never find more than fourteen and that only enlightenment in the Zen sense (*satori*, which is sudden, complete and non-analytical) can produce the complete picture. As D. T. Suzuki puts it in his introduction to Zen:

> Satori is the sudden flashing into consciousness of a new truth hitherto undreamed of. It is a sort of mental catastrophe taking place all at once, after much piling up of matters intellectual and demonstrative. The piling has reached a limit of stability and the whole edifice has come tumbling to the ground, when, behold, a new heaven is open to full survey.

Writing a long-term environmental history is like looking for the fifteenth rock: no matter how many other rocks are inspected and thought about, there is always one which has the capacity to make sense of all the others. But for most of us *satori* is difficult if not impossible and in any case neither approved of nor easily transmissible in the western post-Enlightenment traditions of scholarship. Never mind: there is another Zen story in which a novice meets the famous abbot of another monastery on the forest path. They bow and the novice tremblingly asked the abbot where true Buddha-nature is to be found. 'Walk on, walk on,' said the abbot.

Ignoring the snap-locks

One of the buzzwords of recent times has been 'closure'. In history-writing it might well be desirable that some level of cut-off horizon has been reached. Yet quite obviously we are leaving this story in mid-flow and there is no indication that the year 2000 marked any special stage in the entwined histories of humanity and nature. There is, then, no Conclusion.

Nevertheless there are certain emerging ideas and themes that come out of the previous chapters, some of which relate right back to material introduced in chapter 1. There are a few recurring motifs in different sections and there are some synergies for which there is no obvious place in the chronological structure, all of which deserve some comment, though it is impossible to construct a logical sequence. This, too, is the place to start the hares of 'why' neatly summarised by Kurt Vonnegut (*Cat's Cradle*, New York: Dell, 1981, section 81):

Tiger got to hunt,
Bird got to fly;
Man got to sit and wonder, 'Why, why, why?'

Tiger got to sleep,
Bird got to land;
Man got to tell himself he understand.

But also to abandon this chase in favour of those with greater stamina.

UNDER THE SUN

John McNeill has argued convincingly that the twentieth century was qualita-
tively different from preceding periods and, in the empirical 'progression' of
change, the case is well made.[1] The emphasis here is on ideas and other non-
material aspects of the complexities that are so difficult to convey in sentences
where one word has to follow ineluctably after the other. Some of the novelties
are intensifications of what went before (the environmental changes resulting
from positive feedbacks within capitalism might be one example) and so any
persistent themes which might, for example, be carried through from hunter-
gatherer ecosystems need to be brought out. Equally, any late twentieth-
century synergies or recent exacerbations of extant processes, that have
environmental consequences, need to be examined. We need to look carefully
at the sort of tempting simplicity that, between 1500 and 1800, the drivers of
environmental change were gold, faith and empire and that, after 1900, these
were replaced by meat and motors, with a century's transition between.

Minding our language

The post-structuralist movement has emphasised the slipperiness of language.
The wide use of words such as 'nature' or the way in which 'environment' has
been applied to anything which surrounds, as in, say, 'retail environment', are
casual examples, and the attempts to use a workable but bounded language in
this book have not always lived up to the aims. They should alert us to even
simple instances of where meanings may not be the same for everybody, even
users of the same language. Much use of language is metaphorical and con-
structs a model of reality the accuracy of which is difficult to judge.[2] The
propositions of the natural sciences have been founded on metaphors such as
the atom as miniature solar system, 'wormholes', 'selfish' DNA, and 'elegant'
equations; Dennett goes so far as to describe human consciousness overall as
a war of competing metaphors. Our framing of the world and the language we
use to formulate and test hypotheses, develop tools, and process memory
are all affected by metaphor, especially when we try to cross domain boundaries, like associating an idea with something tangible.[3] The complexities
of the technological world of the late twentieth century placed great

demands on language. Marshall McLuhan (1911–80) was a prophet of the electronic age whose acknowledged main source was James Joyce's interweaving of dream, hallucination and multilingual pun in *Finnegans Wake*. Joyce (1882–1941) claimed to be a music-maker and a great engineer of heaps of other things, including intellectual–emotional complexes of the type encountered in many human–environment interactions, which he called 'feelful thinkamalinks'.[4]

The focus of all kinds of people upon 'environment' is another source of language that carries all kinds of resonances. Some of these are so directed to achieving a practical end that the term 'rhetoric' is often applied, emphasising the persuasive intent. In a negative usage, consider the expansion of towns. In Britain in the interwar period, a threat to the essentially rural image of the nation (especially England) was highlighted by a number of influential writers and artists. Much of this was, by today's standards, unplanned and wasteful of land, expensive to service and consumptive of agricultural land or open space but its extent was not very great. Yet it attracted the phrase 'urban sprawl' which has ever since been aired by any objectors to urban growth beyond some pre-existing boundary, even though such limits were carefully monitored after the 1947 Town and Country Planning Act. Many a map of proposed developments has appeared in neutral colours, only to be transformed by detractors' presentations into something less attractive, such as dark grey or a vicious red.

A positive example is the use of the word 'wilderness'. In the English-speaking context of North America, this was codified in the 1960s to mean landscapes without enduring human presence and, moreover, land which had not been subject to human-induced modifications in the past. Hence, large areas of National Forests, Bureau of Land Management (BLM) tracts and National Parks were subject to designation as wilderness areas and that became the highest category to which protection laws might apply. The cachet was so strong that the term began to creep into other places, even those where the criteria of absolute size, lack of human impact now and in the past, and inviolable remoteness could not possibly apply: the Ardnamurchan peninsula in western Scotland might be wild but it has none of the other features. The remote (by English standards), bleak and mostly nineteenth-century industrial landscape of the North Pennines sometimes attracts the designation of 'the last wilderness'. In all these cases, environmental history was rarely taken into account: in the United States the effects of native North Americans was notably ignored. Then, because 'wilderness' had such positive connotations, it was filtered into other terrain situations, so that people weekending in a busy caravan park in New Hampshire would agree that they were enjoying a 'wilderness experience'. Shifts of this type in experience and language carry the possibility that objects and practices can be assimilated into a culture to the point that a technology or an attitude may be so central as not to attract any opprobrium: it is too wonderful for that and deserves attention and respect rather than critical appraisal. Until the 1970s perhaps, 'globalisation' floating on a sea of cheap oil was an example.[5]

One trap for many non-academic commentators is hidden disparagement. Hunter-gatherers 'camp', for instance, as if their overnight or seasonal shelters and settlements were not embedded in centuries of adaptive cultural practices and aimed, probably, at maximising foraging returns. Similarly, the use of 'nomad' in the context of pastoralism very often implies a sort of random wandering rather than a historically winnowed movement through a set of resources aligned to the terrain and its seasons. In both, the implication is that low levels of material possessions are somehow indicative of a primitive state beyond which we in the industrial economies, thankfully, have progressed.

These are examples of meaning-shift within a language. The whole academic discussion of environment, nature and history is so dominated by English-language media that it is easy to forget that its discourses are also carried on in other tongues and scripts. Not only is there analytical commentary but everyday documentation and discussion, in schools and offices, by the well and in the tavern. In Japanese, for example, the intellectual tradition of Buddhism allows for the concept of *fūdosei*: a relationship which partakes of both the ecological character and the symbolic cultural meaning of a place or environment.[6] 'Nature' in Chinese and Japanese is therefore less of an object in relation to humans (as used throughout this book) but an integral part of a whole: 'not man apart', as the American poet, Robinson Jeffers, put it. The word for nature in Chinese is *shi-zen*; and in Japanese, *tzu-jan*, both of which are literally something like 'self-thusness'. The idea that it is good to have a human-free environment is not therefore a native East Asian one: it is all one consciousness, so to speak.[7] Closer to European experience is the way in which a language-bound phrase can evoke resonances. In interwar Germany, the Nazi party came to espouse certain racial myths which had environmental consequences in terms of the glorification of the outdoors and the status of farmers and foresters. So *Blud und Boden* ('blood and soil') was worked into fascism. Because it was on the surface an environmentally friendly set of policies, occasional commentaries have subsequently accused late twentieth-century environmental advocacy of fascism, which is something of a substitute for rational argument.[8] All these examples point towards the relevance of the linguistics scholar, Benjamin Whorf (1897–1941), in his judgement that 'language is not simply a reporting device for experience but a defining framework for it'. It can therefore stultify by enshrining a outdated world view in print. By contrast, language has the great advantage that it is self-analytical and self-critical. It is inventive whenever new experiences demand new words, and it allows us to move back and forth in (imagined) time. No marvel that T. S. Eliot mused upon '. . . the intolerable wrestle/With words and meanings'.[9]

Postmodernity and environment

The late twentieth century was marked by the appearance in intellectual circles of the cultural movement which became labelled as 'postmodernism' or 'post-modernism'. Postmodernity does not believe in grand narratives, meaning that people have stopped talking to themselves about the contents of the supposedly universal stories and paradigms such as religion, conventional philosophy, capitalism, gender, and the natural sciences, that have defined the stories about culture and behaviour in the past. Instead, they have begun to organise their cultural life around a variety of more local and subcultural ideologies, myths and narratives. Spatial and temporal scales assume a revised importance. Acceptance of postmodernism implies the view that different realms of discourse are incommensurable and incapable of judging the results of other treatments.[10] This may mean that the fragmentation of ideas separates stakeholders from shareholders so that transnational corporations (TNCs) outrank nation states in their influence and control over the environment and its resources. The ideological basis for a 'physical siege' of the non-human world is the encouragement by postmodernism of any number of attitudes towards it. These shelter under the umbrella of 'nature limits, technology enables' in which nature is seen as a mindless force demanding to be tamed. Any respect is a matter of convention or convenience (or, indeed, fear) and there is no distinct ontology of the natural world. Without the findings of the natural sciences, hence, there would be no stopping large corporations eviscerating the world's ecosystems for resources. The social setting may see technology simply as liberation from a pre-industrial societal aspic and ignore the fact that it may provide a lot of solutions for which there is no problem, such as space travel. More importantly, once more, it cuts a 'problem' from its tangle of multiple connections with other realities (no resonances across organ pipes) and the world becomes more like a collection of fragments. To some, the postmodern age in its multiplicity of sympathies offers more wisdom but its setting makes acting on that wisdom more difficult; to others it is simply a not-very-well-disguised form of nihilism.[11]

The ecology of emotion

Not all attitudes and reactions to environment consist of considered words. Many sights and pieces of knowledge provoke simpler responses like awe, fear and pleasure. In any Sunday supplement's travel section, writers will more than once use the term 'stunning' of landscape though it is unlikely that they mean it. It does, though, highlight that emotional reactions are likely in individuals (recall the way to 'sublime' in the eighteenth century) but to communicate them requires words. So the view from the mountain-top may call down silence but saying why it is so to a companion requires words: observe people in an art gallery, where most of them are bursting to tell somebody else what they feel about an exhibit. Likewise with environmental ethics: many people will feel that something is wrong but lack the language to persuade others.

Emotion is difficult to discuss in relation to environment for that very reason: words may be some distance from the original state. But that it provokes action is scarcely in doubt. The campaigns against animal cruelty in the nineteenth and twentieth centuries were scarcely started after a cool assess-ment of the philosophical position of cattle in the cosmos or even the cultural significance of bears in Hokkaido. Instead, a few like-minded people pooled their emotional reactions and began to act. The extreme case has been the campaign against the use of laboratory animals, which has allowed some of its adherents to inflict a great deal of pain and danger on the humans running such facilities.[12] These are examples of a wider scene: that, while abstractions like ecology or economics portray humans as cold and rational, most humans and their institutions live in a world of feelings and moods. Unless there is some emotional response, it is difficult to recognise a stimulus at all. That is not to say that emotion is a raw neurophysiological matter because feelings are also constructed cognitively, which is a way of saying that cultural traditions inform emotional reactions and states: mountains elate some people and frighten others. Some individuals react negatively to all snakes whereas others have learnt which ones are poisonous: emotions are clearly not universal and ethnic differences will never be irrelevant.[13]

Rationality cannot be discounted, however, and it is fair to conclude that most humans are both rational and emotional in their responses to their sur-roundings. A totally self-referential system of being without any relation to the values outside it (the aim of much neoclassical economics) is as far from being functional in the world as an emotionally driven condition which elevates wants to needs and luxuries to necessities; numerous examples with environ-mental relevance have been seen in chapters 3, 4 and 5. Further, a concentra-tion on rationality may produce logical principles that do not lead to action, which requires commitment and motivation.[14] In *Capital* (vol. I, 1867) Marx thought that a possession like a suit was a social hieroglyphic and perhaps the late twentieth century saw some movement towards seeing 'possessions' in terms of being social-environmental hieroglyphics. Examples might be the increasing practice of accompanying flying with planting trees, or the signing of restaurant food with its local sourcing, and less packaging. The growth of the movement to 're-use, repair and recycle' in the high-income economies from the 1970s merits more attention as an idea and in terms of its empirical effects, though the latter has been an everyday matter in poorer countries.[15]

Emotional attachment is not evenly distributed through our environment and E. O. Wilson famously argued [16] that humans had a predilection for an affiliation with other species, which he called biophilia. In its historical setting, it points out that humans have come to love and to fear things which our ances-tors could not have known so that there is at the very least a learned component to it. This leads to the need to investigate the social settings of emotions because our attitudes to environment may well derive from social considera-tions just as much as some quasi-innate reaction to water, spiders or moun-tains. If, therefore, we learn about the world via interest in it then emotions have a part to play in that process: the separation of thought and feeling so implanted by the scientific revolution of the seventeenth century and after is no longer valid. What we learn depends upon the quantity and quality of our personal engagement with it. In historical terms hunter-gatherer children have plentiful contact with soils, plants and animals whereas the child from a run-down urban neighbourhood lives in a setting of concrete, brick, abandoned cars and rats. In the PIE urban economy, most knowledge of nature is likely to come via the television set unless the formal educational process provides other alternatives. (Lest we romanticise the rural, many western farms are rather industrial in their ways with noise and smell, especially those devoted to intensive animal production.) So the historical lesson is one of progressive separation of humans and their environments as they withdraw into cities and industrialised locations which are larger and more completely built up. Nature is not absent from such places (there is still weather, birds learn to adapt, and there are animal scavengers of some kind) but it is a selection which appears most of the time to be mediated by the human-constructed structures.[17]

There is a last historical perspective, in which emotions (in the range in which we currently know them) are pre-social. They are said to be central to the evolution of consciousness and thus an integral part of the being of the species.[18] They are, therefore, not immutable and they can, as mentioned above, be learned. Thus, there may be an emotional content to any changes in consciousness that came along with, for example, the shift from foraging economies and the development of agriculture. Historically, we need not expect that emotional responses to the environment will be the same from one era to the next: those developed from hybrid experiences of visual electronics and occasional visits to tropical beaches will be different from those of the Innuit in the 1920s. Also, while actors and networks are involved in theories of hybrid artifice and artefact construction, emotion is never far from the surface of the page.[19] The question of immediacy is also relevant for, while medical ethics became formalised in many countries after 1950, no such binding developments took place for planetary ethics.[20]

Much of this comes together in the fields of aesthetics which bear on the human surroundings. While pictures may give us a new way of looking, or sensitise us to something not hitherto perceived, they may also reinforce the propriety of the conventional. So, a representation of nature may, in fact, be a self-contained formal system (analogous to a still life) as well as a statement about nature in general in either its own terms or those of a particular culture. Emotional responses to nature itself are usually conceptual rather than causal or sequential and, anecdotally, seem often to be beyond representation. Kant thought that nature was so vast as to exceed any human framing and so it produced awe. That, of course, was then and it would be interesting to interpret 300 years of the arts in terms of the success of 'framing'.[21]

In the dialectic with rationality, a radical position sees that rationality itself is a feeling and is emotionally constituted. This would contradict any divergence between rationality and emotion. Hence it works against any of the socially constituted myths which protect particular interests and ideologies: consider environmental controversies of the PIE period in which the defenders of the status quo were accused of being 'emotional'. Few venture capitalists and developers would agree with David Hume's pronouncement of 1739 that 'reason is, and ought always to be, the slave of the passions'.[22]

Religion

Religions believe that the cosmos and especially life are subject to transcendental power. Everything can then be situated within that framework, in contrast to post-Enlightenment secularism which sees religion as one facet of culture like sport or marriage customs and especially prone to myth-making. Religions have mostly had an attitude to nature, though it is unusually risky to assign them to historical eras because they show considerable spatial and temporal variation. Western Christianity's approval of technology was by no means universal in other faiths.[23]

As the discussion of hunter-gatherers in chapter 2 showed, there was a close relationship with non-material beings, whose transcendence was important and had to be heeded. Any aspect of the world could manifest a presence that might be helpful (in the hunt, for example) or might be an object of fear and avoidance.[24] The world of the non-material might be accessed via an intermediary like a shaman, and the continuity with the past was stressed by the idea of an annual cosmogenesis. A kind of pantheism was also present in the agriculture-based societies of the Classical Mediterranean but this economy was the one in which adherence to a narrower interpretation of the world became important, though we cannot prove that it was causal. In the Middle East, hence, the monotheistic religions of Judaism, Islam and Christianity became established, with their basis of a God who was extra-terrestrial. Their message was enshrined in written form and espoused a doctrine that overrode specific environmental and cultural attachments.[25] Further east, the multiplicity of gods characterised by Hindu religion survived the shift to the worship of the Buddha and the centrality of his writings in a sacred canon.[26] The same is very roughly true in China and Japan in their adoption of Buddhism and in Japan of the polytheistic Shinto. Confucian practice, however, was Earth-bound even if the father figure was all-important. In the breakdown of Marxism as a system of belief in central Asia, strong indigenous religious beliefs have emerged as environmentally tender in character compared with those of the Han Chinese and the Russians. The latter introduced fodder crops and heavy machinery that reduced the need for flock movement and so 38 per cent of the pastures of Inner Mongolia are classified as 'degraded'. This is a term which includes locust invasions as well as the spread of poisonous plants like *Oxytropis glabra* which causes abdominal distension and blindness in domesticated stock. In Tuva and Mongolia, collective institutions with a communist flavour have, however, continued to support movement among pastoralists.[27]

The coming of industrialisation cannot be said to have caused religious shifts in a direct sense. In Europe, the effects of the Reformation and the

Enlightenment preceded widespread use of fossil fuels and were more important within Christianity than burning coal. The successes of science and technology, however, together with the separation of fact and value made space for systems of belief which, though denying sacredness and transcendence, took on some characteristics of the preceding religions.[28] Marxism, for example, had its sacred texts and its saints and the 'conquest of nature' world view which some interpreters have assigned to Christianity as well; both agree on human exceptionalism.[29] The post-industrial outlook of the late twentieth century engendered a number of reappraisals of what was described as a 'disenchanted' world. Attempts to reconnect all things in a holistic fashion almost inevitably resulted in religious interpretations of Lovelock's Gaia hypothesis (see p. 15); other developments have included neo-paganism and ritualistic Green lifestyles. In the mainstream, the World Wide Fund for Nature started gatherings of world faith leaders in 1986 with the intent of declaring a common ground on environmental matters, and leaders of organised groups such as the Anglican and Roman Catholic churches, pronounce from time to time, without noticeable effect. Some Christians even endorse environmental degradation as likely to accelerate the final Armageddon which will precede the Second Coming. Given the tenor of postmodern times, religions appeal largely to signed-up segments and lack the universalism possible in pre-industrial centuries even given the slower communications of those times. Optimists may hope that the intentional and affective (that is, emotional) effects of a supernatural agency might allow or even ensure that humans monitor and accommodate themselves to nature's limits if they expect the species to avoid extinction.[30] (There is among the secular a faith that, somehow, the processes of organic evolution in the form of species extinction do not apply to us, just as many religious people assume that the future of humanity is out of their hands.)

   In some eras, religion could certainly cast a protective ring around non-human entities. Certain animals might not be eaten (the pig and the cow are the most obvious examples) and sacred groves retained their tree cover in periods of intensive agriculture in Greece and India, among many other places.[3] In general, such protection declined with the onset of industrialisation; whether it is possible (even if desirable) to 're-enchant' the world, a suggestion of the late twentieth century, is yet to be seen.[32] The desire to follow Kierkegaard's road of a 'gigantic passion to commit to the absurd' seems limited; even the more limited syncretic ideal of a spirit of compassion that 'extends to the ends of the earth' has a ring of the past and not the future.[33]

Myth, symbol, value

If we think of myth as a condensed and often poetic statement about what is taken to be unquestionably true in a given time and place, its role in environmental thinking and attitudes cannot be ignored. There are various ways in which the myths can be exhibited in symbolic form, sometimes as public ritual. In hunter-gatherers' societies there are obvious connections with the land; propitiation rituals to ensure future food supplies, for example, are common. In many pre-industrial agricultural groups, there are fertility rituals which express the hope of securing future food or water. Worship of images of the corn goddess Demeter in the Classical world is just one sample, transmuted into the blessing of the plough in medieval Europe and partially reclaimed in the shape of the corn dolly. It is difficult to see the carvings of the 'Green Man' in many European churches without concluding that fertility is in some way involved. In industrial times, symbols and myths of power are everywhere. Missionaries in India claimed (in 1862) that rail travel at 30 miles an hour was fatal to the slow deities of paganism.[34] Later, the developing symbols of this era all have environmental connections: the tractor, the aeroplane, clean water, pharmaceuticals, television and above all the private vehicle and tarmacked roads are replete with consequences for nature. Perhaps the greatest of all such symbols is the consumption of meat. The acquisition of hamburger outlets marks the emergence of a country into the 'modern' world (for example, in Russia and China). Meat consumption is divorced from nutrition just as the production of meat is separated from the animal behind the walls of the abbatoir, the packaging plant and the supermarket cabinet. Disguise became important so that some children have no idea of the origin of chicken nuggets or fish fingers.[35] From the 1960s, the centrality of meat came under questioning, ostensibly on health grounds so that new products with a lower fat content came to market, such as venison, ostrich and kangaroo. Some forms of symbolic behaviour persist: the attraction of hunting and fishing for men is important in many countries even where it is ritualised into a kind of drinking contest. It is more likely, though, to be a class-based hangover from the arrogation of sport to an aristocracy than a wired-in piece of Palaeolithic action.

Because the economies of the industrialised world are so dominant, attempts to uncover an overriding mythological narrative for most of post-hunter-gatherer time have been made. The most popular of these is to see many devices, such as the garden, the development of science and the ability to oust pests as attempts to regain the Garden of Eden. If only humans had enough knowledge, they seem to say, they could regain that prelapsarian state, not least by retelling the story in a less patriarchal way so that there emerges a true partnership between humans and the cosmos.[36] It is a nice thought and some environments from time to time typify it: islands are an obvious example, from the eighteenth century on to the marketing of the Caribbean and Indian Ocean islands in the twentieth century; gardens have followed a similar trajectory.[37] The images of agriculture as a profound departure have also spawned attempts at long-term mythologies. The cognate nature of 'culture' and 'agriculture' is central: the primary material theme that emerges from Roman times onwards is that of the conversion of the savage woods (*silvestris*) to gardens and cultivation (*cultura*); the recent inheritor is the Rome

Plow (much used in the Vietnam War) which is a bulldozer with a sharp blade.[38] Later developments have also created mythological narratives in the sense that they are often unexamined: the way in which Malthusian ideas are implicit in, for example, the discourse of 'spaceship earth' is one such.[39] In this context, the natural sciences continue to be important because they are always questioning myths rather than enforcing them in the way that neo-liberal economics, for instance, promotes the myths of a technology-based cornucopia.

In terms of value, there seems little question that humans have moved from a state in which they and their environmental companions had something approaching an equal value to one in which humans are top of the pyramid and the only objects of moral concern. It was perhaps a sign of change that, in the late twentieth century, there was a flurry of environmental philosophy and ethics which advocated the extension of moral concern to non-humans at a legally enforceable level rather than simply the joining of voluntary organisations.[40] But swimming against the tide is scarcely sufficient an image for that outlook: more like standing in the middle of a riverbed undergoing a Mediterranean flash flood from a deforested watershed.

Parts and wholes

When writing and talking about humanity and nature, the usual way of using words is to classify and to formulate smaller units of discourse. The ways of the natural sciences have influenced almost everybody who ventures into the field and they are powerfully aligned to breaking down the world into smaller units the better to be able to model it: hence the popularity of boxes in a flow chart and the relevance of the imagery of organ pipes (pp. 1–2). This mode fits neatly on to a cultural history in which there has been a millennium-aged tendency to categorise phenomena and behaviour into binary pairs. In spite of any evidence to the contrary concerning intermediates, western thought has insisted on delineating male versus female, good versus evil, human and the Other, mind and body and in this context especially, nature and culture. These are often seen as separate entities despite it now being obvious that nature has influenced culture (even when people have preferred not to admit it) just as culture has brought nature within its thrall in many ways. The two are clearly intertwined like the model of DNA's double helix, with the base pairs transferring the effects of both on to the development of the other. The tendency to form binary opposites is evident in this book, where oftentimes the two strands are afforded separate paragraphs and where the two themes of coalescence and fragmentation are juxtaposed. In this type of *Zeitgeist*, the advent of the digital computer with only two states (1/0) was bound to find an easy reception. There, the two are equal but, in most binary pairs, there is a dominant member with a subservient Other.

An enduring interest in this kind of thinking is also manifest. In western thought the notion of there being a perpetual and never-to-be-resolved struggle between good and evil came out of early Zoroastrianism and seems to have been lodged (not without opposition) in our thinking ever since: politicians routinely assert that if you are not with us totally then you are against us. The phrase 'the conquest of nature' arose in the nineteenth century and has certainly not fallen into disuse. Even in the present phase of concern over global climatic change, the effects of nature can be read about as somehow an independent malignity to be overcome rather than avoided by changing ourselves, rather like a heavy smoker complaining that his lung cancer was directly caused by the chief executive of the cigarette company. Nevertheless, the urbanisation of populations makes them unaware that food, for example, has to be produced somewhere and in an ecological and social setting. Into that gap, it is easy to promote such developments as GM crops without sparking a general awareness of the consequences[41] or to excoriate 'food miles' without looking at the total embedded energy of the product.

The attractiveness of different ways of thought is not surprising. In Asia, expressions of total unity in, and relatedness among, the whole of the universe are not rare. They may take the form of everything having a sacredness that needs to be acknowledged, as when Japanese shinto practice invests everything including the lavatory with *kami* or gods; of saying that all will be well if humans simply follow the ways of the universe as water finds its way down the hill (found especially in the *Tao Te Ching*) or in models like the Jewel Net of Indra where a net cast over the universe has faceted jewels in every intersection which in turn reflect every other jewel. There is nothing that resembles the A→→B causation with which we are so familiar, coming from an analytic tradition that demands separate and preferably quantifiable entities. Some of this type of thinking came into the west in the nineteenth century with the translations of Asian texts, some more with the 'alternative thinking' that swept California in the 1960s and 1970s. It is perhaps there as a seed in the ground, full of potential but awaiting the right conditions for germination. In the west, the term holism was coined by Jan Smuts (1870–1950) in the early twentieth century, and the ideas taken up by Arthur Tansley (1871–1955) in his development of the conception of the ecosystem. Indeed, in the last 100 years, more holistic thinking has come through the channel of the discipline of ecology than from any other source. Inevitably, some of the scientific discipline of that thinking has got lost in the attempts to adapt it to a social and political movement, but better that way round than have it ignored: it is easier to trim a beard than to produce hair follicles.

The culmination to date of holistic thinking has been J. R. Lovelock's Gaia hypothesis, part of which was outlined in chapter 1 (p. 15). In a reversal of conventional thinking, Lovelock has argued that life has created many of the conditions on Earth, not that life has simply adapted to the physics and chemistry of a cooling planet. The composition of the atmosphere and the chemical make-up of the oceans would, he has said, be very different if there had been no life on Earth. Ways in which human consciousness can evolve may start to reduce the gap between Darwinian thought and the idea of

purpose for a whole system such as the Earth. In some ways, holistic thinking has been enforced by global climate models (GCMs), with their emphasis on complex and at-a-distance feedback loops and by the way in which the
emitters of carbon compounds contribute to a global homogeneity which
affects us all.

There have, of course, been social studies which attempt a comprehensive
view. The most famous of these is the extension of Malthusian ideas into the
whole people–environment relationship and its transformation into a series
of computer models. Taking existing types of relationship and projecting
them forward produced the notion of 'The Limits to Growth'. The key
element was the nature of exponential growth and the dislocations that it
would produce, especially in resource availability. The social scientists and
economists involved with the topic were especially unhappy and their refuta-
tions of the mostly pessimistic 'Limits' publications continued with the work
of Lomborg which attracted much attention in the early twenty-first century.
The 'Limits' ideas were re-cast in the ideas of an environmental 'Kuznets
Curve'. Its basis was that the economic growth of the conventional kind in fact
reduced the demand for resources as technology became more efficient and
also, therefore, produced less pollution. Evaluation of this viewpoint has later
suggested that it may be limited to a few special cases and that the positional
goods effect is still enhancing demand for resources. It does not seem to apply
to, for example, the profligate use of oil by the United States.[42] Emerging from
all this debate in the later twentieth century came the label of 'sustainability'.
Its origin seems to have been biological, in the shape of the J-curve of a biotic
population which levels off after a period of exponential growth. Since its early
appearances, the term has become rather like 'ecology' in the 1960s: a hurrah-
word with meanings so elastic as to hold up nothing. It is, however, firmly
entrenched in the rhetoric of environmental management and of development. It also became tied in with the notion of resilience. This is a difficult
ecological concept in a world characterised by non-equilibrium states but
hinges around the amount of change which a socio-ecological system can
undergo while still retaining its ecological functions and feed-back mechanisms. An adaptive capacity is obviously part of this set of ideas. The spatial
aspects of it need an acknowledgement that any local sustainabilities must not
rely on a heavy ecological footprint elsewhere.[43] In a general sense, climatol-
ogy seems to have revived a basic environmental determinism, just as its neo-
Malthusian form was declining.

The practicality of 'sustainability' was paralleled by a more mystical view of
humanity as being on the verge of a mass transition of consciousness to a
different kind of world. Its foundations were in part the Gaia hypothesis and
its less scientific interpreters but also the popular versions of particle physics
in which reality depended upon the questions that were asked. This indeterminacy led to ideas about enfolded reality, systems views, autopoietic systems
and wholeness which, with a shot of Californian-style Zen, demanded a move

onwards from neo-liberal economics, Newtonian tramways and Cartesian duality into the experience of sitting on the beach and feeling a total unity with the light and the waves. There were times in the 1960s and 1970s when it felt that the west at least was burgeoning with something new and different which had not yet been born but which was just on the edge of coming into being, rather like some of the slow movements of Mozart piano concertos or some of his opera ensembles. The isomeric and opposite trends to individuation and fragmentation seem not to have been noticed and these at the moment at least appear to have become dominant.[44] The individual line of particulate powder seems to be a more accurate symbol than the shared spliff; John Cage continues the fragmentation of Webern and Schoenberg noticed in chapter 4. In a different sort of language, Isaiah Berlin reminded us that Utopias were generally achieved only be trampling on somebody else and that a kind of equilibrium, necessarily unstable, was likely to prevent mutual destruction; his context was political but it may have a far wider resonance.[45] As Brecht said [*Einverstandnis* (Agreement)],[46]

> . . . it takes a lot of things to change the world:
> Anger and tenacity. Science and indignation . . .
> The understanding of the particular case and the understanding of the ensemble:
> Only the lessons of reality can teach us to transform reality.

One of the lessons seems to be that the ground for agreement and the avoidance of conflict is normally very limited.

Unpredictable woods and pastures

The poet John Milton (1608–74) was a Puritan and so seems an unlikely author of the lines:[47]

> Chaos umpire sits
> And by decision more embroils the fray
> By which he reigns; next him high arbiter
> Chance governs all.

But many a process has unpredictable outcomes, even where modern science is involved. In fact, the natural sciences have developed ways of talking about non-equilibrium states and the ways in which a sensitive dependence upon initial conditions might produce trajectories whose differences grew exponentially through time. Even the most simple models of population change in 'natural' ecosystems show a variety of behaviours that might include a simple return to equilibrium but also cycles, cycles upon cycles, and totally irregular courses. Spatially, these would present themselves as mosaics which exhibited chance and could be partially described by the tools of chaos theory whose popular image was the butterfly effect, in which a butterfly's fluttering in Tokyo results in a storm in New York. This shows some of the gaps in scientific theory and praxis which give rise to predictabilities being overshadowed by the contingencies of detail in a world which humans have made both materially and symbolically. There is, too, the idea of emergent qualities: that a whole can show properties which are not predictable from the component parts. Those of water, it is argued, cannot be forecast from those of hydrogen and oxygen. Much of the thinking along these lines came together in the work of the chemist Ilya Prigogine (1917–2003), a Nobel laureate, who constructed the idea of self-organising systems. These of necessity were structures which dissipated energy and thus created entropy but in their lifetimes also created complexity. Living organisms were possible examples, though the existence of a genetic blueprint argues against such a label. Ecosystems, on the other hand, or the Gaia hypothesis, fit the concept quite well. Purely social systems sit less well, except that they can be so complex that 'self-organising' with all its connotations of apparently random behaviour as well as foreseeable trends (consider global capitalism as an instance) is an applicable term. Management of self-organising systems (the word 'autopoiesis' – self-creating – is sometimes used) begins with minimising entropy and preserving diversity.[48]

No body of theory has yet managed to produce and order historical data from this restless world, and this was true of many historic sciences, such as palaeontology and evolutionary biology, until the late twentieth century when compared with ahistoric fields such as physics, chemistry and molecular biology. Thus, notions such as 'balance' and 'sustainability' need very keen scrutiny.[49]

'The balance of nature'

In either an implicit or a direct form, this phrase appears in from the seventeenth century onwards European languages. It sits upon a history of thinking that encompasses 'nature' as a state to be imitated as a basis for human life. Many human actions upset the balance and, in the words of its Chinese equivalent, 'lose the mandate of heaven'. In contrast, the lessons from science about the last million or so years (and, indeed, beyond them) is that equilibrium is not a normal state. Climatic change, to begin with, imposes all kinds of change upon non-human organisms and ecosystems in which the adjustment of terrestrial biomes to a low-ice planet is still taking place and which has gone on throughout the Holocene, not always at a steady pace. The biological adaptation has proceeded upon an ecological timescale in which, for example, tundra was replaced by birch scrub and then by mixed deciduous forest, or in which woodland climbed up mountainsides, or in which deep deserts like the Sahara had a moist and green phase in the mid-Holocene. This temporal change has been accompanied by an evolutionary timescale in which species have undergone the kinds of adaptation that Darwin and his followers have made a foundation of our worldview. The Holocene has not been regarded as a major period for speciation except in very isolated environments such as islands well separated from other land masses or where humans have created the conditions for hybridisation.[50] Speciation has, until recently, been based entirely upon measurement of anatomical characteristics but, now that direct observation of genetic material is possible, any evolutionary shifts will become much better appreciated. Darwin's basic ideas, as illuminated by Mendelian genetics, have been amplified by what is usually called epigenetic change. The usual definition of this process centres on the study of heritable changes in gene function that occur without a change in the sequence of nuclear DNA. This includes the study of how environmental factors affecting a parent can result in changes in the way genes are expressed in the offspring. Thus, the life of a parent can be expressed in its descendants. One metaphor expresses it thus: 'Just as the conductor of an orchestra controls the dynamics of a symphonic performance, epigenetic factors govern the interpretation of DNA within each living cell'.[51] To which we might add that no two live performances are ever the same. An analogy of this idea in the field of culture enlarges our whole view of how cultural history might have possessed a scarcely heard ground bass upon which the recorded variations dominate the passacaglia of historiography.

Cultural change can also exhibit disequilibria. There are exogenous developments as when an agricultural society, for example, is forced by persistent drought to abandon its territory. Industrial-era equivalents might be the settling-down of itinerant loggers once an old-growth forest has been exhausted of its saleable timber or the removal of miners from the location of a worked-out mineral ore. Cultural inertia is sometimes a factor, as when a community insists on its practices remaining the same even when its economy has been totally transformed: what Marilyn Strathern called in the context of the British BSE epidemic of the 1990s, 'the fatal traditionalism of the English'.[53] Yet, the conflict between a sensitivity towards nature and material demands was certainly present in early modern Europe and has not been transformed in kind, though magnified in intensity.

One consequence of all kinds of disequilibria is the disappearance of economy–ecology relationships at the largest scales. Once agriculture was available then hunter-gatherers disappeared: not everywhere and not always permanently but they were displaced to the margins of cultivated lands in forests, grasslands and the tundra, and by 5000 BC most of the world's population had adopted the newer way of life along with its new attitudes and technologies. Similarly, the penetration of industrialism into solar-based agriculture was even more rapid, aided as it was by some aggressive empire-building. So, a way of life that was normal in AD 1750 had almost everywhere felt the hand of Birmingham, Springfield or Essen within a century. But if asked in 9000 BC, the hunter-gatherers would have said, 'oh yes, we know how to get along fine – we'll be here next spring' and the farmers would have said in 1750, 'we have a famine now and again but we know how to recover, so thanks, we expect to keep this up'. In other words, they both thought they had a sustainable economy.

This is not the book in which to explore the meaning for the future of the 100-odd definitions of 'sustainability'.[53] Viewed historically, most of them are rooted in notions of long-term equilibrium achieved after periods of rapid growth, as in the classic J-curve of a 'natural' population. An upper asymptote and some form of 'balance' are assumed. Yet, the historical evidence suggests that, for a variety of reasons rarely connected with natural phenomena, any economy can make a rapid transition to another: in other words, there has never been such a thing as sustainability in the way current advocates would like to see it. The key reason for the metamorphoses is surely the possession of technology and, for this reason, to leave out of any development of 'sustainability science' the unpredictable effects of, say, biotechnology and nanotechnology seems to be unreal.[54] This brings us to the uncomfortable position that history is not much use in predicting the future, a reason often given as justification for its access to public money. In geopolitics, it may be the case that, for western nations to invade a country in the 1920s and again eighty years later, produces similar chaos but the same is not true of environmental history. The populations of bacteria, the courses of the rivers and the precipitation–evaporation ratios are all different. So disequilibrium is not only observable in the world but its consequences follow through for those who want to project its outcomes into the future. As Gellner says, 'the neolithic and industrial revolutions cannot plausibly be attributed to conscious human design . . . [their social order] simply could not be properly anticipated or planned or willed'.[55] To which we might add 'environmental order' as well. The relationships of disequilbria to indeterminacy are related to energy use. A plentiful supply of energy leads to complexities in society (including hierarchy) and thence to non-linearities, few of which are foreseeable and some of which may exceed the human capacity to adapt.[56] A diminishing supply (since oil production might have passed its peak around the 1980–2000 period) is equally likely to produce non-linear social and environmental effects. It is an interesting thought that capitalism produced relaxed population-resource ratios and that these allowed choices, including post-1789 democracy, which tighter relationships do not. If climate change, biodiversity loss, the spread of HIV and the fear of global pandemics of SARS or avian flu are put into one frame, then the resulting hologram might well be that of the ghost of Thomas Malthus.

The 'nature' of consciousness

All through this chapter runs the nagging question of what constitutes ideas, attitudes and thought. The neurophysiology of the human brain is central but there are also enquiries about transmission of ideas, their modification and their translation into action. The literature from all kinds of scholars on this matter is immense and this is not the book in which to explore it save for one particular aspect that merits a mention: if we accept that there is a holistic entity called 'consciousness', then has it changed during the time of human tenure of the Earth or is it a constant which can adopt new world views from time to time without altering its basic characteristics? (There is, too, the additional complication that only consciousness can talk about consciousness and the possibility that there is 'unconsciousness' in which all sorts of models may be made.)

The historical evolution of consciousness is a difficult topic because it depends more than usual upon inference from incomplete evidence. A complex set of arguments by Mithen[57] locates the origin of today's consciousness in the Middle Palaeolithic, between 100,000 and 30,000 years ago, growing with the evolution of *Homo sapiens sapiens*. The biological function of consciousness is firstly to allow the building and testing of hypotheses in the social sphere: if one can examine one's own mind, then it predicts to some extent what others think as well. Mithen proposes that the Neanderthals did not possess an advanced degree of development and that language is the key to progress. The critical process is the import of non-social information into what had previously been the domain only of social intelligence. Only thus could people consider environment reflexively so that moose became good to think as well as good to eat. If the Middle Palaeolithic was the key time, then what about other transitions such as that from foraging to agriculture? Mithen considers that the evolution of the mind to a recent biological-social format is central, and that all other changes are contingent on local factors. Thus, the 'invention' of agriculture in more than one place shows that the mind deals with certain problems in a certain way but that the details are dependent on what is to hand. This view is the core of one debate about the suitability of the human mind to cope with the environmental pathologies which are detected after industrialisation.[58] The main strand of thought here is that the human mind is still basically a hunter-gatherer or early agricultural mind. This can cope with short-term and limited-space difficulties, but the abilities of today's human beings to deal with the consequences of rapid changes lag behind the facility to create new situations. It is a controversial hypothesis but some evidence for it can be seen in the political approach to environmental instabilities, most days of the week.

Much of the anxiety is condensed in thinking about security. Security is a feature of the conditions of the individual and of society, including the nation state. It involves a sense of well-being derived from protection against harm and injury, together with access to basic needs. Thus, an unpolluted set of surroundings, clean water, food, a stable climate and a comprehensible social matrix are all affected by environmental considerations. Without security, then there is often conflict, whether local violence, civil war or interstate struggles, with poor people and poor nations the most likely to suffer. Exacerbations include rapid population growth that causes not only depletion and degradation of resources and environments but also subdivision of their availability and hence more scope for argument. There are also late-twentieth-century shortages, as with water in the Middle East, and worries about shortages, as with oil. In that period, too, the concerns over climatic change and its likely sudden flips became highly visible.[59]

In historical perspective, insecurities with an environmental dimension are likely to have been a concomitant of human life all along. For foraging groups, famine was probably a recurrent feature, though one which could be ameliorated by social reciprocities and, in any case, an annual cosmogenesis gave hope. Agricultural societies were responsive at larger scales when city states kept grain stores against lean years or enforced periods of fallow for fields and trees. The limitations of their powers were set off as Acts of God, outside their realm entirely. The same attitude came through into the Industrial era when insurance began to be widespread. States found that they could act, as against gross pollution, for example, but might not always choose to, as in the British refusal to help with famines in India and Ireland in the nineteenth century and, indeed, in Bengal as late as the 1940s. Since 1950, truly international action has been possible and has been deployed against famine and in the aftermath of earthquakes and volcanic eruptions. Even (to move outside the time-frame of this book) the great Asian tsunami of 2004 was not regarded as outside either practical help or human mental boundaries, for few except the immediate victims resorted to call it an Act of God. Security is, then, better for some people and worse for a great number, with (as so often) the poor being more vulnerable to increased insecurity, even to global processes such as climate change, let alone the kind of rapid population growth that forces them to live in places vulnerable to environmental instability.

The drive to dominion

The message of Genesis I of the Bible encapsulates an entrenched human trait, no matter how much theological wriggling can be applied to subsequent material: humans are to have dominion over all the living things of the Earth and to get to name them.[60] In two chapters of one 'book' we have the theme of many others: that there is an empirical story of action and a cultural story steeped in language.

The species to which we belong has lived in two worlds during its evolutionary history: an ecological world and a psychological cosmos, but they are not separate, as many examples show. In the spirit of looking beyond the binary divisions to which our culture is prone, the question can be asked, 'are coalescence and fragmentation parts of the same thing?' Coalescence can be seen as a vast market being created by international institutions so that it can be dominated by a few megacorporations that can externalise their costs. Most of them then fall upon environment, with the coalescence extending to all the inhabitants of the world via environmental instabilities. Fragmentation stratifies communities so that either there is a cascading hierarchy of power from those who Know the Truth, or power is so decentralised that everybody feels that, in their small way, they are in control of something: 'For every man alone thinks he hath got / To be a phoenix, and that then can be / None of that kind, of which he is, but he.'[61] Donne could not have known just how obvious that might become in the twentieth century when individualism at many scales became both possible and desirable (thanks to energy accessibility) at the expense of the collective and of any sense of connection to everything else. As one consequence, in the west at least, we seem to be poor at enjoying the Other without possessing and controlling it. At least, power can be tempered by contingency: it is interesting that out of 148 species of mammals with adults weighing over 45 kilograms only fourteen have been domesticated and nine of those were found in south-west Asia; sub-Saharan Africa had fifty-one of the heaviest but none has become really tame. Likewise, two-thirds of the heavily seeded wild grasses that were candidates for domestication grew in western Asia.[62]

So, it comes to dominion. Human consciousness is deeply imbued with the need to control. People, it appears, are not happy unless they have something to dominate, even if it is only a pet parrot. Even cultures which eschew environmental interventions in their literature carry them out in practice, and the brief upsurge of interest in Erich Fromm's ideas of freedom to be, rather than freedom to possess, did not survive the 1970s even in northern California.[63] This urge is easily exploited by those who have the tools to make us want to possess things, especially in terms of positional goods, and so the rich do not make the choice not to have things; the poor of course, do not have choice. So dominion can come through symbolic practices such as using a private vehicle. Even meat can contribute to the definition of national identity: beef is implicated in John Bull's love of freedom, in the centrality of 'steak et frites', and in the supersize steak bigger than the plate.

There is scope, therefore, to investigate the history of humans' sense of identity and the degree to which it depends upon an ability to control other people and to have power over nature. Not least, we might be better off for knowing how much we fear nature. In the west at least, it would be interesting to know (and perhaps not without practical application) whether the men in black clothes are always happiest when cutting down trees, or at the very least ordering others to do so. In other words, the power relations between men and women are part of the whole cultural setting for environmental change.[64] All are perhaps wrapped up in the ability to be literate, because the ability to write and to preserve written accounts have been a source of power within and between societies for thousands of years.[65] Another layer to the ideas of control is 'space' in the sense that the more humans there are, the more control is exerted over both the social and environmental systems of what is seen as a 'full' or at any rate crowded planet.[66]

AT THE YEAR 2000

The year 2000 was the occasion for a great deal of stocktaking, review and advocacy. Most of it provided a great wealth of empirical studies, one of which is quarried here for its information on the 1950–2000 period. There were fewer urges to round up and enclose material on environmental thought, so here the former is followed by a superficial collection of a few of the non-material themes that seem to have become sufficiently crystallised out to be coherently described.[67]

Knowing where we are

It is common currency that environmental change has been faster than ever before during the 1950–2000 period. The Millennium Assessment (MA) jointly carried out for international agencies talks of alteration 'more rapidly and extensively than in any comparable period of time in human history'[68] and applies this precept to 'ecosystem services'. These are the 'natural' systems (no longer unaffected by humans, of course) which supply or affect food, water, disease management, climatic regulation, and non-material aesthetic and spiritual demands. The Assessment calculates that 60 per cent of the 'ecosystem services' are being degraded or used unsustainably, a group which does not include fossil fuels which are non-renewable anyway. The work suggests that the costs of degradation are often shifted from one group to another (notably from the rich to the poor) and from one generation to another. These are judgements about environmental futures and, as such, are not germane to the present study, except in so far as they contain historical resonances, too. Whenever there has been exploitation of nature by a particular stratum of a society, then the costs have been borne elsewhere whenever possible. Medieval aristocracies allowed the preservation of game for hunting to diminish the diets of local peasantry; industrial barons made sure they lived upwind of the smoke and stink that their factories visited on the working classes. Storing up trouble for the future is not new, either: wastes from eighteenth- and nineteenth-century industrial plant still influence the development of cities directly as unreclaimable land and radioactive material from careless atomic establishments keep appearing despite assurances to the contrary over the years. By contrast, though, some practices store up wealth and gain, as with tree-planting or with continued attention to the fertility of fields, achieved by judicious cropping and manuring. It may be that the burning of fossil fuels has allowed energy to be turned into knowledge of how to do without them. What has been different in the last fifty years, it is argued, is the extent and intensity of metamorphosis which, in turn, leads to a high degree of irreversibility. We could say that all changes through history conform to Heraclitus' dictum that you cannot step into the same river twice, but we can agree that systems taken beyond their renewability limits or used to push other ecosystems to a new state (perhaps stable, perhaps fluctuating) can be regarded as irreversibly changed. Further, study of the past suggests that not all changes occur on a gradual timescale that allows slow adjustment: there may be sudden changes in the physical parameters of an ecosystem (be it a 'natural' one or one with a high degree of human influence) or in the abilities of a given society to cope with the changes it encounters.[69] The probability that any global environmental change will first have a differential impact according to place and power, and then cascade on to everybody, seems high.[70] The ecological footprint is already like a mountain path, in the sense that separate bootprints have coalesced into a bare track which is open to erosion and which gets wider as people use the edges more and more.

The timing of major accelerations in the rate of change of human-driven processes is variable: carbon dioxide concentrations increase markedly from 1800, but much-manipulated ecosystems (including urban growth) and the loss of tropical forests are 1900-onwards changes. Energy usage underwent a number of quickenings but those after 1950 are critical. The nineteenth and twentieth centuries involved a much more rapid response to medical emergencies, such as pandemics, and also to famine. The perception of poverty and a feeling of pessimism about the human-environment relationships came about mostly after 1950. The period 1950–2000 was also a time of enhanced measurement and dissemination of knowledge about all kinds of ecosystem change leading to amplifications and shifts in our understandings and interpretations of them, several of which are informed by environmental histories (Tables 6.1 and 6.2). None is more graphic than the International Panel on Climate Change finding that the globe's atmosphere is in a state never before experienced. The rapid metamorphoses of the last 300 years have been highlighted by international programmes such as the LUCC (Land-Use and Cover Change) and PAGES (PAst Global chanGES) programmes of the International Geosphere–Biosphere Programme (IGBP). Much other work of the kind discussed in chapters 2 to 5 has made for a far more complex but, clearly, much more accurate picture than would have been possible twenty years ago.[71] The kind of assumptions (still occasionally to be found) that all ecosystems were in a pristine condition until hit by intensive agriculture, or that all grasslands were virgin until converted to cultivation or ranching, and that tropical forests were basically 'natural' environments, have all been replaced; this is possibly the most important long-term finding of historical studies. Not the least of the implications is that land-use change is an integral and significant component of the surface-energy budget of the planet as well as of the carbon cycle.[72] Beyond such studies, another major shift in knowledge has brought the realisation that climatic shifts are often non-linear on all scales and, hence, episodic and abrupt, with multiple equilibria. In such a world of chaos and complexity, prediction becomes a much more difficult enterprise, even when good accounts of long-term trends exist.[73]

An economics-based view of such impacts also has lessons for historical interpretation. As an ecosystem enters a wider market (as when coalescence brings about trade opportunities, for example) then its non-market benefits

TABLE 6.1 Changes in the understanding of land-cover and land-use changes

Previous interpretations	End-of-twentieth-century understandings
Rate of land-cover conversions indeterminate	Importance of last 300 years
Concentration on tropical forests	Acknowledgement of effects on grasslands, open woodlands, wetlands
Assumption of pristine condition in a previous era	Millennial depth of manipulation
Changes are permanent	Trajectories are reversible, not forecastable and in constant state of flux
Spatial homogeneity	Local and contingent fragmentation is important
Population growth the main cause	Responses to changes in economy and policy trigger events in both physical and cultural spheres
Localities are key	Hotspots of change are important but local influences interplay with global and worldwide processes which may be either attenuated or amplified
Impact on carbon cycle a main focus	Wider considerations such as human health, biodiversity, albedo, hydrology, methane, nitrogen oxides and many more
Impact on humans dependent on magnitude of biophysical change	Impact dependent upon vulnerability of settlements and economies

Source: modified (with permission) from E. F. Lambin and H. J. Geist, 'Global land-use and land-cover change: what have we learned so far?' *Global Change Newsletter* **46**, 2001, 27–30.

Note: 'Previous' is not defined closely but most of them could be found in the mainstream thinking of the 1960s and 1970s: the end-of-twentieth-century understandings have come up quite fast in the 1980s and 1990s. The driving force underlying many of them has been research into climatic change and modelling, but concern with biodiversity loss and the life-support of poor people have also been key features. Compare with the pivotal role of the 1980s in Table 6.2.

are often lost or degraded, even if in the long term they may be higher than the market values. Had the forests of Indonesia been harvested for renewable products rather than logged off then their income-producing potential would be much higher and there would be fewer watershed disasters. A long-term and relatively low-level output from a system (and especially those depending upon biological resources) will last longer and be more stable than intensive reaping followed by abandonment: this was a lesson learnt in the twentieth century by the managers of federally owned forests in the United States in the face of an aggressive private sector. The consequences of short-term exploitation are often high: the collapse of the Newfoundland cod fisheries in the early 1990s cost tens of thousands of jobs and a monetary loss of $2 billion; in the United Kingdom in 1996 alone water pollution and eutrophication cost $2.6 billion in clear-up charges, and, in Italy, algal blooms (again, caused by eutrophication) in 1989 deprived aquaculture of $10 million and

TABLE 6.2 Shifts in attitude in recent decades

Pre-1980s	1980s	1990s
Culture is natural	Nature is cultural	Nature and culture have a reciprocal relationship
Humans react to environment	Humans are pro-active in the environment	Humans are interactive with the environment
Environment is dangerous to humans	Humans are dangerous for the environment	With care, neither is dangerous; without it, both are dangerous
Environmental crises hit humans	Environmental crises are caused by humans	Environmental crises are caused by interactions between humans and environment
Adaptation	Sustainability	Resilience
Technofixes	No new technology	Minimal and careful use of technology

Source: adapted (with permission) from S. E. van der Leeuw, ' "Vulnerability" and the integrated study of socio-natural phenomena', *IHDP Update* **2**, 2001, 6–7.

Note: this Table is a companion to Table 6.1. In a sense it acts as a context for 6.1 by providing more general ideas within which 6.1 sits. But some of the broader concepts of this table have fed back into the more specific notions of 6.1. Some of the headings of 6.2's column 1 are nineteenth century in their formulation, others like 'environment is dangerous' are much more recent.

tourism of $11.4 million. One dichotomy made clear in the 1970–2000 period was the one between those who believed in regulatory approaches to environmental 'problems' and those who believed equally fervently in the ability of 'free' markets to produce solutions via the routeways of neo-liberal economics. It is, of course, possible that some of the triumphs of the 'free market' were, in fact, diffuse responses to environmentalist publicity and campaigning. The simplicities of single-cause changes in human–environment relationships have to be accompanied by those that acknowledge complexity and are place-specific.[74]

Any discussion of the last fifty years has to recall that economic forms of hybridity have been common. Early agricultural societies carried on hunting; advanced solar-powered agriculturalists adopted some powered tools but not others (think of the slow rate of introduction of the tractor into Europe); and 'alternative' movements look askance on bagged fertilisers and hook up small wind turbines to their power supplies. A late-twentieth-century Thai village, for example, may produce its staple foods by solar agriculture except that some of the equipment used is industrial in origin. The population will migrate seasonally to a city to earn some cash but their way of life is also touched upon by government development programmes. So, if there is enough forest for some hunting, then they encompass most of the lifeways of the last 10,000 years. They are, though, closely connected to their environment and so are able to respond quickly to any perceived loss of capability of the local ecosystems to process wastes or yield useful materials. At a wider spatial scale, when such feedbacks are retarded or decoupled, then the kinds of problems listed by the Millennium Ecosystem Assessment (MA) authors become significant. The Thai experience also points to the drawbacks in simple tables of energy use as a surrogate for environmental impact, such as Table 1.1 in chapter 1. Such data can be made more sensitive to 'ground truth' but they still conceal massive qualitative differences: even with an allowance for an 80:20 split in the late twentieth century between the low-income- and high-income-economy sectors of the human population, the 1,800 years of pre-industrial agriculture seem still to have had the highest long-term impact but it was nearly all in worldwide patches whereas the recent impacts are truly global, involving the atmosphere as well, to a measurable degree. Our understanding of the nature and magnitude of environmental changes is itself always in flux, not least when the variety of both temporal and spatial scale is considered (Tables 6.1 and 6.2). One difficulty for most normal humans is that understanding environmental change requires measurement *and* judgement, science *and* emotion, interpretation *and* being there, plus the interaction of all of these.[75] No wonder that the demands and uncertainties of modern living have been labelled as 'liquid life' so that a quantifiable concept of 'risk' has emerged since the nineteenth century as a simplifying aid.[76]

At a sub-global scale, one of the surprising findings was a correlation between human population density and species richness. This was first shown for the tropics: human population densities lined up with terrestrial vertebrates and higher plants in sub-Saharan Africa, and bird endemism with human settlement in the tropical Andes.[77] Extended to Europe, similar relationships apply for plants and mammals but not for birds. In Britain, solar energy availability correlates with human population density which, in turn, correlates with total species density, that is, richness.[78] In this analysis there seems to be no place for history or fossil fuels. Looked at more widely in time and space, however, these findings (though not so discouraging to environmentalists as might be imagined) are the outcomes of a decline in worldwide avian abundance. This seems to have gone down by 20 to 25 per cent since agriculture became widespread: Oceania alone lost 8,000 species. In Britain, breeding wild birds numbered fewer than chickens, and there was one cat for every seventeen wild birds. The end of the twentieth century yielded the information that there were then fewer than fifteen birds per person worldwide.[79]

This reinforces the idea (chapter 5) that humans come less and less into contact with wild forms, and that our knowledge of the 'worldwide' and the 'global' is inevitably mediated by technology; unmediated contact (if such a thing is possible) is local, small scale and contingent.[80] 'The living moment' restores meaning to local flows but has no sense of precision or global connections. It is parallel to a reminder, however, that small-scale communities (whether agricultural or foraging) were capable of species extinction, though usually by a combination of predation, biotic introduction and the alteration of vegetation rather than by predation alone. In terms of purpose, few such societies deliberately managed their resource base with the aim of 'conservation' and so effects on biodiversity were consequential impacts rather than management aims.[81]

ROLLING SMITHY-SMOKE

The poet Philip Larkin is famous for his dyspepsia but there is also a lot of affection shown, for instance, in his account of an agricultural show where '. . . time's rolling smithy-smoke / Shadows much greater gestures;'[82] The smoke of detail needs pulling back a little to see how some of the ideas in chapter 1 have fared, in longer perspectives than in the Tables of this chapter. The notion of a 'Golden Age', for example, is alive and well in many spheres but not in environmental history where the evidence of permanent change is now so massive, and its consequences so pervasive, that no era of the past can be considered rationally as golden. (Rationality is never the whole story, of course.) Similarly, the idea of the sublime, a learned behaviour from the eighteenth century in the west, has been replaced by a more existentialist set of attitudes in which we are on our own in a secular world. Thus, 'progress' is not guaranteed, especially when the twentieth-century assumption that it equals 'growth' is subject to questioning, albeit without much effect at the level of those driving economies. The technological fix is another area in which the pen is deployed to little effect against the sword. There are now so many words that those with the power to deflect or redirect the sword run the high risk of getting lost: what now constitutes a greater gesture in the way of *Capital* or *The Origin of Species*? The flows of the cosmos, so important to foragers and to the Chinese, are now understood to be too vast (especially since the universe seems to be still expanding) to be reckoned with: Gaia is big enough. After a difficult start, the Gaia idea seems to get more support from conventional science every year and, though many natural scientists would prefer another name, a holistic approach to feedback systems is an aspiration for many modellers. Focus on climatic change can be seen as a resurgence of environmental determinism. Fatalism is also alive and well, as are many binary concepts. Though recognised and analysed by many intellectuals, they provide a useful shorthand in the 'practical' world. An either/or presentation is always more impressive than a both/and alternative; metaphorical black and white (absent from the Powerpoint repertoire) more impressive than anything with intermediate shades in it, even if greying is more important in practice.

All this points to a realisation that ideas are mutable: they weave in and out of the material world and so cannot escape being influenced by it even if there is neither environmental nor technological determinism. In a new world there is new thinking, though there is never any guarantee that the two will be synchronous, for one may run ahead of the other, as happened with Darwin's and

Russel Wallace's ideas in the nineteenth century and with biotechnology today. Yet, some of the themes of our first chapter remain important throughout the 10,000 years of history examined here. Material linkages and their interweaving non-material equivalents are as important as ever though their scales have changed in spatial terms and in degrees of intensity. That the activities of the 1950–2000 years have affected the composition of the atmosphere and the surface of the Earth as never before is agreed by even those commentators who do not agree on the climatic and economic consequences. When we look at the whole picture – temporally and spatially – then one process can surely dominate all the others: if there had been very slow or zero population growth then everything mentioned in these pages would have been different in some way or other.[83]

Indra's internet?

In his concept of the noosphere, V. I. Vernadsky anticipated the degree of worldwide instant communication which many (but by no means all) humans can use. One of its consequences is that an event in one place can be instantly transformed into opinion in another: a tiger is shot in India and later that day ten more people in Luxemburg decide to join a wildlife-protection group. Another upshot, however, is that with very little effort we can find another three hundred environmental stories to which we could respond in some way, even if only with indifference. Yet a dead tiger is not simply an inert beast: it is one part of a worldwide (indeed perhaps global) network of ideas and capabilities in which the fact of its death (irreversible though it is) has to be accompanied by an examination of the context of ideas in which it occurred. One consequence is that we may know about many factors which influence environmental history but find it difficult to work them into broader treatments.[84] Historical events and processes happen within Newtonian principles and the laws of thermodynamics but after that there is whole world of chance and contingency. These preclude prediction because even an identical starting point would give different results if it was rerun. As Stephen Jay Gould summarises it, 'The law of gravity tells us how an apple falls, but not why that apple fell at that moment, and why Newton happened to be sitting there, ripe for inspiration.'[85]

One factor in the communications net is that of authority. What do we know, do we believe, do we act upon? In the times of the foragers, there were many instances of a continuous flow of authority between the humans and their spirit equivalents, with humans usually at the receiving end of instructions. Access to the knowledge of the other world was often mediated by a special person such as an elder or a shaman. In an agricultural era, authority is locally contingent to begin with but expands with conquest and empire. It tends to be hierarchical, with an off-planet Ultimate source, mediated as a mystery by specially endowed individuals (mostly but not entirely male), typically with restricted access to a written set of instructions.

The industrial era saw the written word as print taken to the far ends of the Earth by trade, conquest and empire-building, and the beginnings of electronic transfer by telegraph and telephone. Also important was the agency of the visual image following the development of photography. The dominance of western science and technology (and hence of 'the scientist') was spread by all these means. In the 'post-industrial' time since 1950 the dominance of science and technology has been ever greater, not least via the visual media of television and film. The downsizing of technology seems to have given greater influence to the small screen. Its combination with a person in a white coat and an on-screen computer seems to be the mid-twentieth century onward's equivalent of the newspaper and before that the sermon and the shaman.

Above all, the natural sciences offer us a narrative of authority to which we have to respond. We accept that the findings of the science are provisional but, because they are self-critical, are regularly tested and can be measured against their predictive value, then some reaction is essential. It is above all the natural sciences which laid the foundations for the greater intensity of environmental concern after 1950. Their authority has, nevertheless, been challenged from time to time. One attitude is a refusal to accept the 'facts' of scientific investigation for reasons such as:

- The people publishing the research are biased towards a particular finding because they are on a funding/promotional gravy train that favours their outlook. This is levelled against manipulators, who are often accused of concealing evidence of environmental effects, and against environmentalists, who are said to have 'their own agenda', whatever that may be.
- There is an obviously political motive, often on the part of national governments. The refusal to accept as definitive the fisheries science in Canada or in the European Union is based in a fear of the fishing industry; the refusal to sign up to the Kyoto Protocol is a refusal to join symbolically in a world endeavour even if it may be an inadequate set of measures.
- There is a rooted world view which regards any limits imposed by 'nature' as inconvenient: some social scientists would like to regard the laws of thermodynamics as temporary hindrances to the perfectibility of humankind, or the size of the human population as irrelevant to any environmental matters: 'that old Malthusian thing'.
- The local and the contingent are overlooked in the discourses of science, politics, technocratic attitudes and moral discussions which make for worldwide and global evaluations. The intricate enmeshing of the environmental and the social at local scales is difficult to relate to bigger pictures, especially when abstraction and reflexivity are dominant. One outcome is an 'informed bewilderment'.[86]

Setting aside the critical validity of the natural sciences does not impede the embrace of technology as a solution to perceived difficulties. It will, the advocacy runs, provide more of something, in the way remote sensing and (to a lesser extent) the geographic information system (GIS) open up the possibilities for more roads, more quarries and bigger irrigation projects based on larger dams. It will make the world cleaner because atomic power emits no carbon dioxide (not from the plant once it is running, at least[87]), some carbon dioxide can be piped under the oceans and stored and, in any case, hydrogen fuels can replace the carbon economy. Low-income economies can be encouraged to omit the carbon-fuelled stage of development and go straight to renewables.

There is resistance to this hyper-technological view of the world in both rich and poor countries. But in spite of 50 years of devoted environmental activism and thinking it has not been all that influential; from time to time radical leaders seem to throw up their hands, as when James Lovelock endorsed nuclear power in 2004. Some parts of the world still look like 'nature' even if an environmental historian knows that there has been human influence upon them. But their transition to control seems very likely: they will become 'environment' and then part of an energy-dense network of urban settlements the surroundings of which leave no room for contingency.

This book is about history, not futures, but a few last remarks may have some relevance to coming decades, taking their imagery from the pipe-organ at Lübeck. It looks as if the organist is becoming increasingly constrained to play from the one score, with his freedom to improvise being restricted to a few of the less important pipes. Some observers wonder if the resonances will, in fact, cause the whole machine to disintegrate; if so then we cannot complain that it was not built to cope with such pressures and that we never knew that the structure was shaking loose. But, equally, we were sure for many years that we knew how to make repairs and not to ask for impossible volumes. Somewhere in history, those restraints were overwhelmed by a mythology that triple forte was always best. There's none so deaf as those that will not hear, we could add. Frederich Schiller (who wrote the famous 'Ode to Joy') thought in 1786 that 'Die Weltgeschicte is der Weltgerichte' (The world's history is the world's judgement), without always remembering that it was made with a great deal of chance and contingency, was always changing and was distinctly unpredictable; who can say what the results of developments in biotechnology, electronics and nanotechnology will be? In human history, business was rarely as usual and this seems even less likely to be the case in future. We have now learned that our actions can bring about fluctuations of almost unimaginable amplitudes and that it is possible to reduce both cultural and natural diversity. If we wish to turn away from that trajectory then there has to be mutual coercion, mutually agreed upon: the organist has to decide upon a theme and to play it, even if some variations are allowed and the instrument moves from hand-blowing to solar power. At the same time, many of us may hear for ourselves as individuals the unharmonious resonances coming from more than one pipe, and decide to lower the volume.[88]

NOTES

1. J. McNeill, *Something New under the Sun. An Environmental History of the Twentieth Century*, New York: W. W. Norton, 2000/London: Allen Lane, 2000.

2. There are examples of metaphors in 'tribal' groups in N. Bird-David, 'Tribal metaphorization of human–nature-relatedness. A comparative analysis', in K. Milton (ed.) *Environmentalism. The View from Anthropology*, London and New York: Routledge, 1993, 112–25.

3. D. Dennett, *Consciousness Explained*, Boston MA: Little, Brown, 1991/London: Penguin Books, 1993, summarised in the source of much of this paragraph: S. Mithen, *The Prehistory of the Mind*, London: Thames & Hudson, 1996/Phoenix, 1998, pp. 245–6.

4. M. McLuhan, *The Medium is the Massage* [*sic*], New York: Random House, 1967; James Joyce, *Finnegans Wake*, London: Faber & Faber 1939, Part 4, episode 15. There are many later editions and the whole text is online at more than one location.

5. B. Latour, *Aramis or the Love of Technology*, Cambridge MA: Harvard University Press, 1996, trans. C. Porter. Originally published in French, 1993. The example of oil is mine, not his.

6. A. Berque, *Le Sauvage et L'artifice: les Japonaises devant la Nature*, Paris: Gallimard, 1985.

7. H. Tellenbach and B. Kimura, 'The Japanese concept of "nature" ', in J. B. Callicott and R. T. Ames (eds) *Nature in Asian Traditions of Thought: Essays in Environmental Philosophy*, Albany NY: SUNY Press, 1989, 163–82; D. E. Shaner, 'The Japanese experience of nature', ibid. 183–209. See also the special issue on the state of the field of 'nature in Asian traditions' in *Worldviews. Environment, Culture, Religion* **10** (1) 2006.

8. A. Bramwell, *Ecology in the 20th Century. A History*, New Haven and London: Yale University Press, 1989.

9. B. L. Whorf (ed. J. B. Carroll), *Language, Thought and Reality*, London: Chapman & Hall, 1956/Cambridge MA: MIT Press, 1956, reprinted by MIT in 1974 and available from them on CogNet subscription service. Whorf's arguments have been controversial because they are very difficult to test without using language. Empirical tests have mostly been about the naming of colour. Language's historical functions are discussed in W. H. McNeill, 'A short history of humanity', *New York Review of Books*, 29 June 2000, 9-11; The Eliot quotation is from 'East Coker', the second of his *Four Quartets*. The whole of this Quartet is of relevance to any historian.

10. This paragraph is extracted from the opening material in the online Wikipedia at http://en.wikipedia.org/wiki/Postmodernism#Uses_of_the_term accessed on 16 January 2006. The basic document is generally held to be F. Lyotard, *The Postmodern Condition. A Report on Knowledge*, Manchester: Manchester University Press, 1984 and amplified by F. Jameson, *Postmodernism or, the Cultural Logic of Late Capitalism*, London and New York: Verso, 1991. An application to the present discussion is in M. Gandy, 'Crumbling land: the postmodernity debate and the analysis of environmental problems', *Progress in Human Geography* **20**, 1996, 23–40.

11. In order of topic: A. E. Gare, *Postmodernism and the Environmental Crisis*, London and New York: Routledge, 1995; M. Soulé and G. Lease (eds) *Reinventing*

Nature? Responses to Postmodern Deconstruction, Washington DC: Island Press, 1995; Z. Bauman, *Postmodern Ethics*, Oxford and Cambridge MA: Blackwell, 1993. See also M. E. Zimmerman, *Contesting Earth's Future. Radical Ecology and Postmodernity*, Berkeley, Los Angeles and London: University of California Press, 1994. Nihilism is especially strong in A. Gare, *Nihilism Incorporated. European Civilization and Environmental Destruction*, Bungendore NSW: Eco-Logical Press, 1993. There is a very interesting analysis and a set of prescriptions in C. Birch, 'Eight fallacies of the modern world and five axioms for a postmodern worldview', *Perspectives in Biology and Medicine* **32**, 1988, 12–30.

12. The campaigners can be so hostile that, in Britain, new legislation was enacted in 2005 to make some of their activities criminal offences.

13. D. P. Tolia-Kelley, 'Affect – an ethnocentric encounter? Exploring the "universalist" imperative of emotional/affectual geographies', *Area* **35**, 2006, 213–17.

14. K. Milton, *Loving Nature. Towards an Ecology of Emotion*, London and New York: Routledge, 2002; N. R. Anderson, *Emotion, Belief and the Environment*, New York and Oxford: Oxford University Press, 1996. Both authors are anthropologists and we need to be grateful that they have ventured into this difficult area of study.

15. As an example, see M. Pacheco, 'Recycling in Bogota: developing a culture for urban sustainability', *Environment and Urbanization* **4**, 1992, 74–9; a more general paper is P. J. H. van Beukerin and M. N. Bowman, 'Empirical evidence on recycling and trade of paper and lead in developed and developing countries', *World Development* **29**, 2001, 1717–37.

16. E. O. Wilson, *Biophilia. The Human Bond with other Species*, Cambridge MA: Harvard University Press, 1984.

17. The leaflet accompanying the vast and wonderful installation ('The Weather Project') by Olafur Eliasson in the Turbine Hall at Tate Modern in London in 2003–4 tells us that '47 per cent believe that the idea of weather in our society is based on culture' and '53 per cent believe that it is based on nature'. As worded, the two propositions are not necessarily a binary pair, but we can see what they (the 'Tate Weather Monitoring Group') mean.

18. A. R. Damasio, *Descartes' Error: Emotion, Reason and the Human Brain*, New York: Putnam, 1994.

19. B. Braun and N. Castree, *Remaking Reality. Nature at the Millennium*, London and New York: Routledge, 1998, Part 3.

20. The efforts of, for example, UNEP, UNFPA, and international development agencies should not be denigrated. But they rarely produce binding outcomes, partly because of the diversity of cultures and purposes which they consider.

21. There are relevant discussions in, for example, R. W. Hepburn, *'Wonder' and Other Essays*, Edinburgh: Edinburgh University Press, 1984; A. Berléant, 'The historicity of aesthetics. I.' *British Journal of Aesthetics* **26**, 1986, 101–11 and II, ibid. 195–203. See also A. Carlson, *Aesthetics and the Environment. The Appreciation of Nature, Art and Architecture*, London and New York: Routledge, 2000.

22. My friend and mentor the late Dan Luten used to say, when confronted with such an accusation by Californian developers, 'the love of money is emotional, too'. Many terms that are difficult to define, such as 'balance' and 'nature', are links between the scientific community and the wider emotional responses of so many people. See K. Richards, 'Psychobiogeography: meanings of nature and motivations for a democratized conservation ethic', *Journal of Biogeography* **28**, 2001, 677–98. The Hume quotation is from *A Treatise upon Human Nature*, bk 2, pt 3.

23. But see, for example, the socio-ecological links in J. Goudsblom, 'Ecological regimes in the rise of organized religion', in J. Goudsblom, E. Jones and S. Mennell, *The Course of Human History. Economic Growth, Social Process, and*

Civilization, Armonk NY and London: M. E. Sharpe, 1996, 31–47 and chs 1–3 of M. Oelschlaeger, *The Idea of Wilderness*, New Haven and London: Yale University Press, 1991. Technology and medieval Christianity are dealt with by L. T. White, *Medieval Technology and Social Change*, Oxford: Oxford University Press, 1964.

24. K. Milton, 'Nature and the environment in indigenous and traditional cultures', in D. E. Cooper and J. Palmer (eds) *Spirit of the Environment: Religion, Value and Environmental Concerns*, London and New York: Routledge, 1998, 86–99.

25. S. Afran, *In Gods We Trust. The Evolutionary Landscape of Religion*, New York: Oxford University Press, 2004.

26. K. Vatsyayan, 'Ecology and Indian myth', in G. Sen (ed.) *Indigenous Vision. Peoples of India Attitudes to the Environment*, New Delhi: Sage, 1992,157–80. [Title is *sic*]

27. E. Erdenijab, 'An economic assessment of pasture degradation', in C. Humphrey and D. Sneath (eds) *Culture and Environment in Inner Asia*, vol. 1: *The Pastoral Economy and the Environment*, Cambridge: White Horse Press, 1996, 189–97; Chen Shan, 'Inner Asian grassland degradation and plant transformation, ibid. 111–23; C. Humphrey and D. Sneath (eds) *The End of Nomadism? Society, State and the Environment in Inner Asia*, Durham NC: Duke University Press/Cambridge: White Horse Press, 1999.

28. B. Szerszynski, *Nature, Technology and the Sacred*, Oxford: Blackwell, 2005.

29. The more that is learned about genetics and ethology, the less this type of exceptionalism (i.e. the notion that humans are totally different from other species) seems to hold up.

30. S. Afran op. cit. 2004, p. 272.

31. M. D. S. Chandran and J. D. Hughes, 'Sacred groves and conservation: the comparative history of traditional reserves in the Mediterranean area and in South India', *Environment and History* **6**, 2000, 169–86.

32. J. D. Proctor, 'Resolving multiple visions of nature, science and religion', *Zygon* **39**, 2004, 637–57; A. McGrath, *The Reenchantment of Nature. The Denial of Religion and the Ecological Crisis*, New York: Doubleday/Galilee , 2003.

33. Compassion and quotation from the closing pages (396–9) of K. Armstrong, *The Great Transformation. The World in the Time of Buddha, Socrates, Confucius and Jeremiah*, London: Atlantic Books, 2006. The dissent is mine.

34. A Revd J. Cummings, quoted in M. Adas, *Machines as the Measure of Men: Science, Technology and Ideologies of Western Dominance*, Ithaca NY and London: Cornell University Press, 1989.

35. A. Franklin, *Animals and Modern Cultures. A Sociology of Human–Animal Relations in Modernity*, London: Sage, 1999.

36. C. Merchant, 'Reinventing Eden: Western culture as a recovery narrative', in W. Cronon (ed.) *Uncommon Ground. Toward Reinventing Nature*, New York and London: W. W. Norton, 1995, 132–59; *idem*, *Reinventing Eden: the Fate of Nature in Western Culture*, London and New York: Routledge, 2004.

37. R. H. Grove, *Green Imperialism: Colonial Expansion, Tropical Island Edens and the Origins of Environmentalism, 1600–1860*, Cambridge and New York: Cambridge University Press, 1995. Aldous Huxley's novel about heaven (little known in contrast to his vision of hell in *Brave New World*) is called *Island* (1962). Its paradise is destroyed by the discovery of oil.

38. R. Waswo, *The Founding Legend of Western Civilization: from Virgil to Vietnam*, Hanover NH/London: University Press of New England, 1997.

39. M. Leach and J. Fairhead, 'Challenging Neo-Malthusian deforestation analyses in West Africa's dynamic forest landscapes', *Population and Development Review*, **26**, 2000, 17–43.

40. The seminal work was C. Stone, *Should Trees Have Standing? Toward Legal Rights for Natural Objects*, Los Altos, CA: William Kaufmann, 1974. In, 2006, Spain considered a law which gave 'legal person status to great apes (*Guardian*, 7 June 2006, page 9 of *SocietyGuardian*).

41. C. B. Herrick, ''Cultures of GM': discourses of risk and labeling of GMOs in the UK and EU', *Area* **37**, 2005, 286–94.

42. The computer models were developed by J. Forrester, *World Dynamics*, Cambridge MA: Wright-Allen Press, 1971; the key work was D. H. Meadows, D. L. Meadows, J. Randers and W. W. Behrens, *The Limits to Growth*, London: Earth Island, 1972, with its counterblast H. S. D. Cole et al. (eds) *Models of Doom. A Critique of the Limits to Growth*, New York: Universe Books, 1973 (in the UK it was called *Thinking about the Future*). The riposte is D. L. Meadows, D. H. Meadows and J. Randers, *Beyond the Limits*, London: Earthscan, 1992; then comes B. Lomborg, *The Skeptical Environmentalist: Measuring the Real State of the World*, Cambridge: Cambridge University Press, 2001. The later discussion of types of curve can be seen in D. I. Stern, M. S. Common and E. B. Barbier, 'Economic growth and environmental degradation; the environmental Kuznets Curve and sustainable development', *World Development* **24**, 1996, 1151–60. American society is built on the profligate use of oil according to D. Nye, *Consuming Power. A Social History of American Energies*, London and Cambridge MA: MIT Press, 1998.

43. 'Resilience' was the focus of an issue of the *IHDP Update*, Bonn: IHDP, 02/2003. For a more formal treatment see C. Perrings, 'Ecological resilience in the sustainability of economic development', *Economie Appliquée* XLVIII, 1995, 121–42.

44. Contrast F. Capra, *The Turning Point. Science, Society and the Rising Culture*, New York: Simon & Schuster, 1982/London: Fontana 1983, with R. Bennett, *The Fall of Public Man*, New York: Knopf, 1977/London: Faber & Faber, 1986 and Penguin Books, 2002.

45. I. Berlin, 'The decline of Utopian ideas in the West', in H. Hardy (ed.) *The Crooked Timber of Humanity*, London: Fontana 1991, pp. 20–48. There is only a limited treatment of the environment in F. E. Manuel and F. P. Manuel, *Utopian Thought in the Western World*, Oxford: Blackwell, 1979.

46. *Poems 1913–56*, London: Methuen, 1987.

47. *Paradise Lost*, Part I, line 907 et seq.

48. I. Prigogine and I. Stengers, *Order out of Chaos. Man's New Dialogue with Nature*, London: Flamingo Books, 1984; F. Capra, *The Web of Life*, New York: Random House, 1997.

49. Many of these ideas are brought together in I. G. Simmons, 'History, ecology, contingency, sustainability', in J. Bintliff (ed.) *Structure and Contingency. Evolutionary Processes in Life and Human Society*, London and New York: Leicester University Press, 1999, 118–31.

50. R. L. H. Dennis, 'Butterfly habitats, broad-scale biotope affiliations, and structural exploitation of vegetation at finer scales: the matrix revisited' *Ecological Entomology* **29,** 2004, 744–52.

51. The cultural interface with biology is explicated in E. O. Wilson, *Consilience. The Unity of Knowledge*, London: Little, Brown, 1998. The nature–nurture dichotomy and the role of epigenetic change is the topic of M. Ridley, *Nature via Nurture*, London: Harper Perennial 2004, which is written in accessible language, as was most of the website http://epigenome.eu in November 2006, from whence comes the quotation.

52. M. Strathern, *After Nature. English Kinship in the Late Twentieth Century*, Cambridge: Cambridge University Press 1992, p. 198.

53. J. Holmberg and R. Sandbrook, 'Sustainable development: what is to be done?', in J. Holmberg (ed.) *Policies for a Small Planet*, London: Earthscan, 1992, 18–38. Much of the original writing is collected in the seventy-two papers edited with an introduction by M. Redclift, *Sustainability*, London and New York: Routledge, 2005, Critical Concepts in the Social Sciences, 4 vols.

54. Influential statements, such as those of Kates et al. (see below), are sufficiently general so as not to exclude 'complex self-organizing systems' but seem to me to lack a sense of dealing with rapid and radical technological change. They are not in any sense 'wrong' but perhaps do not go far enough. R. W. Kates et al., 'Sustainability science', *Science* **202**, 2001, 641–2.

55. E. Gellner, *Plough, Sword and Book. The Structure of Human History*, Chicago: University of Chicago Press, 1988, p. 20.

56. R. N. Adams, *The Eighth Day. Social Evolution as the Self-Organization of Energy*, Austin TX: University of Texas Press, 1988; J. Gray, *False Dawn. The Delusions of Global Capitalism*, London: Granta Books, 1998.

57. S. Mithen, *The Prehistory of the Mind*, London: Thames & Hudson, 1996.

58. R. Ornstein and P. Ehrlich, *New World New Mind*, London: Methuen, 1989.

59. This paragraph relies on N. Myers, *Ultimate Security. The Environmental Basis of Political Stability*, New York and London: W. W. Norton, 1993; P. Roberts, *The End of Oil*, London: Bloomsbury, 2003, ch. 10; F. Dodds and T. Pippard (eds) *Human and Environmental Security. An Agenda for Change*, London and Sterling VA: Earthscan, 2005, Part 2; and a focused number of *IHDP Update*, 'Conflict and cooperation', Bonn: IHDP, 03/2004.

60. Genesis 1: 29 and 2: 19.

61. *An Anatomy of the World. The First Anniversary. (c.*1600)

62. W. J. Burroughs, *Climate Change in Prehistory*, Cambridge: Cambridge University Press, 2005, p. 278.

63. E. Fromm, *The Fear of Freedom*, London: Kegan Paul, 1942, reprinted down to 1980. The relinquishing of the desire to control is best achieved by experiencing a relation to the eternal, says M. E. Zimmerman, *Contesting Earth's Future. Radical Ecology and Postmodernity*, Berkeley, Los Angeles and London: University of California Press, 1994, p. 373.

64. In the west, wearing black has for centuries been associated with powerful men, especially in the nineteenth century when it became a kind of uniform. The tree-feller is William Gladstone (1809–98, Prime Minister of Great Britain at various times) and the commentary on power and clothing (in the west) is J. Harvey, *Men in Black*, Chicago: University of Chicago Press, 1996. See also M. French, *Beyond Power: Men, Women and Morals*, London: Cape, 1985; L. C. Zelezny, P.-P. Chua and C. Aldrich, 'Elaborating on gender differences in environmentalism', *Journal of Social Issues* **56**, 2000, 443–57.

65. J. Goody, *The Power of the Written Tradition*, Washington DC and London: Smithsonian Institution Press, 2000.

66. The 'full planet' idea has always been associated with the books of Herman Daly, such as *Economics, Ecology, Ethics. Essays towards a Steady-State Economy*, San Francisco: Freeman, 1980. A recent exploration is in J. A. Swaney, 'Are democracy and common property possible on our small earth?' *Journal of Economic Issues* **37**, 2003, 259–88.

67. There is a very broad and penetrative historical treatment of the social facets of world systems in J. Friedman, 'General historical and culturally specific properties of global systems', *Review* **15**, 1992, 335–72.

68. Millennium Ecosystem Assessment 2005, accessed at www.millenniumassessment.org in August 2005: 'Summary for Decision-Makers', p. 1. The meaning of

'comparable period' is not immediately clear but I take it to mean a fifty-year interval.

69. Examples of 'choice' in the face of environmental adversity are the main theme of J. Diamond, *Collapse. How societies choose to fail or survive*, New York: Viking Penguin 2005/London: Penguin Books 2006.

70. See the careful discussion of this by D. Liverman, 'Vulnerability to global environmental change', in J. X. Kasperson and R. E. Kasperson (eds) *Global Environmental Risk*, Tokyo: UNU Press/London: Earthscan, 2001, 201–16.

71. There is a good summary in *IHDP Update* 03/2005 with links to appropriate websites.

72. R. A. Pielke et al., 'The influence of land-use change and landscape dynamics on the climate system: relevance to climate-change policy beyond the radiative effect of greenhouse gases', *Philosophical Transactions of the Royal Society of London* **A 360**, 2002, 1705–19.

73. J. A. Rial et al., 'Nonlinearities, feedbacks and critical thresholds within the Earth's climate system', *Climatic Change* **65**, 2004, 11–38. An excellent summary account of both biophysical and social trends is given in R. W. Kates and T. M. Parris, 'Long-term trends and a sustainability transition', *Proceedings of the National Academy of Sciences* [of the USA] 100, 2003, 8062–67. There is a contextual discussion in S. R. Dovers and J. W. Handmer, 'Uncertainty, sustainability and change', *Global Environmental Change* 2, 1992, 262–76. A treatment in tune with the present book is in M. Fischer-Kowalski and H. Haberl, 'Sustainable development: socio-economic metabolism and colonization of nature', *International Social Science Journal* **50**, 1998, 573–87.

74. E. F. Lambin et al., 'The causes of land-use and land-cover changes: moving beyond the myths', *Global Environmental Change* 11, 2001, 261–9.

75. This is a free adaptation of a quotation from T. O'Riordan in H. Haberl, S. Batterbury and E. Moran, 'Using and shaping the land: a long-term perspective', *Land Use Policy* **18**, 2001, 1–8 at p. 3. O'Riordan's original is in his introduction to T. O'Riordan (ed.) *Environmental Science for Environmental Management*, London: Prentice Hall, 2000, p. 15.

76. Z. Bauman, *Liquid Life*, Cambridge: Polity Press, 2005; U. Beck, *Risk Society. Towards a New Modernity*, London: Sage, 1992, first published in German 1986.

77. A. Balmford et al., 'Conservation conflicts across Africa', *Science* **291**, 2001, 2616–19; L. Fjeldså and A. Rahbek, 'Continent-wide conservation priorities and diversification processes', in G. M. Mace, A. Balmford and J. Ginsberg (eds) *Conservation in a Changing World*, Cambridge: Cambridge University Press, 1998, 139–60.

78. M. J. Araújo, 'The coincidence of people and biodiversity in Europe', *Global Ecology and Biogeography*, **12**, 2003, 5–12; K. L. Evans and K. J. Gaston, 'People, energy and avian richness', *Global Ecology and Biogeography* **14**, 2005, 187–96.

79. K. J. Gaston and K. L. Evans, 'Birds and people in Europe', *Proceedings of the Royal Society of London* **B 271**, 2004, 1649–55; K. J. Gaston, T. M. Blackburn and K. K. Goldwijk, 'Habitat conversion and global avian biodiversity loss', *Proceedings of the Royal Society of London* **B 270**, 2003, 1293–300.

80. D. Abram, *The Spell of the Sensuous*, New York: Vintage Books, 1997.

81. E. A. Smith and M. Wishnie, 'Conservation and subsistence in small-scale societies', *Annual Review of Anthropology* **29**, 2000, 493–524; D. K. Grayson, 'The archaeological record of human impacts on animal populations', *Journal of World Prehistory* 15, 2001, 1–49.

82. 'Show Saturday', in Collected Poems (ed. A. Thwaite), London: The Marvell Press/Faber & Faber, 1988, 199–201.

83. A nice subject for some counterfactual history writing: a change from Napoleon winning at Waterloo or a different outcome to the War of American Independence.

84. Many of the creative arts fall into this category, but think also of our sonic environment: P. A. Coates, 'The strange stillness of the past: towards an environmental history of sound and noise', *Environmental History* **10**, 2005, 636–65. Even in a more conventional context, consider the diversity of material in the proceedings of the 2nd international conference of the European Society for Environmental History: L. Jeleček, P. Cromý, H. Janů, J. Miškovský and L. Uhlířová (eds) *Dealing with Diversity*, Prague: Charles University Faculty of Science, 2003.

85. S. J. Gould, *Wonderful Life*, London: Penguin Books, 1989, p. 278. The preceding sentence is an adaptation of his statement on contingency in the same paragraph.

86. This theme is explored for the Classical times to nineteenth century period by R. Collins, *The Sociology of Philosophies. A Global Theory of Intellectual Change*, Cambridge MA and London: Belknap Press of Harvard University Press, 1998, esp. ch. 15; M. Castells, *The Information Age. Economy, Society and Culture*. Vol. III. *End of Millennium*, Oxford: Blackwell, 1998. Baumann (op. cit. 1993) says something very similar.

87. There is no consensus as to how much carbon dioxide is produced in the early stages of the nuclear fuel cycle and in the decommissioning.

88. Not in the volume of talking to ourselves, clearly. The years 2006 and 2007 saw a remarkable burst of book-length treatments in the field of environmental history, most of which became available after this book was completed. The world scale gets attention but there seemed to be an interesting concentration on the regional and continental scales. An outstanding volume rooted firmly in a scientific world-view is F. Oldfield, *Environmental Change: Key Issues and Alternative Perspectives*, Cambridge: Cambridge University Press, 2005.

Further reading

The chapter notes lead to a very wide variety of source material but most of it is from books and journals. Every new book is, however, undertaken in the context of what went before and which has remained in the memory or the filing system of the author. Yet between the writing of this paragraph and its appearance in print there will almost certainly be published some other relevant book-length works. So this section will attempt very briefly to suggest a few follow-on sources for readers interested in general treatments of human impact and management in the environmental sphere, which have a historical dimension.

The books that attempt a worldwide scale of treatment are dominated by North American writers, and the names of John McNeill, Donald Hughes, John Richards and Donald Worster stand out. Any list must include:

D. Worster (ed.) *The Ends of the Earth: Perspectives on Modern Environmental History*, Cambridge: Cambridge University Press, 1989.

J. D. Hughes, *The Environmental History of the World: Humankind's Changing Role in the Community of Life*, London and New York: Routledge, 2001.

J. R. McNeill, *Something New Under the Sun: An Environmental History of the World in the Twentieth Century*, London: Penguin Books, 2001.

J. F. Richards, *The Unending Frontier: An Environmental History of the Early Modern World*, Berkeley and Los Angeles; University of California Press, 2003.

I have tried to put these and other works into context in:

I. G. Simmons, 'The world scale', *Environment and History* **10**, 2004, 531–6.

A very approachable text on land cover and land use through time is:

A. M. Mannion, *Dynamic World. Land-Cover and Land-Use Change*, London: Arnold, 2002.

A taster of the variety of work being pursued under the banner of environmental history is seen in:

Leoš Jeleček et al. (eds) *Dealing with Diversity*. Proceedings of the 2nd International Conference of the European Society for Environmental History, Prague 2003, Prague: Charles University Faculty of Science, 2003.

Books and journals are now supplemented by an immense array of web materials. Most of these are concerned with today's anxieties, though global orientation usually includes attempts at a historical setting even if only of the last fifty to a hundred years. The PAGES programme in Bern should be consulted, along with the IPCC, the IHDP and UNEP. The Worldwatch Institute's annual printed *State of the World* will convey any reliable data that can be found and is backed up by web material. The websites of the two major scholarly organisations in the field (ASEH and ESEH) are also worth scanning for bibliographical guides. The journals *Environment and History* and *Environmental History* carry the field further, as do those of a more ideas-based nature such as *Environmental Values* and *Worldviews; Environment, Culture and Religion*.

Glossary

Algal blooms   Sudden surges in growth of algae in aquatic environments, often colouring or clogging the water; bacterial decay then uses up much of the oxygen in the water and causes the death of, for example, fishes.

Anoxia   Without oxygen. Hence a toxic and usually fatal environment for animals. Most often used for a zone of the water column that has suffered pollution by over-loading of nutrients such as phosphorus and nitrogen.

Biodiversity   The variety of species per unit area or, more specifically, the totality of genes, species, and ecosystems of a region. This may comprise all or some of genetic diversity, species diversity and ecosystem diversity.

Biological productivity   The amount of living tissue per unit area per unit time added during periods of growth of plants and animals. Usually expressed in units such as kg/ha/yr, and the weight of tissue is usually dry weight.

Biomass   The quantity of living matter per unit area at one moment in time, usually given in units such as kg/ha or t/ha. The weight of biomass is usually given as the dry weight.

Browse   Plant matter within the reach of a herbivorous animal but not growing in the soil, that is, twigs and leaves of a shrub, not grasses.

Deep Ecology   An environmental philosophy based on the idea of the intrinsic value of non-human components of the cosmos; the world is not our oyster, we share it with the oysters. Associated especially with the Norwegian thinker Arne Naess (1912–) and much influenced by Buddhism, Mahatma Ghandi and Spinoza.

Desertification   The loss of plant biomass from an area with low and probably irreg-ular rainfall. Climatic change (possibly cyclical) and grazing intensity are both ele-ments of its cause and there is often controversy about which is the most important at any one time.

Dinoflagellates   Simple plants, often part of the marine plankton. If human activity floods the sea with phosphates, dinoflagellates may form a bloom and their exudates colour the water forming a Red Tide, which is often toxic to shellfish and thence to humans fond of oysters or mussels.

Discourse   A way of thinking and writing which is institutional because it delimits what can be said about a particular topic and the language which is used. Thus, it affects what can be accepted as 'truth'.

Dualism The idea that there are two fundamental classes of things or categories which are opposed to each other, like 'man' and 'nature', or 'economic' and 'uneconomic'. 'Good' and 'bad' are also examples.

Ecofeminism The movement that saw women as being guardians of environmental values, because they were essentially caring and nuturing and not exploitive. Coined in 1974, the core ideas were much thinned out by the 1990s.

Ecotone The transition zone between two ecological formations, such as the reedswamp between dry land and open water in a lake, or an area of scrub between a woodland and open grassland.

Ecumene The parts of the planet Earth that are inhabited by humans.

Endemic Species of plants and animals that have a very limited and specific distribution. Islands and mountain chains are often high in endemics that have evolved there and had little chance to spread: the subspecies of finch and tortoise found on different islands in the Galapagos are famous endemics owing to Charles Darwin's notice of them in his thinking about evolution.

Energy The ability to do work, coming in forms such as chemical, potential and kinetic. It is measured as force times distance: the joule is one newton applied through one metre. The transformation of energy from one form to another entails the loss of the ability to do work, in the form of heat. (See also Entropy.)

Energy density The amount of energy stored in a unit mass of a resource (often in joules per kilogram): a good comparator for foods and fuels.

Energy intensity The cost of a product or service in energy terms. Aluminium is energy intensive (230–340 MJ/kg), whereas steel (20–50 MJ/kg) is not.

ENSO El Niño-Southern Oscillation, a set of interacting parts of a single global system of coupled ocean–atmosphere climate fluctuations caused by oceanic and atmospheric circulation. ENSO is the most prominent known source of variability in weather and climate around the world on a time scale of 3 to 10 years. Its effects are felt in all the major oceans and in South America, Africa and Oceania.

Entropy A difficult concept in energy studies which, in environmental contexts, usually refers to the production of disorder in a system through time. The presence of life seems to be contrary to the formation of entropy but is achieved only at the cost of the dissipation of potential energy to heat.

Equids Horse-like animals, including, for example, zebras.

Feedback loop A mechanism in a machine or a living system which keeps it stable.

Gaia hypothesis Associated with the chemist, former NASA scientist, and environmentalist, James Lovelock (1919–), this postulates that the planet functions as kind of super-organism which tends to maximise the conditions for life, though not necessarily the human form. The concept is controversial among many biological scientists because it appears to have elements of teleology in it.

High forest   Apparently mature forest, with many trees growing close enough for their canopies to overlap or have only a very narrow space between them.

Holism   The notion that the whole is more than the sum of its parts; a system can only be understood as a whole and not by breaking it down into subsystems. An important idea in the science of ecology.

Hydrocarbon   A chemical compound consisting only of carbon and hydrogen. Many hydrocarbons are combustible in normal conditions.

Infield   In pre-industrial agriculture, a field that is usually near a settlement and is cultivated in most years, carefully manured and regularly allowed periods of fallow.

Intrinsic value   The idea that non-human entities have their own value and not simply that placed on them as elements for human usage; the opposite state is called 'instrumental value'.

Kant, Immanuel (1724–1804)   German philosopher and geographer whose ideas have had a lasting impact upon environmental thinking as well as in many other spheres. The importance of rationality in formulating morality is central.

Land ethic   Associated with Aldo Leopold (1886–1948), American zoologist, who formulated it as 'A thing is right when it tends to preserve the integrity, stability, and beauty of the biotic community. It is wrong when it tends otherwise.' Along with Henry David Thoreau (1817–62), his ideas are often central to American thinking about environmental matters.

Malthusian   Pertaining to Thomas Robert Malthus (1766–1834), demographer and political economist. He thought that the potential of population growth to outstrip resources was ever-present. His influence is still strong.

Mendelian   Pertaining to Gregor Mendel (1822–84), the father of modern genetics. Using sweet peas, he demonstrated that inherited characteristics could be predicted on a probabilistic basis.

Mutualism   The connection between two organisms which is to the benefit of both. It may be essential to both species or they may be able still to exist in its absence. The closest form of mutualism is symbiosis.

Nietszche, Friederich (1844–1900) A philosopher of complex ideas but often quoted as an 'existententialist': the freedom of the individual to exist, act and formulate morality is central and more important than, for example, systems of rationality.

Nomadic pastoralism   An economy based on herding domesticated or semi-domesticated animals which involves a seasonal round of movement between pastures and/or water sources. Commonest in dry or mountainous regions.

Normative   Normative statements tell us how things ought to be, how to value them, which things are good or bad and which actions are right or wrong. Clearly many statements made in an environmental context are of this kind.

Objectivity   The suggestion that humans can step back and view a phenomenon as it really is, without bias. The natural sciences most frequently make such a claim.

Outfield   In pre-industrial agriculture, a field taken in from wild vegetation and used for a limited period before its fertility declines and it is allowed to revert to an earlier condition.

Ovicaprids   In archaeology, bones of either sheep or goats but which cannot be determined any further.

PCB   Polychlorinated biphenyls. A class of organic compounds used in many industrial processes which are very stable and break down very gradually in the environment. Thus they have the potential to be long-lived pollutants.

Pelagic   The open sea: the zone away from the coast and above the sea-floor.

Photosynthesis   The way in which green plants 'fix' solar energy into chemical form. In a simple expression it can be represented by the equation:

Carbon dioxide + Water + Light energy $\rightarrow$ Glucose + Oxygen + Water

Phytolith   A microscopic spicule of silica found in the stems, roots and leaves of, for example, grasses. Its size and shape may be specific to a species or other taxonomic group and hence useful in environmental archaeology.

Phytomass   As in biomass, but confined to plant material.

Pollarding   A form of tree management: the trunk is cut off just above the browsing level of domesticated animals and the young sprouting stems harvested for fodder or other uses.

Positivism   The belief system underlying the natural sciences, especially the need for observation, the formulation and testing of hypotheses, and the logical structure of any statements made.

Prometheus   In Greek mythology, the Titan who stole fire from the gods and gave it to humankind. In the course of conflict with Zeus, he was chained to a rock and an eagle came and ate out his liver, which grew again every night. Thus, his presumptions (*hubris*) were punished.

Pyrophytes   Plants which are not damaged by fire; some even need it for successful reproduction.

Qanat   An irrigation system from south-west Asia in which water from the foot of mountains is led to fields in a series of parallel tunnels with vertical air shafts. An Arabic word, pronounced 'kanat'.

Saqiya   A water-lifting device powered by a draught animal, such as a donkey, mule or camel, used in pre-industrial irrigation systems. An Arabic word.

Sentientism   The idea that moral precepts apply only to 'sentient' creatures such as mammals and birds.

Shaduf   A counterbalanced pole used as a lifting device in pre-industrial irrigation systems. The power source is a human. An Arabic word.

Shaman   An intermediary between the natural and spirit worlds who can travel between worlds in a state of trance. Commonest in, but not confined to, hunter-gatherer societies.

Shredding   A form of tree management in which the side branches are all lopped to provide animal fodder or usable wood.

Tao   A Chinese word, usually translated as 'The Way'. The source and guiding principle of the universe. Adherence to it is necessary for harmony in the world. Often pronounced 'Dao'.

Teleology   The proposition that a process is purposeful and has a goal or predetermined end. A main characteristic of Darwinian thought is that evolution has no teleology: it is open-ended.

Theriomorphy   In which a spirit takes the form of an animal or even in which a human-other animal hybrid is formed.

Transhumance   This occurs when domestic animals are taken to a different environmental zone for a season and some of the human population follow them: for example, to use mountain pastures only available in the summer.

Xerification   Drying out. An early stage in desertification, when the overall phytomass is diminished and more drought-tolerant plants replace those requiring more water.

Zoonoses   Diseases transmissible to humans but originating in wild animals. Anthrax and Ebola virus are examples.

Acronyms

ACLS	American Council of Learned Societies
ATC	Air Traffic Control
BSE	Bovine Spongiform Encephalopathy ('mad cow disease')
CEO	Chief Executive Officer
CHC	Chlorinated Hydrocarbons
DNA	Deoxyribonucleic Acid
ENSO	El Niño-Southern Oscillation
FAO	Food and Agriculture Organization [of the United Nations]
GCM	Global Climate Model (or General Circulation Model)
GIS	Geographical Information Systems
GM	Genetically Modified
GNP	Gross National Product
HEP	Hydro-Electric Power
HGV	Heavy Goods Vehicle
HIE	High-Income Economy
HIV/AIDS	Human Immunodeficiency Virus/Acquired Immunodeficiency Syndrome
IPCC	Intergovernmental Panel on Climate Change
LIA	Little Ice Age
LIE	Low-Income Economy
MNC	Multi-National Company (see also TNC)
NGO	Non-Governmental Organisation
PIE	Post-Industrial Economy
SARS	Severe Acute Respiratory Syndrome
TNC	Trans-National Company (same as MNC)
UNEP	United Nations Environmental Programme
WWF	World Wide Fund for Nature

Index